Science and Technology in Modern China, 1880s-1940s

Science and Technology in Modern China, 1880s-1940s

Edited by

Jing Tsu and Benjamin A. Elman

BRILL

LEIDEN | BOSTON

This paperback was originally published in hardback under ISBN 978-90-04-25853-2 as volume 27 in the series *China Studies*.

The copyediting, proofreading, and production costs for this volume have been provided through a publishing subvention provided by a Mellon Foundation Career Achievement Award for Benjamin Elman.

Cover illustration: 1910 issue cover of *Fiction Times* [*Xiaoshuo shibao*].

The Library of Congress has cataloged the hardcover edition as follows:

Science and technology in modern China, 1880s-1940s / edited by Jing Tsu and Benjamin A. Elman.
     pages cm. -- (China studies, ISSN 1570-1344 ; volume 27)
  Includes bibliographical references and index.
  ISBN 978-90-04-25853-2 (hardback : acid-free paper) -- ISBN 978-90-04-26878-4 (e-book) 1. Science and state--China--History--19th century. 2. Science and state--China--History--20th century. 3. Science--China--History--19th century. 4. Science--China--History--20th century. 5. Technology--China--History--19th century. 6. Technology--China--History--20th century. I. Tsu, Jing, editor of compilation. II. Elman, Benjamin A., editor of compilation.

  Q127.C5S332 2014
  509.51′09041--dc23

                    2013048160

This publication has been typeset in the multilingual 'Brill' typeface. With over 5,100 characters covering Latin, IPA, Greek, and Cyrillic, this typeface is especially suitable for use in the humanities.
For more information, please see brill.com/brill-typeface.

ISBN 978 90 04 26841 8 (paperback)
ISBN 978 90 04 26878 4 (e-book)

Contents

List of Figures

Notes on Contributors

Iwo Amelung

is professor of Chinese Studies at Goethe University, Frankfurt am Main. He has studied Sinology and history at Göttingen University and in Bonn. His Ph.D. is from the Free University, Berlin, with a dissertation on the control of the Yellow River during the late Qing dynasty. He has served as the managing director of the European Centre for Chinese Studies at Peking University and was a visiting professor at the Chinese Academy of Sciences. He is interested in the history of late imperial China and also focuses on the history of the Republican era. He has worked extensively on the reception of Western knowledge in late imperial China, including the historiography of science and technology in modern China.

Benjamin A. Elman

is Gordon Wu '58 Professor of Chinese Studies, professor of East Asian studies and history, and former chair of the Department of East Asian Studies at Princeton University. He received his Ph.D. from the University of Pennsylvania. He works at the intersection of several fields, including history, philosophy, literature, religion, economics, politics, and science. He is currently studying cultural interactions in East Asia during the eighteenth century, in particular the impact of Chinese classical learning, medicine, and natural studies on Tokugawa, Japan, and Chosŏn, Korea. Elman's recent books include *Meritocracy and Civil Examinations in Late Imperial China* (2013) and *A Cultural History of Modern Science in China* (2009).

Fa-ti Fan

teaches at Binghamton University, State University of New York. He is the author of *British Naturalists in Qing China: Science, Empire, and Cultural Encounter* and many articles on a wide range of topics in the history of science and of modern China. He is completing two book projects, one on science in Republican China and the other on earthquakes in socialist China.

Danian Hu

is the deputy chair of the Department of History at the City College of New York. He received his Ph.D. from Yale University. He teaches courses on East Asian science and technology, China's nuclear weapons program, and Chinese history. His current research concerns scientific developments in modern China and Sino-Japanese scientific exchanges in the twentieth century. He is the author of *China and Albert Einstein: The Reception of the Physicist and His Theory in China, 1917–1979* (2005), which has also been published in 2006 in a revised and expanded Chinese edition.

Joachim Kurtz

joined the Heidelberg University Asia and Europe in a Global Context Cluster as professor of intellectual history in summer 2009. Earlier he was an associate professor of Chinese at Emory University and a Research Group director at the Max Planck Institute for the History of Science in Berlin. He has held visiting positions at the Institute for Advanced Study in Princeton and the École des Hautes Études en Sciences Sociales in Paris. His research focuses on cultural and intellectual exchanges between China, Japan, and Europe, with special emphasis on practices of argumentation, logic, political theory, rhetoric, translation studies, historical semantics, and the history of the book. He is the author of *The Discovery of Chinese Logic* (2011).

Eugenia Lean

is an associate professor of modern Chinese history at Columbia University. She is the author of *Public Passions: The Trial of Shi Jianqiao and the Rise of Popular Sympathy in Republican China* (2007), which was awarded the John K. Fairbank Prize by the American Historical Association. Her current project, "Manufacturing Modernity: Chen Diexian, a Chinese Man-of-Letters in an Age of Industrial Capitalism," focuses on polymath Chen Diexian, a professional writer/editor, science enthusiast, and pharmaceutical industrialist, to examine the intersection among vernacular science, commerce, and ways of authenticating knowledge and things in an era of mass communication.

Thomas S. Mullaney

is associate professor of modern Chinese history at Stanford University and received his Ph.D. from Columbia University. He is the author of *Coming to Terms with the Nation: Ethnic Classification in Modern China* (2011) and principal editor of *Critical Han Studies: The History, Representation and Identity of China's Majority*, which examines China's majority ethnonational group. His current book project on the Chinese typewriter examines China's nineteenth- and twentieth-century development of a character-based information infrastructure encompassing Chinese telegraphy, typewriting, character retrieval systems, shorthand, Braille, word processing, and computing.

Hugh Shapiro

is associate professor of modern Chinese history at the University of Nevada. He earned his B.A. and M.A. from Stanford and his Ph.D. from Harvard. He has held visiting appointments at Princeton University, Nichibunken in Kyoto, National Taiwan Normal University in Taipei, and the Institute for Advanced Study in Princeton. His archival research and fieldwork in China, Taiwan, and Japan focus on the comparative history of disease, especially on the diverse ways that biomedical and neuropsychiatric systems impinge on the imagination of bodily and psychological experience in modern China. He coedited *Medicine across Cultures: History and Practice of Medicine in Non-Western Cultures* (2003). He received the Li Ching Prize for the History of Chinese Science and won his university's highest teaching award and its research mentor award.

Grace Shen

is assistant professor of history at Fordham University. Her current research interest is studying the connections between geology and modern China. She stresses geology's link to Chinese nationalism in her recent book from the University of Chicago Press entitled *Unearthing the Nation: Modern Geology and National Identity in Republican China, 1911–1949*. Shen explores the multiple valences of the land itself as a territory, resource, physical environment, and native place, and she uncovers the ways that models of science and nation converged in geological research.

Shen Guowei

is a member of the faculty in the Department of Foreign Language Studies at Kansai University in Osaka, Japan. He completed his graduate studies in the Doctor's Degree Program in the Graduate School of the Division of Letters at Osaka University in 1991. He was promoted to professor at Kansai University in 2001. His research fields focus on Chinese and Japanese linguistics, the history of scientific concepts in China and Japan, and translation studies. He received the Stephen C. Soong Translation Studies Memorial Award in 2011 from the Research Centre for Translation at The Chinese University of Hong Kong.

Jing Tsu

is Professor of Modern Chinese Literature and Culture at Yale University and holds a Ph.D. in Chinese studies from Harvard University. She has been a Fellow at the Harvard Society of Fellows (Junior Fellow), the Radcliffe

Institute for Advanced Study at Harvard, the Center for Advanced Study in
the Behavioral Sciences at Stanford, and the Institute for Advanced Study at
Princeton. She has published *Sound and Script in Chinese Diaspora* (2010) and
*Failure, Nationalism, and Literature: The Making of Modern Chinese Identity,
1895-1937* (2005). Her current book project, *The Alphabetic Mind in Chinese*,
examines modern scientific thinking in China as a basic project of reinvent-
ing the power of writing and thinking itself across divergent contexts.

Introduction

Jing Tsu and Benjamin A. Elman

In the past several years, China historians and literary scholars who study the culture and institutions of modern science have developed a strong interest in comparative methodology in intraregional and global histories. Critical interests are merging in new ways in an effort to rethink the impact of nationalism, the ways in which ideas and technology interact and circulate, and new intersections between local episteme and global knowledge. This is, in part, due to a felt need to move beyond the "belated" hypothesis of the development of science in China (the "Needham paradigm") based on the often-reified "Western" frame of comparison. How to go about this, however, has been a question of some debate and dissensus. This volume results from a two-day symposium that was convened in New Haven, Connecticut, in January 2010, to address these methodological crossroads and changing historical viewpoints.[1] The conference brought together junior and senior scholars of history, history of science, and literature in an attempt to reevaluate the meaning and practice of "scientism" in nineteenth- and twentieth-century China. Revisiting the foundations of modern intellectual history as well as charting new directions in the study of science and technology in relation to culture, we identified the following themes:

- Intellectual History and Humanism
- Language, Translation, and the Technology of Scripts
- Scientism and National Empires of Knowledge
- Lexical Changes and Epistemological Shifts
- Professionalizing Science
- Dissemination, Circulation, and Competing Networks
- Popular Science and Empirical Practice in Everyday Life

---

1 We thank the Council on East Asian Studies and the Department of East Asian Languages and Literatures at Yale University for sponsoring this project, which was also supported by the East Asian Studies Program at Princeton University. The Andrew W. Mellon Foundation has provided a timely publication subvention for the conference volume.

It became clear to the editors that one of the major lacunae in the study of "modern science" in China was the late Qing and Republican era, from the 1890s to 1949. Wedged between a period of prolific missionary translation activities accompanied by domestic technological buildup in the nineteenth century, on the one hand, and the state-driven agenda of the Communist period starting in the mid-twentieth century, on the other, the setting for science during this transitional period comprised new ideas, failed attempts, and renewed efforts. The experimental nature of these adjustments in part led to a flatly interpreted narrative concerning the alleged lack of adequate growth in science and technology in modern China.[2] This assumed story line was generally taken at face value. In recent years, however, this view has become particularly suspicious, especially in light of the crumbling of the overall premise of deficiency that is associated with the Needham paradigm. For more than three-quarters of a century, Joseph Needham and *Science and Civilisation in China* have supplied the dominant framework for interpreting the limitations of science, medicine, and technology in premodern, "imperial China."

At the same time, above and beyond reversing long-standing presuppositions, our collective examination makes new inroads into a much more varied topography, primarily during the Republican era. In each case under examination, the adapted uses of scientific knowledge range from creative appropriation to disarticulated, small-scale efforts. The result does not always follow planned institutional goals. Nor do these incremental acts necessarily respond, in a unified way, to overarching local, national, or global exigencies of scientific modernity. Together, these different threads weave distinct webs of established, as well as informal, networks. Each locus of activities extended beyond the local frontier while shaping international and global scales of contestation. An analysis of their dynamics disputes a narrative of deficiency and the failure to modernize drawn from the interpretation of late imperial developments. More pertinent to the modern period, an examination of their highly differentiated circumstances forestalls the a priori dominance of any generalizing framework in favor of new conceptual and empirical syntheses.

Beginning during the discussions at the conference and continuing afterward, the editors and participants decided that the papers presented at Yale, and the additional contributions solicited later, could serve as an important first step in readdressing the place of the Republic of China in the history of modern science, medicine, and technology before 1949. Accordingly, at the

---

2   While the failure narrative was literally deployed in this way, it paradoxically also acquired an expansive, rhetorical utility in projecting the prospect of national, racial, and cultural rejuvenation in the late Qing and Republican periods. See Tsu 2005.

March 2010 Association for Asian Studies Annual Meeting in Chicago, the editors consulted with Albert Hoffstädt from E. J. Brill in Leiden. A blueprint for the present volume was in the making. We agreed that the Yale conference papers would move beyond our earlier fixations on the failures of "imperial China" (see further below) and its impact on understanding modern science in China. We wanted to readdress the unbalanced historiography of the late Qing and Republican periods by finally acknowledging the limitations of uncritically applying premodern labels (disguised as "debates") to the remarkable growth of science and technology, now called "technoscience" (*keji* 科技) in China, during the twentieth and twenty-first centuries.[3] In this light, habitually invoked concepts and categories for comparison such as "logic," "freedom," "natural law," "despotism," "classical Greece," "ideograph," "linear perspective," "capitalism," "Westernization," and so on would require a new set of historical sensibilities as well.

Our discussions also helped us realize how murky our understanding of the history of science, medicine, and technology in Republican China had become when viewed through the refracted shadows of the premodern historiography on science in imperial China before 1850. The Yale conference brought together the leading lights in the modern field, whose new and ongoing work make this new turn in understanding "science in Republican China 1911–1949," possible. This volume is thus not only one of the first to move beyond past historiography about late imperial China, but also one of the first interdisciplinary efforts to explore the contemporary implications of science during the two "republics": the ROC since 1912 and the PRC since 1949.

Just as past limitations are taken into account in these pages, the desired new perspective also brings its own challenges. To be sure, a paradigm shift will not be accomplished within the confines of a single volume. At the same time, our attempt resonates with similar stirrings in other corners of China studies. We join the recent chorus of scholars who are engaged in rethinking the customary methods used in comparing China and the West, which have generally downplayed the primacy of intraregional influences and other local-global circuits of transmission. The focus, instead, on large differences between grand civilizations has traditionally set the terms of comparison from the time of Needham to even more recent works on empires that seek to rebrand the "Asian Age."[4] Lately, everywhere a "China theory," or something with a Chinese

---

3 Kurtz 2011 provides an example of how the new literature approaches topics such as "logic" as a factor in the history of science in China.

4 Frank 1998.

twist or characteristics, is propagated to stand as an alternative developmental and historical model to Euro-American versions of capitalism and modernity.

While it has been instructive to reverse the terms of asymmetry for heuristic purposes in this way, the editors also hope to encourage further steps by suggesting a global context for reevaluating China's scientific modernity. Among other things, the paths that were not taken in the history of the West are no longer recapitulated as examples of belatedness or lack of development, confined to non-Western contexts. Instead, we consider these other historical scenarios as active shifts in the standards of comparison, thereby highlighting the multiple possibilities of divergence in materializing the various shapes and scales of modernization. This includes, rather than defines itself against, that which has been traditionally referred to as "Western." While the past is prelude to the present and future, our currently changing "present" standpoint on the "past" may not have much of a "future" in a reassessment of the global history of modern science.[5]

Part of the methodological challenge stems from the evolution of the field of China studies itself. Premodern historians, for instance, have tended to reify the perennial fields of "Chinese Science" (CS) and "Traditional Chinese Medicine" (TCM). Looking back from the present in reverse teleology, they opt to essentialize in order to explain the recurrent natural philosophies that informed premodern intellectual life in imperial China.[6] Working in fields that existed before the rise of modern science even occurred in either Europe or China, such studies have anachronistically stressed the "failure" of these fields (CS and TCM) to become modern in imperial China on their own.[7] There have been notable exceptions. Premodernists like Nathan Sivin, for instance, have artfully exposed Needham's ideological blind spots in relation to Daoism (misreading who ancient Chinese scientists were), Marxism (overdetermining Chinese despotism), and capitalism (inflating the limits of Chinese commerce). Needham's assumptions, Sivin shows, were so far off the mark that they compromised Needham's "grand titration" of East and West.[8]

Despite such suggestive advances beyond Needham, premodernists have also spent much of their time researching issues in earlier natural studies while continuing to explain why modern science, technology, and medicine arrived so late in China. Recent efforts by Geoffrey Lloyd and Nathan Sivin in their

---

5   Sivin 1978, for example, provides an incisive review of Elvin 1973 by attending to the book's use of evidence pertaining to science and technology.

6   Sivin 1995a.

7   Elman 2006.

8   Sivin 1995b; see also Sivin 2005. Compare the more radical critique of Needham in Hart 1999.

book *The Way and the Word: Science and Medicine in Early China and Greece* (2002) are a case in point. They reaffirm the millennial differences in mentality between the ancient Chinese (as monolithic and uniform in their imperial worlds of public discourse) and the decentralized city-state Greeks (as inherently disputatious). Earlier, Frederick Mote had contended in his *Intellectual Foundations of China* that there existed a deeply metaphysical "cosmological gulf" between ancient China and the West that had continued into modern times.[9] Like Mote's book, Lloyd and Sivin's *The Way and the Word* reestablished incommensurability between the "ancient" Chinese and Greeks as a background gloss to explain why modern science and medicine appeared first among modern Europeans in the classical-Greek-oriented West.[10] Such asymmetrical grounds of comparison have been uncritically applied to the evaluation of other cultural and linguistic forms. On the level of language, for example, the perceived differences between "alphabetic" and "ideographic" writing systems, representing Greek and Chinese civilizations, respectively, have for centuries been made to support the claims about the West's unique capacity for abstract thinking.[11]

The "Needham Question"—Why did a divided Europe, and not imperial China, develop modern science first?—until recently remained preeminent. This question was paralleled by continuing scholarly efforts in other fields to explain why China did not develop capitalism or democracy before Europe.[12] We are, however, now entering a new era that explores modern science in Republican China in more active, rather than simply receptive, terms. What was once vaguely described as appropriation can be seen in stronger relief by examining the specific contexts of agency and mobilization. Increasingly, we are able to address modern science in Republican China from a comparative point of view and include it in the story of global science, instead of excluding it on the view that its predecessors in China had failed to materialize it in a particular way. Our volume thus recasts the roots of modern science in the late Qing and its continued development in early Republican China, rather than building a house of cards on the abstractions that are associated with the Olympian vision of the premodernists and that start with ancient Greece and China.

---

9   Mote 1989, 12–25, 95–96.
10   Needham 1969, 150–168; and Lloyd and Sivin 2002, passim. Lloyd 1990 has usefully recognized the danger of anachronistically positing perennial "mentalities" to explain long-term historical developments in mind-sets. Cf. Kuhn 2000.
11   Tsu 2010.
12   See Pomeranz 2000; Jones 1987; and Needham 1959, 150–168.

As part of the broad history of China's modern transformation, it will continue to be useful to make references to the advent of Western science. One sees great promise of further, innovative interpretations when one moves beyond identifying similarities and differences, impact and response. Rather than unfairly homogenizing both Chinese and Western contexts in the process by naming what one has in terms of what the other does not, the possibility of their comparison is considered anew in the following pages. Westernization has been a historically necessary, but in itself insufficient, explanation for how science developed in modern China. In a similar way, we no longer can afford to deflate the place of science, or the platform it provides, in China's increasing participation and growing stakes in the contemporary world. China's plans to send expeditions to the moon and Mars in the twenty-first century are partly a response to the shock of heavy-handed Western and Japanese imperialism since 1850. It is therefore important that the role of modern science, technology, and medicine in contemporary China is properly considered in its varied social, political, and cultural dimensions—and not only by historians of science.

While China's quest for modernization has often been described in terms of progress and developmental models, it is better analyzed as a complex process through which China sought to reboot itself in an unfamiliar global order. In this regard, the demonstrability and reproducibility of knowledge across contexts were of exceptional importance. It is impossible to appreciate the significance of science in modern China without taking into account how well it can be communicated to, and grafted onto, existing knowledge structures and infrastructures. China's experience of modernity has also been marked by a general distrust of knowledge, both local and foreign. Though empiricism, experience, and experiment are more commonly associated with questions of truth value, they were just as often implicated in debates about believability and utility. Could science, then, be relied upon to cross boundaries? Or would it work only where complicated systems were in place to support it? How could the Chinese be confident about their new understandings when appearances and well-established systems had failed before?

The role of the Chinese state in modern science continues to be decisive. But a state-centered approach to modern science, however useful, would be incomplete without an equal emphasis on the global and comparative issues involved in the mastery of modern science in different social and cultural strata. We have also overlooked the advent of early Chinese practitioners and intellectuals who acted as spokespersons for modern science, before the "scientist" became an accredited informant. At the same time, we need to rethink our reliance on the early, at times crude, political rhetoric and philosophical theory enunciated by Chinese publicists of science since the 1919 May Fourth Movement. While they championed the cause of a "Mr. Science" made in Japan that was called *kexue* 科學, they left few directives and theoretical foundations on which further pillars could be built. The problem is how best to incorporate these different emphases into a more conceptually coherent, while empirically differentiated, picture.

Similarly, we must problematize post-Mao efforts to distinguish Chinese socialism from scientific progress. Most Euro-American and Chinese accounts have indicted Maoist mass science and its rhetoric of science's role in class struggle as a smokescreen for power politics. We have elided what socialist ideals were about during the Great Proletarian Cultural Revolution from 1966 to 1976. Although the victimization of many scientists during this period and the role of Maoist ideology in leading some Chinese scientists to oppose relativity in the name of dialectical materialism, for example,[13] are important issues in the unmasking of Maoism after 1976, the broader aspects of understanding why mass science appealed to many Chinese and some Euro-Americans in the 1960s force us to question the easy separation of scientific practice from social and political agendas.[14] More researchers in socialist laboratories will reveal the peculiar nature of socialist rhetoric and communist institutions in forging myths about science that enhance its revolutionary status in China and elsewhere in the increasingly postsocialist world. After all, liberal capitalist ideals have informed our own Euro-American notions of modern science as the sine qua non for the rise of the middle classes via science and engineering since the industrial revolution.

In sum, the account above, taken as a whole, suggests a number of ways that a comparative history of science can lead us in new directions. First and foremost, historicizing the Western scientific revolution in a global context makes it possible to compare other, non-Western approaches to modern science without reducing such efforts to simple reception history.[15] Second, differential studies that wield appropriate concepts and categories for comparing precise historical situations are mandatory. In particular, case studies can successfully integrate scientific contents and the historically dynamic contexts as the key to moving from the local to the global and back again. We should explore Chinese interests in modern science as scientists there articulated and practiced them, rather than speculate about why they did not act the way Americans and Europeans expected them to act. Future research on the active careers of mod-

---

13   Wang Zuoyue 2007.
14   Schmalzer 2007.
15   Shen 2007.

ern Chinese scientists, both individually and as a group, will allow us to super-sede past accounts of the passive reception history of modern science in China by publicists such as Liang Qichao.

To this end, each of the essays in this volume opens up a potential venue that invites further study and debate. They are organized into two groups: (1) critical historical frameworks and (2) new research directions. As part of the first group, Benjamin Elman's opening essay introduces the terminological changes in the concept of science from the late Qing to the early Republican period. The essay rethinks the historiography of modern science in China in light of the influence of Meiji Japan on the development of modern science there after the 1894–1895 "First" Sino-Japanese War. Meiji-style science, Elman argues, trumped the Western sciences that Christian missionaries and Chinese reformers had enunciated during the "Self-Strengthening Movement" after 1865. In order to move toward a clearer understanding of the place of science in modern China, Elman contends, we need more case studies that success-fully integrate the technical contents of scientific assimilation with the histori-cally dynamic contexts in which those contents were deployed. This may require us to move from the local to the global and back again.

At the Yale conference, Wang Hui presented an essay on Yan Fu's translation of Thomas Henry Huxley's (1825–1895) *Evolution and Ethics* (*Tianyan lun* 天演 論). Although we could not include it in this volume, we should mention here that Wang noted in his essay that, early on, some readers considered Yan Fu a follower of the metaphysics and epistemology of the *Changes Classic* (*Yijing* 易 經). They associated John Maynard Smith's *On Evolution* with nativist dis-course, not Darwinian inspired theories. Wang Hui contended that Yan Fu, read in this light, can be understood as a Westernizing scholar who tried to cre-ate a universal naming system (*mingxue* 名學) to rationalize the conflicting ideas of evolution, induction, individualism, collectivism, and so forth into an ordered structure. The world of the *Changes* that Yan Fu read into Spencer and Huxley was not a Social Darwinist world of the "survival of the fittest" and "nat-ural selection." What Yan was using Spencer and Huxley for was elaboration of a cosmic principle that was compatible with Chinese discourses of change based on the investigation of things (*gewu* 格物) and Western notions of induction. For our conference, what was important was the fact that Yan Fu's notion of evolution was more than just a reflection of Spencer and Huxley. He used the logic of the *Changes* to articulate a universal human order that would explain human society (*qun* 群), ethics, and politics and that would escape the "competition for survival" that was the source of political chaos.

Focusing on the historiography of science and technology of the modern period, Iwo Amelung provides a panoramic view of the cultural and intellec-

tual context in which science acquired its intellectual and disciplinary significance. Emphasizing the driving force of national consciousness against the
larger background of modernization, Amelung examines how science and
technology were discussed and interpreted through the lens of national salvation. A key point is Chinese intellectuals' and educators' reclassification of traditional texts in order to reposition scientific knowledge within a worldview
that was more familiar to them. Subsequently institutionalized as part of
"National Studies," the integration also entailed curricular changes in academic
institutions, further reinforcing its legitimacy.

While new classifications made it easier to reproduce scientific knowledge,
their artifice is far from neutral. Taking up the issue of how taxonomies themselves constituted the very battleground on which traditional and new systems
of thought were amalgamated, Joachim Kurtz analyzes the issue in the intellectual context of an important, if eclectic, National Essence proponent, Liu
Shipei. Focusing on this single figure to cross different intellectual terrains,
Kurtz demonstrates how Liu used the core disciplines of modern Western
humanities—psychology, ethics, and sociology—to reinvent the relevance of
ancient Chinese learning.

Together, these four essays (three included in this volume) place science in
its broader institutional, intellectual, and epistemological context. Reorienting
the basic issues at stake, they pave the way for the next set of essays, which take
off in new directions, thereby building different contexts for new research
questions. If science in modern China was heatedly debated among leading
figures and institution builders, it was often the lesser-known inventors, educators, amateurs, and other unofficial practitioners who helped to bring it to life
in the popular consciousness. Turning to the middle strata of knowledge dissemination through its commercial, industrial, educational, media, and
regional players, the next set of essays shifts our focus from core intellectual
circles to scientism at large. They provide different reference points for the
issues outlined in the first set of essays and raise new questions in little-known
areas of research.

While knowledge dissemination through translation and popular culture
has been widely studied, one hitherto neglected issue is how the general communication between high and popular culture, involving foreign and indigenous words as well as concepts, was also intimately tied to the changing
materiality of language. Shen Guowei's, Jing Tsu's, and Thomas Mullaney's
essays address the role of language in translation, typewriting technology, and
telegraphy, respectively. Departing from the general reliance on translation as
a blanket term for a space of autonomy and creative appropriation, Shen
explains the semantic nuts and bolts of the notoriously difficult translations of

the late Qing intellectual and reformer Yan Fu. Shen focuses on Yan's complex negotiation with Chinese and Western philological and philosophical traditions, with Japanese as a crucial intermediary. He provides a detailed study of how commensurate meanings are made and unmade in the hands of this pivotal figure during this epistemic transition.

While the contestation between languages in contact is most visible on the level of translated meanings, Jing Tsu argues that the material transformation of the Chinese language in the late nineteenth through mid-twentieth centuries opened up a new, unprecedented playing field. The very physical medium of the Chinese script had to negotiate its own terms of scientific modernity. Turning our gaze toward a hitherto little-studied history of the materiality of the Chinese script, Tsu examines the technological crossing of alphabetic and "ideographic" writing systems since the late nineteenth century that culminated in the invention of a pathbreaking Chinese-language typewriter in the 1940s. This particular invention intersected with several international and linguistic battles that were being waged between civilizational writing systems, global language wars, typewriting technology, late Qing language reform, and the Cold War. It responded in ways that fundamentally changed the linguistic conditions of civilizational strife from the inside out.

Looking at China's entrance into the family of nations, Thomas Mullaney's piece shows how attempts were made to put the Chinese language on equal footing with Western languages through the international codification of telegraphy in the nineteenth century. The favoritism toward the English language shown by the institutionalization of the Morse code, and Chinese telegraphers' strategic maneuvers in response, shed light on the disparate, but no less historically significant, efforts to combat China's disadvantaged entry into global communication. Mullaney explains how these incremental if underestimated technological innovations unfolded and later even became an object of jokes and derision. The study shows that China was already enmeshed in a new global order in the nineteenth century, which required the Chinese to adopt a set of linguistic technologies and symbolic systems structured around alphabets rather than the Chinese script.

If science and technology in China were often enmeshed in questions of sovereignty, their commercial and educational aspects involved a different set of problems. From brand names to academic rivalry, the spread of authoritative knowledge relied on middlemen, entrepreneurs, and professors, as well as scientists, to bring about a widely shared sense of scientific modernity and materialism. Eugenia Lean discusses how science manuals intersected practices of editorship and entrepreneurship as novel knowledge and commerce created the need for authentication. She focuses on the maverick cultural figure of Chen Diexian, who, among other things, was also a fiction writer. Industrial knowledge, as Lean shows, facilitated the transfer between specialist knowledge and popular consumption. Taking a different angle, Fa-ti Fan analyzes how the cultural authority of science was established in a debate about spontaneous generation in the 1930s. Its level of controversy drew in specialists and spectators alike, making the bounds of experimental science itself an institutionally and epistemologically contested affair. Both essays are concerned with how the authority of science was established in cultural and public discourse.

While debates about the public digestibility of science raged on, Danian Hu takes us back to the familiar intersection between Western propagators and the Chinese audience. Instead of missionaries, however, Hu looks at the collaborative efforts between Western and Chinese academic institutions through the career of the British physicist William Band at Yenching (present-day Beijing) University. Band was interested in making physics available for solving China's social problems, such as the services it could offer to the Rural Reconstruction Movement in the 1930s. His case exemplifies how science was positioned to make tangible social and political contributions that were specific to China's needs.

Harking back to the issue of language, but in the specific context of the geological sciences, Grace Shen looks at how the circulation of scientific authority depended on the access to print sponsorship. Such sponsorship provides a concrete barometer for understanding the degree of interaction, collaboration, and competition between Chinese geological societies and the international scientific community. Maintaining Chinese authority over the development of the discipline in China's hinterland as well as urban areas, on the one hand, and sustaining a friendly line of exchange with foreign channels, on the other, mapped out a distinct local-global nexus for science building.

Finally, Hugh Shapiro's essay turns the rationality of science back on itself by probing the limits of neurosis, insanity, state power, possession, self-mutilation, and cultural taboos against mental illnesses in 1930s Beijing. Using the case of a thirty-year-old woman who communicated her symptoms almost exclusively in the performative style of traditional Chinese opera, Shapiro provides a rare glimpse into the explicit and implicit social and cultural contexts in which mental illness was evaluated, authenticated, and co-opted by perceived norms as well as by the voices of modern medical science. By demonstrating the extent to which mental illness pushed the bounds of both traditional and biomedical prescriptions of psychic expression, Shapiro blurs the distinction between rational science and accepted cultural norms for rationality.

In terms of overall findings, the essays in this volume break much new ground, going well beyond the earlier Needham-inspired limits on the history of modern science in China that Elman's paper describes. Iwo Amelung, for example, contends that the emergence of the historiography of Chinese science and technology needs to be understood as part of a number of larger developments, namely the reception and appropriation of Western knowledge since the late imperial era. Similarly, Joachim Kurtz reconstructs the key elements of Liu Shipei's reinvention of ancient China's intellectual history in Euro-American terms and analyzes the violent conceptual transformations that foreshadowed the many reformulations of the histories of Chinese science and thought published throughout the Republican period. Shen Guowei suggests that Yan Fu made only a cursory effort to use the Japanese concept of "science" (*kexue*) as a core concept of the modern age. Yan himself thought deeply about the full meaning of science, the necessity of adhering to the scientific method, and the nature of scientific terminology. For Yan, however, the fundamental chasm between Chinese and Western scholarship, and the attitudes of traditional society toward "science" and "art," remained in place.

In light of the revolution in communications at the turn of the twentieth century, Jing Tsu shows how machine translation eliminated the possible distortions caused by multiple approximate and overlapping semantic meanings between languages. It also created a different translation, in which language itself was submitted to another level of permeability. The ideograph/word was divided into smaller units of denomination, thereby revamping the basic assumption of scale in a word-to-word translation. For Thomas Mullaney the Morse code connected the Qing Empire to the rapidly expanding international telegraphic network. The product of this convergence was the Chinese telegraph code of 1871, in which approximately seven thousand common-usage Chinese characters were assigned a series of nonrepeating, four-digit numerical codes.

Eugenia Lean addresses the commercializing forces inherent during the early Republic whereby long-standing practices of collecting and compiling knowledge were combined with new forms of industrial and commercial pursuits to popularize and legitimate the incorporation of modern science into daily life and industrial endeavors. Fa-ti Fan shows how science, especially the scientific experiment, gained and maintained its epistemological and cultural authority in Republican China. Scientists formed a group identity and guarded their institutional and intellectual status whereby science became part of a public discourse. Likewise, Danian Hu shows how William Band's mentoring of a large group of top native physicists (including several distinguished women) fostered the study of theoretical physics while making contributions to the indigenous Rural Reconstruction Movement. The transnational contributions made by individual Western scientists grew out of the special role played by mission colleges.

Grace Shen traces how local scientists deployed language, sociability, and publications first to enroll outsiders in their cosmopolitan vision of Chinese geology and then to carve out a "private" sphere for domestic debate. Both aspects reflected the Chinese understandings of the international scientific economy and their shifting place within it. Hugh Shapiro points to the ambiguous interplay between healing, theater, possession, gods, actors, and patients in the formation of a nascent neuropsychiatry in urban China. The hybridity of competing vocabularies for understanding individual experiences and for analyzing a person's interiority using the clinical terms of psychiatry empowered a sacrosanct-curative-theatrical zone where performance, religious practice, healing, escape, and anxieties about madness comingled.

Offering new frameworks, orientations, issues, and research proposals, these eleven essays draw from the history of science in China, as it has been known and not known, in order to launch the particular intersection between early twentieth-century science and the history of Chinese science as a new starting point. The fate of science in modern China is, after all, far from a uniform or conclusive tale. How it will intersect with and challenge scholarship on modern China is a question we leave open for others to explore.

Bibliography

Elman, Benjamin 2006. *A Cultural History of Modern Science in China.* Cambridge, MA: Harvard University Press.

Elvin, Mark 1973. *The Pattern of the Chinese Past.* Stanford, CA: Stanford University Press.

Frank, Andre Gunder 1998. *ReOrient: Global Economy in the Asian Age.* Berkeley: University of California Press.

Hart, Roger 1999. "Beyond Science and Civilisation: A Post-Needham Critique," *East Asian Science, Technology, and Medicine* 16: 88–114.

Jones, E. L. 1987. *The European Miracle: Environments, Economies, and Geopolitics in the History of Europe and Asia.* 2nd ed. Cambridge: Cambridge University Press.

Kuhn, Thomas 2000. "Commensurability, Comparability, Communicability," in Thomas Kuhn, *The Road since Structure.* Chicago: University of Chicago Press.

Kurtz, Joachim 2011. *The Discovery of Chinese Logic.* Leiden: E. J. Brill.

Lloyd, G. E. R. 1990. *Demystifying Mentalities.* Cambridge: Cambridge University Press.

Lloyd, G. E. R., and Nathan Sivin 2002. *The Way and the Word: Science and Medicine in Early China and Greece*. New Haven, CT: Yale University Press.

Mote, Frederick 1989. *Intellectual Foundations of China*. 2nd ed. New York: Knopf.

Needham, Joseph 1959. *Science and Civilisation in China*. Vol. 3, *Mathematics and the Sciences of the Heavens and the Earth*. Cambridge: Cambridge University Press.

_____ 1969. *The Grand Titration: Science and Society in East and West*. Toronto: University of Toronto Press.

Pomeranz, Kenneth 2000. *The Great Divergence: China, Europe, and the Making of the Modern World Economy*. Princeton, NJ: Princeton University Press.

Schmalzer, Sigrid 2007. "On the Appropriate Use of Rose-Colored Glasses: Reflections on Science in Socialist China," *Isis* 98, 3: 571–583.

Shen, Grace 2007. "Murky Waters: Thoughts on Desire, Utility, and the 'Sea of Modern Science,'" *Isis* 98, 3: 584–596.

Sivin, Nathan 1978. "Imperial China: Has Its Present Past a Future?," *Harvard Journal of Asiatic Studies* 38: 449–480.

_____ 1995a. "State, Cosmos, and Body in the First Three Centuries BC," *Harvard Journal of Asiatic Studies* 55, 1 (June): 5–37.

_____ 1995b. "Taoism and Science." http://ccat.sas.upenn.edu/~nsivin/taos.pdf.

_____ 2005. "Why the Scientific Revolution Did Not Take Place in China—or Didn't It?" http://ccat.sas.upenn.edu/~nsivin/scirev.pdf.

Tsu, Jing 2005. *Failure, Nationalism, and Literature: The Making of Modern Chinese Identity 1895–1937*. Stanford, CA: Stanford University Press.

_____ 2010. *Sound and Script in Chinese Diaspora*. Cambridge, MA: Harvard University Press.

Wang Zuoyue 2007. "Science and the State in Modern China," *Isis* 98, 3: 558–570.

Toward a History of Modern Science in Republican China

Benjamin A. Elman

Abstract

Despite the recent increase in the number of teachers of the history of science and medicine, historians of "Chinese science" until recently have spent much of their time researching issues in premodern natural studies and, usually, trying to explain why modern science, technology, and medicine arrived so late in China. The "Needham Question"—Why did a divided Europe, and not imperial China, develop modern science first?—until recently remained preeminent. This question was paralleled by scholarly efforts in other fields to explain why China did not develop capitalism or democracy before Europe. We are entering a new era that explores modern science in contemporary China in more active, rather than simply receptive, terms. Increasingly, we are able to address modern science in China from a comparative point of view and include it in the story of global science. The earlier lack of studies of modern science in China was not due to the burden of historiography alone, however. Historians used the potential sources for modern Chinese science, when available, to focus on individual Chinese scientists or representative scientific institutions in the Republic of China (1911–) and the People's Republic of China (1949–), rather than exploring the larger problems of how science has been practiced in the modern context of nationalism, state-building, and socialism.

The Historiography of Modern Science in China

Most Western accounts have described how British imperial expansion during the eighteenth century collided with a Sinocentric Qing state unsympathetic to scientific knowledge. But this view should be amended. We should not read the Qianlong emperor's (r. 1736–1795) famous 1793 letter to George III gainsaying Western gadgets as the statement of a Manchu dynasty out of touch with reality. The emperor did not categorically reject Western technology. His court simply contested the originality of the astronomical instruments—a replica of the solar system, for example—that the Macartney mission brought to China. Qianlong, on the other hand, showed great interest in the model warship equipped with cannon that Macartney presented. Unaware of the industrial revolution to come in Europe, the emperor had widely employed European Jesuits as astronomers, architects, and cannon-makers, who advised him against accepting the English demands.

Once the Qing calendar functioned properly with Jesuit help, the emperor was not inclined to think Macartney's planetarium so fabulous. Later emperors who faced irresistible English military firepower in the aftermath of the Opium War (1839–1842) were dealing with a different set of technological circumstances. Chinese had incorporated algebra and geometry and made natural studies a part of classical studies in the eighteenth century, but the continued development of science and technology in Europe required Chinese to depend on the modern sciences introduced by Protestant missionaries in the new historical conditions of the post-Napoleonic age.

In light of the important place mathematics and astronomy occupied in Qing dynasty evidential studies (*kaozheng xue* 考證學), it is remarkable how quickly—not overnight to be sure—the Chinese people adapted to the needs of science and technology, again under the umbrella of the "investigation of things and extension of knowledge" (*gewu zhizhi* 格物致知). With the introduction of the differential and integral calculus in the mid-nineteenth century, for which the Chinese could not find an ancient, native precedent, Li Shanlan (1811–1882) and other Chinese mathematicians admitted that although the "four unknowns" notation (*siyuan shu* 四元術) was perhaps superior to Jesuit algebra, the Chinese had never developed anything resembling the calculus. Moreover, after the Opium War, the most influential Chinese mathematicians no longer were devoted exclusively to the revival of ancient Chinese mathematics. They merged European and Chinese mathematics into a new synthesis.

Even after the Opium War, missionary inroads in China remained limited. Protestant missions principally funded the new translations, newspapers, and schools that introduced modern science in the 1850s. The massive Taiping conflagration from 1850 to 1864 was led by anti-Manchu and anti-gentry discontents who took advantage of a demographic catastrophe when the total population reached about 450 million. It left a swath of destruction in South China that significantly changed the tenor of things, once the peasant rebellion was quelled using new Western armaments. From the 1860s on, the impetus for science and technology shifted from the Protestants to the reforming Qing state and its new Western-oriented policies and institutions.[1]

Dr. Benjamin Hobson (1816–1873) was among the key pioneers in the late 1840s and early 1850s. After moving to Hong Kong, Hobson, an English medical missionary, pioneered a series of medical and science translations coauthored with Chinese for his premedical classes in Guangzhou. Hobson prepared the *Treatise of Natural Philosophy* (*Bowu xinbian* 博物新編, 1851), associating

---

1 Biggerstaff 1961.

science with the Chinese tradition of "broad learning about things" (*bowu* 博物). The missionary community preferred calling science "the investigation of things and extension of knowledge" in their scientific translations for the Inkstone Press (Mohai shuguan 墨海書館).[2]

Research on Western Anatomy and Traditional Chinese Medicine

Hobson also produced a series of other works to educate his students. His *Summary of Astronomy* (1849) and the *Treatise on Physiology* (1851) were also designed for his medical students. The *Treatise on Physiology* presented modern anatomy. The missionaries believed that medicine was at a low ebb in China. Yet while Hobson was translating Western medical works into classical Chinese, the heat factor tradition, which dealt with fever-inducing illnesses and had emerged in the seventeenth century, was growing increasingly prominent in South China, where the missionaries were often assigned. Regional traditions dealing with southern infectious diseases and northern cold damage disorders continued to evolve in the nineteenth century. In the process, heat factor illnesses became a new category. The mid-nineteenth century emergence of a medical tradition stressing heat factor therapies coincided with the introduction of Western medicine in the treaty ports, particularly Guangzhou, Ningbo, and Shanghai.[3]

Chinese accepted anatomy when they could assimilate it within their focus on conduits of *qi* 氣 in the body. Moreover, Song physicians had mapped acupuncture and moxibustion therapy onto the skeletal body, and the internal organs had also been drawn and modeled. Chinese medical efforts to treat southern infectious illnesses paralleled the gradual emergence of tropical medicine during the late nineteenth century when the British Empire increasingly populated the tropics with its own physicians. These networks of doctors and their medical reporting system from Africa to India and South China in turn addressed interregional infectious diseases such as malaria. Colonial physicians cumulatively sent back information about epidemics and infectious illnesses to London, the metropole of global medicine.[4]

Chinese increasingly acknowledged the need to synthesize Chinese and Western medicine. They linked cold damage disorders to the specific illness that Westerner physicians identified as typhoid fever. Germ theory was added to discussions of warm versus cold factor illnesses. Chinese physicians began to explain the wasting of the body's natural vitality in terms of tuberculosis

---

2 Wright 2000.
3 Hanson 2001.
4 Anderson 1996.

(= wasting disease) and gonorrhea (= depletion illness). Western public health procedures also began to be enacted in the coastal treaty ports.[5]

Unlike Ming-Qing astronomy, which was completely reworked in the seventeenth and eighteenth centuries by the introduction of Western techniques, traditional Chinese medicine did not face a serious challenge from Europe until the middle of the nineteenth century. Except for smallpox inoculations, quinine therapy for malaria, and a number of herbal medicines unknown in China, the European medicine brought by Jesuit or Protestant missionary physicians was not superior in therapeutic results until a relatively safe procedure for surgery combining anesthesia and asepsis was developed at the turn of the twentieth century.[6]

The translations Hobson prepared led some literati to question traditional Chinese medicine in the nineteenth century, however. Xu Shou (1818–1884), John Fryer's (1839–1928) collaborator, was one of the first scholars to complain that while literati had integrated Western and Chinese mathematics, they paid little attention to the strengths of Western medicine. Xu called for a similar synthesis of Western experimental procedures, linking chemistry and Chinese strengths in *materia medica*. Outside the missionary hospitals and clinics in the treaty ports, Hobson's translations were not popular due to Chinese distaste for surgery. Hobson's works introduced invasive surgery for childbirth drawn from the anatomical sciences that had evolved in Europe since the sixteenth century. Although anatomy could pinpoint childbirth dysfunctions in women in the uterus, such procedures were dangerous even by Western standards until modern surgery integrated sterilization techniques with anesthetization procedures to make local interventions safer.[7]

From Western Medicine to Modern Science in China

Hobson's work represented the first sustained introduction of the modern European sciences and medicine in the first half of the nineteenth century. His 1849 digest of modern astronomy, for instance, presented the Copernican solar system in terms of Newtonian gravitation and pointed to God as the author of the works of creation. Thereafter, Newtonian celestial mechanics based on gravitational pull was increasingly presented in Protestant accounts of modern science. A natural theology also informed Hobson's *Treatise of Natural Philosophy*, which was the first work to introduce modern Western chemistry. Along with the fifty-six elements, the textbook presented God as ultimate cre-

---

5   Rogaski 2004.

6   Chang 1996.

7   Wu 1998.

ator behind all the myriad changes in things. Although later changed, Hobson's chemical terminology the names of gases in Chinese and outlined the chemical makeup of the world; Hobson's scheme supplanted the four-elements theory of the Jesuits and challenged the Chinese notion of the five phases.[8]

By including sections on physics, chemistry, astronomy, geography, and zoology for his Chinese medical students, Hobson unexpectedly attracted the interest of literati unsuccessful in the civil examinations. Fryer described a group of Chinese literati investigators who earlier had met to study Jesuit works on mathematics and astronomy. They used Hobson's *Treatise* to catch up with findings since the Jesuits. This group, which included Xu Shou and Hua Hengfang (1833–1902), also carried out experiments. After fleeing the Taiping rebels in the early 1860s, they were invited by the leader of the victorious Qing armies, Zeng Guofan (1811–1872), to work in the newly established Anqing Arsenal. Hua began translation projects with Alexander Wylie (1815–1887) and Joseph Edkins (1823–1905), while Xu worked on constructing a steamboat based on Hobson's diagrams.[9]

The Role of Treaty Ports and Modern Science in Shanghai

Among treaty ports, Shanghai by 1860 was the main center of foreign trade, international business, and missionary activity. The London Missionary Society Press in Shanghai became the most influential publisher of Western learning after 1850. It published translations from a distinguished missionary community. These missionaries worked with outstanding Chinese scholars who moved to Shanghai after failing to gain a place via the imperial civil examinations. In the 1850s, Protestant journals that published in Chinese, such as the *Shanghae Serial* 六合叢談 (*Liuhe congtan*) at Inkstone Press, introduced new fields in the Western sciences. Beginning with the *Shanghae Serial*, the literati notion of investigating things (*gewu* 格物) moved from encompassing classical learning and natural studies to designating a specific domain of knowledge within the natural sciences again called "investigating things and extending knowledge" (*gezhi* 格致). Through the Protestant translation work of Wylie, Li Shanlan, and others for the *Shanghae Serial*, the investigation of things increasingly demarcated the new Western natural sciences. A scientist was now called "someone who investigated things and extended knowledge."

A talented missionary printer and translator, Alexander Wylie produced the *Shanghae Serial* monthly in 1857 and 1858, before it suddenly stopped. Wylie made some remarkable inquiries into Chinese science and mathematics with

---

8 Andrews 1994.
9 Bennett 1967.

the help of Li Shanlan. Through this interaction, Li successfully completed the transition from the traditional craft of algebra to the modern calculus. Wylie's and Li's 1859 translation of John Herschel's (1792–1871) *The Outline of Astronomy* (1851) grew out of their early collaboration. The astronomy of the Cambridge-educated Herschel moved away from that of the late eighteenth-century Newtonians, who had stressed geometrical demonstrations over algebraic processes.

Wylie and Li stressed modern algebra as a mathematical language for the natural sciences. They related it to traditional Chinese mathematics by substituting it for procedures solving equations with a single unknown or four unknowns. Wylie emphasized that Chinese "quadrilateral algebra" (*siyuan* 四元, "four unknowns," procedures) was superior to the Jesuits' elementary algebra and acknowledged that Western scholars had not studied the two traditional methods. Nevertheless, Li and Wylie also refuted the theory that the science of algebra had originated in China.

In the 1860s, the Qing government employed many missionaries as translators to work with the Chinese in the Qing dynasty's Jiangnan Arsenal in Shanghai. Like the Jesuits, who had changed their focus from proselytizing among Chinese, the Protestants were committed to the gospel of science in China because they also thought its success in government would redound to Christianity. Many Chinese literati saw in Western learning and the modern sciences an alternative route to fame and fortune. Literati whom the Protestants had trained in the sciences began to establish links with the ruling dynasty by serving as advisors and translators after the devastations of the Taiping Rebellion. Many Chinese who had worked for Inkstone Press in Shanghai moved from the Protestant missions to the dynasty's arsenals and new schools. In this milieu, some Chinese grasped modern evolution long before the 1890s, and others became pioneering translators of Western medical works.

During this era, conservative Manchu officials, such as Woren (d. 1871), and traditionalist literati attempted to derail foreign learning in official schools such as the Beijing School of Foreign Languages. Literati who feared that Western learning would subvert state orthodoxy produced several major anti-Christian tracts in the nineteenth century. Reformers neutralized them in the 1870s, however, and they were finally routed in the aftermath of the Sino-Japanese War.

The dynasty's pursuit of Western technology began in earnest when Yung Wing (Rong Hong 1828–1912), a Cantonese who graduated from Yale University in 1854, represented Zeng Guofan in buying all-purpose machinery in Europe in 1864. In 1863, Yung had advised Zeng to launch an ironworks in Shanghai. The Nanjing Arsenal quickly produced fuses, shells, friction tubes for firing cannon, and small cannon for the Anhui Army. New machinery was added in 1867–1868, along with some British mechanists. By 1869, Nanjing was producing rockets and trying to forge larger guns.

In 1866, the Hunanese general Zuo Zongtang (1812–1885) suggested creating a modern shipyard in Fuzhou, Fujian, to build and operate Western-style warships. The regents of the Tongzhi emperor (r. 1862–1874) quickly authorized the proposal. When Zuo was sent on military campaigns to Chinese Turkestan (Xinjiang) to put down rebellions, Shen Baozhen (1820–1879) became the director general of the Fuzhou Shipyard in 1867. Depending on French knowhow, Fuzhou quickly became the largest and most modern of all the Chinese military defense industries established in the 1860s and 1870s. It also had the largest gathering of foreign employees. Until the Sino-French War of 1884–1885, Fuzhou remained a major center of French interests.[10]

Subsequently, in 1866–1867, the court approved a proposal to add a Department of Mathematics and Astronomy to the Beijing School of Foreign Languages. The goal was to teach students about modern science through instruction in chemistry, physics, and mechanics. The addition of mathematics and astronomy in particular was unsuccessfully opposed by Woren while he was a Hanlin academician and imperial tutor. Woren's failure encouraged Chinese literati to accept appointments in the Beijing School. A special civil examination in mathematics, however, was successfully opposed in the 1870s, but Li's mathematics examinations at the School of Foreign Languages were influential.

Industrialization in the Jiangnan Arsenal and Fuzhou Shipyard

The Qing government established the Jiangnan Machine Manufacturing General Bureau, usually called the Jiangnan Arsenal, in Shanghai in 1865 to administer the industrial works and educational offices. At its crest, it contained four institutions: (1) the Translation Department; (2) the Foreign Language School; (3) the school for training skilled workmen; and (4) the machine shop. In addition, the Jiangnan Arsenal had thirteen branch factories. By 1892, it occupied seventy-three acres of land, with 1,974 workshops and a total of 2,982 workers. The arsenal acquired 1,037 sets of machines and produced forty-seven kinds of machinery under the watch of foreign technicians who supervised production. From 1868 to 1876, shipbuilding in the Jiangnan Arsenal was highly productive: eleven ships were built in eight years. Ten were warships. Five of these had wooden hulls; the other five, iron hulls. All parts of each ship, including the engine, were built at the arsenal. When compared to

---

10   Pong 1994.

the warships built following French models at the leading Japanese dockyard in Yokosuka in the 1870s, the higher level of shipbuilding technology at the Jiangnan Arsenal was attained earlier.[11]

The second major industrial site for shipbuilding and training in engineering and technology was the Fuzhou Shipyard. When Zuo Zongtang submitted his 1866 memorial to establish a complete shipyard at Fuzhou, he expected that after five years he could eliminate the need for foreign experts. In return for their support, neighboring provinces would receive naval protection from the Southern Fleet based at Fuzhou. Zuo and his successor Shen Baozhen relied mainly on French expertise for Fuzhou. Once the Qing established the shipyard, however, the Fujian maritime customs left the venture in a perpetual financial bind. At its peak the shipyard employed 3,000 workers. When construction was completed, the force dropped to 1,900, with 600 in the dockyard   800 in workshops, and 500 manual laborers. The shipyard had more than forty-five buildings on 118 acres set aside for administrative, educational, and production purposes.

In terms of scale, the Fuzhou Shipyard was the leading industrial venture in late Qing China. For organizational efficiency, a modern tramway with turntables at important workshops and intersections served the whole plant. Nineteen ships, with 80- to 250-horsepower engines, were planned. Of these, thirteen would be transport ships with 150-horsepower engines. Sixteen ships were finished from 1869 to 1875. Of these, ten transports with 100-horsepower engines and one corvette as a showpiece with a 250-horsepower engine were realized. After 1874, the shipyard sent graduates to Europe, especially England and France, for advanced training.[12]

Why have we undervalued such pre-1900 industrial achievements? The answer lies principally in the fact that, during the Sino-Japanese War from 1894 to 1895, the Japanese army and navy decisively defeated the armed forces of the Manchu Qing dynasty. Since then, Chinese and Japanese patriots and scholars have assumed that Meiji Japan (1868–1911) was vastly superior to Qing China in modern science and technology prior to 1894. Actually, prior to the war, many contemporary observers thought that the Qing army and navy were superior, even if only in sheer numbers. After 1895, each side rewrote their histories to validate triumphant Japan or lament the defeated Qing. For the Chinese and the Manchus, the Sino-Japanese War turned the Qing era of Self-Strengthening reforms from 1865 to 1895 into an alleged scientific and technological catastrophe. Thousands of Chinese students who studied modern

---

11   Meng 1999.
12   Pong 1994.

science and medicine in Meiji Japan quickly assimilated the Japanese terminology for the modern sciences under the Meiji neologism for "science" as "organized fields of learning" (*kexue* 科學; Japanese, *kagaku*).[13]

The decisive Qing defeat in the Sino-Japanese War energized public criticism of the dynasty's allegedly inadequate policies. The unexpected naval disaster and the way it was presented as Japan's technological victory shocked many literati and officials. A greater respect for Western studies emerged in China. Technology alone had not been the key determinant, however. Japan, for example, could not match China's two major battleships. But Japan proved superior in naval leadership, ship maneuverability, and the availability of explosive shells.

Enhanced by the capture of twelve Chinese warships and seven torpedo boats during hostilities, the Japanese navy added significant tonnage to the Meiji fleet. Moreover, Japanese industrialization accelerated after the Qing dynasty was forced to pay a considerable indemnity to the Meiji regime. The Japanese government used the 1895 Qing indemnity of 200 million taels of silver and later Boxer indemnities as a windfall to bankroll a massive rearmament program to address the Russian expansion in northeast China. Korea and Taiwan were ceded to Japan and became productive colonies. The indemnities meant that the money could not be used to augment the Qing dynasty's reconstruction projects. Qing reparations amounted to 450 million silver taels plus interest. This sum was never fully paid, but an estimated 669 million taels were transferred from China to the foreign countries involved. The Jiangnan Arsenal and Fuzhou Shipyard never fully recovered from the indemnities that they had to pay for over two decades. Before the war the Qing government had been unable to integrate development so that innovative institutions reinforced each other, and so the added weight of Japanese and European imperialism after 1895 tipped the scales against Qing reforms initiated in 1865.

Although the late nineteenth-century naval battles that China lost are still used to demonstrate the failure of the Self-Strengthening reforms initiated after the Taiping Rebellion, the rise of the new arsenals, shipyards, technical schools, and translation bureaus should be reconsidered in light of the increased training in military technology and education in Western science available to Chinese after 1865. If we repopulate this impressive list of factories with the human lives and literati careers they contained, then we can trace more clearly the post-Taiping successors to the native mathematical astronomers of the eighteenth century. A new group of artisans, technicians, and engineers emerged between 1865 and 1895 whose expertise no longer depended on

---

13   Elman 2003.

the fields of classical learning monopolized by the customary scholar-officials. Increasingly, they were no longer subsidiary to the dynastic orthodoxy or its old-fashioned representatives.

We should not underestimate the significance of the schools and factories launched within the Jiangnan Arsenal in Shanghai and the Fuzhou Shipyard. The arsenals, machine shops, and shipyards provided the institutional venues for an education in science and engineering. They also trained the architects, engineers, and technicians who later provided the manpower for China's increasing number of public and private industries in the early twentieth century.[14]

Past accounts of China's failures in science and its dynastic losses on modern military battlefields are instructive, but their overblown rhetoric about the reasons for that failure has overshadowed acknowledgment of the more contingent conditions that placed China at the mercy of Europe and Meiji Japan beginning in the 1890s. Above I have addressed the quieter story of long-standing Chinese interests in the natural world, medicine, the arts and crafts, and commerce under the umbrella of "investigating and extending knowledge" (*gezhi*). These endeavors set the stage for the interaction with European science, technology, and medicine under the influence of Japanese *kagaku*.

The Influence of Meiji Japan on Modern Science in China

In the late nineteenth century, an increasing familiarity with Western learning exposed the Chinese to the limits of traditional categories for scientific terminology. Increasingly, the claim that Western learning derived from ancient China was unacceptable. In the revival of traditional positions after the Sino-Japanese War, which represented the third stage of the Chinese-origins argument, younger literati perceived a latent conservatism that obstructed the introduction of modern science and technology rather than facilitated it. Hence, those students who studied abroad after 1895 began to question the use of investigating things and extending knowledge (*gezhi*) as a traditional trope of learning to accommodate modern science.

Instead, to make a complete break with the Chinese past, many turned to Japanese terminology for the modern sciences. The Japanese neologism *kagaku* 科學 (pronounced *kexue* in Chinese, lit., "knowledge classified by field"), for example, was perceived as a less-loaded term for science than "investigating things," which had so many semantic links to classical learning and the

---

14   Elman 2006, chaps. 6-7.

Song Learning conventions (often called Neo-Confucianism in the twentieth century) still in place as the curriculum for the civil examinations until 1904. By 1903, state and private schools increasingly borrowed from Japanese translations to enunciate the modern classifications of the social sciences (*shehui kexue* 社會科學), natural sciences (*ziran kexue* 自然科學), and applied sciences (*yingyong kexue* 應用科學).

Before 1894, Japan had imported many European books on science from Qing China, particularly after 1720, when the shogun Yoshimune relaxed the Tokugawa prohibition of all books related to Christianity. Many had been translated during the Ming and Qing after the Japanese expelled the Jesuits for their meddling in the late sixteenth-century civil wars there. Ricci's *mappa mundi*, Chinese translations of Euclid's geometry, and Tychonic astronomy, for example, made their way to Tokugawa Japan.[15] The Japanese also avidly imported eighteenth-century Chinese terminology for Sino-Western mathematics. Physics, chemistry, and botany books, imported from Europe via the Dutch trading enclave in Nagasaki harbor in the early nineteenth century, were translated into Japanese from Dutch.[16]

In addition, the translations on science prepared under the auspices of Protestant missionaries such as Daniel Macgowen (1814–1893) and Benjamin Hobson in the treaty ports were immediately coveted by the Meiji government. Prominent translations into Chinese of works dealing with symbolic algebra, calculus, Newtonian mechanics, and modern astronomy quickly led to Japanese editions and Japanese translations of these works. Macgowen's 1851 *Philosophical Almanac* and Hobson's 1855 *Treatise of Natural Philosophy* came out in Japan in the late 1850s and early 1860s. Four other of Hobson's medical works from 1851 to 1858 came out in Japan between 1858 and 1864.[17]

In early Meiji times, many Japanese scholars still preferred Chinese scientific terms over translations derived from Dutch Learning scholars. The Chinese name for chemistry (*huaxue* 化學), for example, replaced the term *chemie* (*semi* セミ in Japanese) derived from the Dutch. Similarly, the impact of Jiangnan Arsenal publications can be seen in the choice of Chinese terminology for metallurgy (*jinshi xue* 金石學, which also meant "study of bronze and stone inscriptions") in Japanese publications. The Chinese characters were later changed in Japan and reintroduced to China using a new term for "mining" (*kuangwu xue* 礦物學).

---

15   Horiuchi 1994, 119–155. See also Jiang 2003.
16   Kobayashi 2002.
17   Wang Yangzong 2000.

Japan's Iwakura mission visited Shanghai in September 1873 at the end of their journey to Europe and the United States and took a tour of the Jiangnan Arsenal on September 4. They described the shipyard, foundry, school, and translation bureau there in very positive terms. The mission noted how the shipyard had been operated by British managers initially. The latter were aided by Chinese who had trained abroad. The account added that "now the entire management of the yard is in the hands of Chinese" and concluded: "This one yard would be capable of carrying out any kind of work, from ship repair to ship construction."[18]

When the diplomat Yanagihara Sakimitsu (1850–1894) visited China, he purchased many of the Chinese scientific translations. On his third visit, in 1872, for instance, he bought twelve titles on science and technology in thirty-one volumes from the Jiangnan Arsenal. These included works on chemistry, ship technology, geography, traditional mathematics, mining, and Chinese trigonometry (*gougu* 勾股). The Japanese government continued to buy arsenal books until 1877. In 1874, Yanagihara received twenty-one newly translated books from China. Despite the influence of Dutch Learning and of translations from China, and even though the Japanese began teaching modern Western science on a large scale in the 1870s, the Chinese did not borrow many scientific terms from Japan before the Sino-Japanese War.

Unlike the Chinese translations that were readily transmitted to and disseminated in Japan, Tokugawa authorities kept translations of Dutch Learning secret. While much has been made of the contributions of Dutch Learning to Japanese science during the Tokugawa period, the Yokosuka Dockyard remained dependent on French engineering advisors until the 1880s and British technical aid in the 1890s. There is no evidence that Dutch Learning per se enhanced the Yokosuka enterprise or determined the course of Meiji science and technology. Moreover, the impact of Dutch Learning, while important among samurai elites in the late eighteenth and early nineteenth centuries, was not sufficient to launch in Tokugawa Japan a technological revolution based on Newtonian mechanics and French analytical mathematics.

Indeed, the concrete advantages that Dutch Learning provided in the rise of modern, industrial science during the Tokugawa-Meiji transition remain undocumented. Japan's overwhelming triumph in the Sino-Japanese War created an environment in which most accounts since 1895 have simply assumed that Dutch Learning gave Tokugawa Japan a scientific head start over the Qing dynasty.[19]

---

18   Kunitake 2002, 352.

19   Wang Yangzong 2000, 142. Cf. Wright 1998, 671; and Masini 1993, 91–92.

Japanese Science in China after 1895

From 1896 to 1910, Chinese translated science books that Japanese no longer worked with foreigners to translate. By 1905, the new Qing Ministry of Education was staunchly in favor of science education and textbooks based on the Japanese scientific system. Instead of the West being represented by Protestant missionaries such as William Martin (1827–1916) and John Fryer, Japan now mediated the West for Chinese literati and officials.[20] After the Sino-Japanese War, reformers were encouraged to study in Japan. Kang Youwei (1858–1927) promoted Meiji Japan scholarship in his *Annotated Bibliography of Japanese Books* (*Riben shumu zhi* 日本書目志) and in his reform memorials to the Guangxu emperor (r. 1875–1908). He recommended 339 works in medicine and 380 works in the sciences (*lixue* 理學), which now replaced as reference sources the formerly popular prize essays from the 1894 Shanghai Polytechnic essay competition. The Guangxu emperor's edict of 1898 encouraged study in Japan.[21]

As a publicist while in exile in Japan, Liang Qichao translated Japanese materials into Chinese at a fast clip. In addition to his antiquarian interests, Luo Zhenyu (1866–1940), for example, published the *Agricultural Journal* (*Nongxue bao* 農學報) from 1897 to 1906 in 315 issues. The articles were mainly drawn from Japanese sources on science and technology. Luo also compiled the *Collectanea of Agricultural Studies* (*Nongxue congshu* 農學叢書) in 88 works, with 48 based on Japanese books. Du Yaquan edited journals in 1900 and 1901 that translated science materials from Japanese journals. These were the first science journals edited solely by a Chinese. The massive translation by Fan Diji in Shanghai of a Japanese encyclopedia took several years. When it appeared in 1904, the encyclopedia contained over 100 works, with 28 in the sciences and 19 in applied science.

Post-Boxer educational reforms of 1902–1904 were also crucial in the transformation of education in favor of Japanese-style science and technology. The last bastion of modern science as Chinese science (*gezhi*) remained the civil examinations, where the Chinese-origins approach to Western learning remained obligatory. After the examination system was abolished in 1904, Japanese science texts finally became models for Chinese education at all levels of schooling. In 1886–1901, for instance, Japan officially approved eleven different texts on physics. Eight of those, which were produced after 1897, were translated for Chinese editions. In 1902–1911, twenty-two different physics texts were approved in Japan, and seven were translated into Chinese.

---

20   Wang Yangzong 2000, 139–144, and Masini 1993, 104–108.
21   Zhao 1897. See also Wang Yangzong 2000, 144-145; and Reynolds 1993, 48, 58–61.

Similarly, in chemistry, from 1902 to 1911, seventy-one Japanese texts were translated into Chinese. Most were produced for middle schools and teachers colleges. Twelve middle school chemistry texts were produced in Japan between 1886 and 1901. Of these, six were translated into Chinese. Eighteen Japanese middle school chemistry texts were produced between 1902 and 1911. Five were translated into Chinese. Japanese scientists were also invited to lecture in China. Chinese also translated more technical physics and chemistry works from the Japanese. Iimori Teizō's (1851–1916) edited volume *Physics* (*Wuli xue* 物理學; *Butsurigaku*) was translated into Chinese at the Jiangnan Arsenal from 1900 to 1903. The translators were aided by the Japanese educator Fujita Toyohachi (1869/70–1929). Iimori's influence on Chinese physics grew out of this project.[22]

Chinese also compiled updated Sino-Japanese dictionaries such as the 1903 *New Progress toward Elegance* (*Xin Erya* 新爾雅), which modernized ancient Chinese lexicons. By 1907, when Yan Fu was in charge of the Qing Ministry of Education's committee for science textbooks, he approved the use of Japanese scientific terms. We should not underrate the historical importance of Japanese translations for the development of modern science in China. Japanese translations were much more widely available in China than those produced earlier by the Jiangnan Arsenal had been. In addition, the new Japanese science textbooks contained newer content than the 1880s arsenal and missionary translations, which were already outdated by European standards in the 1890s. The introduction of post-1900 science via Japan, which included new developments in chemistry and physics, went well beyond what Fryer et al. had provided to the emerging Chinese scientific community.[23]

Chinese presses also published in greater numbers the translations of Japanese texts, which were easier to read because only Chinese compiled them. Moreover, the quality of the translations from works by Japanese scientists was better than that of the earlier science primers since Chinese translators themselves could understand the Japanese originals. In addition, the Japanese texts were available to a new and wider audience of students in the new public schools and teachers colleges that the Qing government established after 1905 as part of its education reforms. The Imperial University in Beijing also invited Japanese professors to join its faculty.[24]

Finally, to make the new translations more easily understood than standard classical translations, Chinese translators helped produce a new literary form

---

22   Wang Yangzong 2000, 146–147. See also Cong 2007.
23   Masini 1993, 145–151.
24   Weston 2004, 50–52.

for presentation of the sciences, which contributed to the rise of the vernacular for modern Chinese scholarly and public discourse. Among urbanites, especially in Beijing and Shanghai, the first decade of the twentieth century provided the basic education in modern science via Japanese textbooks for the generation that matured during the New Culture Movement of 1915 and the May Fourth era after 1919.[25]

The Delayed Emergence of Physics as a Technical Field in China

When we compare the development of modern physics in Meiji Japan and Qing China, we find that scholars in both countries had started to master Western studies in the early and mid-nineteenth century. The Translation Department at the Jiangnan Arsenal had produced Chinese books on physics beginning in the 1850s in China, and the Dutch Translation Bureau in Tokugawa Japan had provided such works in Japanese beginning in 1811. Although the introduction of Dutch Learning in the seventeenth and eighteenth centuries enabled an earlier start in Japan, the materials on physics in the Protestant translations produced in China after 1850—quickly transmitted to Japan—made those earlier studies out of date. Moreover, the Primer Series produced in the 1870s and early 1880s in China remained superior overall to its Meiji counterparts until the 1890s.[26]

Despite the range of science translations in Qing China through the 1880s, physics textbooks were not available in China until they were first published in Japan. This lack had much to do with the way the Protestant missionaries such as Martin and Fryer had introduced the physical sciences to literati audiences after 1860. Rather than a unified field of physics—or natural philosophy, as it was often called by Euro-American specialists until the 1860s—missionary translators first introduced the disaggregated branches of physics. Accordingly, mechanics (*lixue* 力學 or *zhongxue* 重學), optics (*guangxue* 光學), acoustics (*shengxue* 聲學), electricity (*dianxue* 電學), and thermodynamics (*rexue* 熱學) were presented as independent fields in China. By presenting the subfields of physics independently, the translators made it difficult for Chinese later to appreciate the unity of physics. Moreover, introducing the branches first made it more complicated later to reach a consensus on a more general term for physics.

Often physics was equated with investigating things (*gewu*). Others preferred calling physics "investigating things and extending knowledge" (*gezhi*), which frequently overlapped vaguely with the general term for science and

25   Wang Yangzong 2000, 147–150.
26   Wang Bing 1994.

created substantial misunderstanding. Edkins's 1886 *Science Primers* associated "investigating the materiality of things" (*gezhi zhi xue* 格致之學) with physics. In 1895, the school of physics in the Beijing Foreign Language School changed its name from the Hall for Investigating Things (Gewu guan 格物館) to the Hall for Investigating and Extending Knowledge (Gezhi guan 格致館). Unlike the Japanese, who developed independent translation techniques, the Chinese remained dependent on their Protestant informants into the 1890s. This dependency placed severe limits on what the Chinese alone could translate. Overall, the Western translations prepared by Macgowen, Hobson, and Martin in China dealt with physics in very general, textbook terms and never produced useful handbooks.[27]

The Qing state also was slower in reforming its educational system. Meiji Japan's new educational system was established in 1868; Qing education reforms were not comparable until 1902. A Japanese Ministry of Education (Mombusho 文部省) was created in 1871; its Qing counterpart was not established until 1905. Similarly, Tokyo University was founded as Japan's key modern teaching institution in 1877; the Imperial University of Beijing did not exist until 1898. Courses in physics had already started in 1875 in Japan, when the Tokyo school that evolved into Tokyo University shifted from foreign-language lectures by Europeans to lectures in Japanese by returned students who had studied physics abroad. The first Japanese students trained in physics in Japan graduated in 1883.

Chinese science faculties were not established at the Imperial University of Beijing until 1910, but even then only classes in chemistry and geology were taught. Physics was added in 1912. Of 387 students recruited in the sciences, only 54 received diplomas in 1913. Beijing first recruited Japanese science teachers to the university in 1902, but they left in 1908–1909 after their six-year contracts expired. From 1898 to 1911, only 200 students were trained in the sciences at the Imperial University, and the initial absence of faculties of mathematics and physics remained a serious problem in training scientists. The science curriculum was formalized in terms of requirements at the high school level beginning in 1911. In Japan, there were few students of physics when compared with the more popular fields of law and medicine. Between 1882 and 1912, however, Tokyo University graduated 186 in physics.[28]

Japan's educational system had a head start in editing and translating physics textbooks. China, by comparison, lacked textbook materials to teach physics at all levels of the educational system. Similar delays occurred in other

---

27   Smith 1978; and Amelung 2004.
28   Bastid 1988.

technical fields such as chemistry and geology. By 1873, Japanese taught physics in the new Meiji schools, and by 1877, Tokyo University had a physics program. By comparison, the Beijing School of Foreign Languages asked only occasional physics questions on examinations from 1868, which were based on Martin's elementary *Natural Philosophy*. The subfields of physics were taught separately as mechanics, hydraulics, acoustics, pneumatics, heat, optics, and electricity. In addition, the military and arsenal schools also taught some physics, especially its subfields.

Meiji educators produced physics textbooks in the 1870s, but none were available in China until the 1890s. Although Japanese relied on Protestant translations from China initially, the Ministry of Education ordered Katayama Junkichi (1837–1887) to compile an official physics textbook when physics (*butsuri* 物理, *wuli*) became a specialized discipline. Katayama's textbook was added to the Japanese curriculum in 1876 and republished many times. Moreover, Japan invited Western scientists to Japan. K. W. Gratama (1831–1888) began to serve in the Chemistry Bureau in 1869. He was succeeded by H. Ritter (d. 1874). Later, Iimori Teizō completed his edition of *Physics* by consulting the works on physics published by the German J. Müller.

In the late 1890s, the Qing recognized the need to translate physics textbooks. As a result of the 1898 reforms, the government decided to copy the Meiji model for education and create a public school system for science education rather than simply rely on schooling in the arsenals, shipyards, and factories. Full implementation of this program was not feasible until the civil examination system was scrapped, and the new school system replaced it in 1904–1905. The Sino-Japanese War had taught the Qing government that relying on arsenals to modernize was insufficient.[29]

Because there were few science textbooks in China and none that dealt chiefly with physics, Chinese immediately translated Japanese texts such as Iimori's *Physics*. In the early twentieth century, direct Chinese translations of the best physics texts by the most famous Japanese physicists were the most efficient means to prepare textbooks for the new Qing school system. This policy also guaranteed that Chinese would no longer rely on Western informants for specialized translations in important fields such as physics. But China's dependence on Japan was reconsidered after 1915, when Japan's policies toward the Republic of China became increasingly predatory.

Although high-level education in physics began at the Beijing Imperial University in 1912, the best-trained physicists studied in the United States and Japan: Li Fuji (b. 1885) studied in the United States; He Yujie (1882–1939), Xia

---

29   Wang Bing 1994. Cf. Reynolds 1993, 65–110, 131–150.

Yuanli (1884–1944), Li Yuebang (1884–1940?), and Hu Gangfu (1892–1966) studied in Japan. When the Imperial University was reorganized as Beijing University in 1912, it had formal divisions between the humanities and the sciences, with the latter including the fields of mathematics, chemistry, and physics. An independent physics department was not created until 1917, however. The greater availability of physics texts in the school system after 1905, however, did provide for wider knowledge of the field in China than had been the case before 1900.[30]

Japan also had a lead over China in physics research, the unification of technical terminology, and research associations by 1900. For instance, Japanese scholars started publishing in physics in the 1880s. Over two hundred articles in the various subfields of physics had appeared by the end of the Meiji era in 1912. Moreover, several Japanese physicists had emerged who were approaching Western levels of expertise in physics. Translators chose the official Meiji designation for the term "physics," *wuli xue* 物理學, in 1872. Terminology in Japanese physics achieved a final unification with the 1888 publication of an official list of technical terms with foreign counterparts. The committee for systematizing the translation of terms for physics, which was formed in 1885, was led by three of the first Japanese graduates in physics from Tokyo University. Scholars unified terms for 1,700 items from English, French, and German, which they then translated into Japanese and published. Chinese started using the Japanese term for "physics" in 1900, when a Japanese book by that name was published in China. Before, the term had usually referred to the principles of things as part of the traditional fields of natural studies.[31]

Academics created the first mathematics society in Tokyo in 1877, with fifty-five members. In 1884, ten of its seventy-five members specialized in physics. When the Tokyo Mathematics-Physics Society was formed in 1884, it had eighty-two members, twenty-five of whom were physicists. The latter changed its name in 1919 to the Japan Mathematics-Physics Society, which survived as an organization until it separated into two parts in 1948. Smaller specialized groups in physics were also formed in Japan in the 1880s. China was also later than Japan in training physicists and organizing associations. The Chinese had to study physics abroad, and the research institutes for physics at the Academia Sinica, the Beijing Institute, and the Qinghua Institute were not formed until 1928–1929. Although Chinese words for physics terms were unified in 1905, they were not finalized until the 1920s. Moreover, the Chinese Science Society and its journal were not founded until 1915, and that took place abroad in the

---

30   Wang Bing 1994.
31   Wang Bing 1999.

United States at Cornell University. Physicists did not form the Chinese Physics Society until 1932.[32]

The belief that Western science represented a universal application of objective methods and knowledge was increasingly articulated in the journals associated with the New Culture Movement after 1915. The journal *Science* (*Kexue* 科學), which the newly founded Science Society of China created in 1914, assumed that an educational system based on modern science was the panacea for all China's ills because of its universal knowledge system. Meiji Japan served as the model for that panacea until 1915, when Japanese imperialism, like its European predecessor, forced Chinese officials, warlords, and intellectuals to reconsider the benefits of copying Japan.[33]

Toward Republican Science

Despite the late Qing curriculum changes described above, which had prioritized science and engineering in the new public schools since 1902 and in private universities such as Qinghua (Tsing Hua 清華), many Chinese university and overseas students were by 1910 increasingly radical in their political and cultural views, which carried over to their convictions about science. Traditional natural studies became part of the failed history of traditional China to become modern, and this view now asserted that the Chinese had never produced any science. How premodern Chinese had demarcated the natural and the anomalous vanished when both modernists and socialists in China accepted the West as the universal starting place of all science.[34]

After 1911, many radicals such as Ren Hongjun linked the necessity for Chinese political revolution to the claim that a scientific revolution was also mandatory. Those Chinese who thought a revolution in knowledge required Western learning not only challenged classical learning, or what they now called Confucianism (Kongjiao 孔教), but also unstitched the patterns of traditional Chinese natural studies and medicine long accepted as components of imperial orthodoxy.[35]

As Chinese elites turned to Western studies and modern science, fewer remained to continue the traditions of classical learning (Han Learning) or Song Learning moral philosophy (Neo-Confucianism) that had been the basis for imperial orthodoxy and literati status before 1900. Those who still focused on traditional learning, such as Gu Jiegang (1893–1980) in Beijing and others elsewhere, often did so by reconceptualizing ancient learning in light of

---

32   Wang Bing 1994.

33   Sheng 1995, 11–12.

34   Chen Yuanhui et al. 1991, 608–650. Cf. Geertz 1975.

35   See Elman 1997.

"doubting antiquity" and applying new, objective procedures for historiography that they derived from the sciences. Thereafter, the traditional Chinese sciences, classical studies, and Confucianism survived as vestigial native learning in the public schools established by the Ministry of Education after 1905. They have endured as contested scholarly fields taught in the vernacular in universities since 1911.[36]

The Great War from 1914 to 1919 acted as a profound intellectual boundary between those modernists who still saw in science a universal model for the future and the "New Confucian" (Xinru 新儒) traditionalists, such as Zhang Junmai (Carson Chang 1886–1969), who showed renewed sympathy for distinctly Chinese moral teachings after the devastation visited on Europe. The former reformer and now scholar-publicist Liang Qichao, who was then in Europe leading an unofficial group of Chinese observers at the 1919 Paris Peace Conference, visited a number of European capitals. The group witnessed the war's deadly technological impact on Europe. They also met with leading European intellectuals, such as the German philosopher and Zhang Junmai's teacher Rudolf Christoph Eucken (1846–1926) and the French philosopher Henri Bergson (1859–1941), to discuss the moral lessons of the war.[37]

In his influential *Condensed Record of Travel Impressions while in Europe* (*Ouyou xinying lu jielu* 歐遊心影錄節錄), Liang Qichao related how the Europeans they met regarded the First World War as a sign of the bankruptcy of the West and the end of the "dream of the omnipotence of modern science." Liang found that Europeans now sympathized with what they considered the more spiritual and peaceful "Eastern civilization" and bemoaned the legacy in Europe of an untrammeled material and scientific social order that had fueled the world war. Liang's account of the spiritual decadence in postwar Europe indicted the materialism and the mechanistic assumptions underlying modern science and technology. A turning point had been reached, and the dark side of "Mr. Science" had been exposed. Behind it lay the colossal ruins produced by Western materialism.[38]

Bibliography

Adas, Michael 2004. "Contested Hegemony: The Great War and the Afro-Asian Assault on the Civilizing Mission Ideology," *Journal of World History* 15, 1 (March): 31–63.

---

36   Luo 2000; and Weston 2004, 83.
37   Ding 1972, vol. 2, 551–574. For African and Indian critiques, see Adas 2004.
38   Liang 1972, vol. 7, 10–12. See also Chow 1960, 327–329; and Grieder 1970, 129–135.

Amelung, Iwo 2004. "Naming Physics: The Strife to Delineate a Field of Modern Science in Late Imperial China," in Michael Lackner and Natascha Vittinghoff, eds., *Mapping Meanings: The Field of New Learning in Late Qing China*. Leiden: E. J. Brill 381–422.

Anderson, Warwick 1996. "Immunities of Empire: Race, Disease, and the New Tropical Medicine 1900–1920," *Bulletin of the History of Medicine* 70, 1: 94–118.

Andrews, Bridie 1994. "Tailoring Tradition: The Impact of Modern Medicine on Traditional Chinese Medicine 1887–1937," in V. Alleton and A. Volkov, eds., *Notions et perceptions du changement en Chine*. Paris: Collège de France, Institut des Hautes Études Chinoises 149–166.

Bailey, Paul 1990. *Reform the People: Changing Attitudes towards Popular Education in Early Twentieth Century China*. Edinburgh: Edinburgh University Press.

Bastid, Marianne 1988. *Educational Reform in Early 20th-Century China*, translated by Paul J. Bailey. Ann Arbor: University of Michigan China Center.

Benedict, Carol 1993. "Policing the Sick: Plague and the Origins of State Medicine in Late Imperial China," *Late Imperial China* 14, 2 (December): 60–77.

Bennett, Adrian 1967. *John Fryer: The Introduction of Western Science and Technology into Nineteenth-Century China*. Cambridge, MA: Harvard University Research Center.

Biggerstaff, Knight 1961. *The Earliest Modern Government Schools in China*. Ithaca, NY: Cornell University Press.

Chang Chia-feng 1996. "Aspects of Smallpox and Its Significance in Chinese History." Ph.D. dissertation, London University.

Chen Yuanhui 陳元暉 et al., comps. 1991. *Zhongguo jindai jiaoyu shi ziliao huibian: Xuezhi yanbian* 中國近代教育史資料彙編: 學制演變. Shanghai: Jiaoyu chubanshe.

Chow, Tse-tsung 1960. *The May 4th Movement: Intellectual Revolution in Modern China*. Cambridge, MA: Harvard University Press.

Cong, Xiaoping 2007. *Teachers' Schools and the Making of the Modern Chinese Nation-State 1897–1937*. Vancouver: University of British Columbia Press.

Crozier, Ralph 1968. *Traditional Medicine in Modern China: Science, Nationalism, and the Tensions of Cultural Change*. Cambridge, MA: Harvard University Press.

Dagenais, Ferdinand 2001. "Organizing Science in Republican China (1914–1950)," paper presented at the "Ideology and Science Symposium," Center for Chinese Studies, University of California, Berkeley.

Ding Wenjiang 丁文江 1972. *Liang Rengong xiansheng nianpu changbian chugao* 梁任公先生年譜長編初稿. 2 vols. Taibei: Shijie shuju.

Elman, Benjamin 1997. "The Formation of 'Dao Learning' as Imperial Ideology during the Early Ming Dynasty," in Theodore Huters, R. Bin Wong, and Pauline Yu, eds.,

Culture and the State in Chinese History. Stanford, CA: Stanford University Press 58–82.

____ 2003. "Naval Warfare and the Refraction of China's Self-Strengthening Reforms into Scientific and Technological Failure," *Modern Asian Studies* 38, 2: 283–326.

____ 2006. *A Cultural History of Modern Science in China*. Cambridge, MA: Harvard University Press.

Epler, D. C. 1980. "Bloodletting in Early Chinese Medicine and Its Relation to the Origin of Acupuncture," *Bulletin of the History of Medicine* 54: 337–367.

Geertz, Hildred 1975. "An Anthropology of Religion and Magic, I," *Journal of Interdisciplinary History* 6, 1 (Summer): 71–89.

Grieder, Jerome 1970. *Hu Shih and the Chinese Renaissance: Liberalism in the Chinese Revolution 1917–1937*. Cambridge, MA: Harvard University Press.

Hanson, Marta 2001. "Robust Northerners and Delicate Southerners: The Nineteenth-Century Invention of a Southern Medical Tradition," in Elisabeth Hsu, ed., *Innovation in Chinese Medicine*. Cambridge: Cambridge University Press 262–291.

Horiuchi, Annick 1994. *Les mathématiques japonaises à l'époque d'Edo (1600–1868): Une étude des travaux de Seki Takakazu (?–1708) et de Takebe Katahiro (1664–1739)*. Paris: J. Vrin.

Hsiao, Kung-chuan 1975. *A Modern China and a New World: Kang Yu-wei, Reformer and Utopian 1858–1927*. Seattle: University of Washington Press.

Kobayashi, Tatsuhiko 2002. "What Kind of Mathematics and Terminology Was Transmitted into 18th-Century Japan from China?," *Historia Scientiarum* 12, 1: 1–17.

Kunitake, Kume, comp. 2002. *The Iwakura Embassy 1871–73*. Chiba, Japan: Japan Documents. Distributed in North America by Princeton University Press, Princeton, NJ.

Kuriyama, Shigehisa 2002. *The Expressiveness of the Body and the Divergence of Greek and Chinese Medicine*. New York: Zone Books.

Kwong, Luke S. K. 1996. *T'an Ssu-t'ung 1865–1898: Life and Thought of a Reformer*. Leiden: E. J. Brill.

Lachner, Michael, et al., eds. 2001. *New Terms for New Ideas: Western Knowledge and Lexical Change in Late Imperial China*. Leiden: E. J. Brill.

Lean, Eugenia 1996. "The Modern Elixir: Medicine as a Consumer Item in the Early Twentieth-Century Press," M.A. degree paper, University of California, Los Angeles.

Lei, Hsiang-lin 1999. "When Chinese Medicine Encountered the State: 1910–1949." Ph.D. dissertation, University of Chicago.

Liang Qichao 梁啟超 1972. *Yinbing shi zhuanji* 飲冰室專集. Taibei: Zhonghua shuju.

Liu, Lydia 1995. *Translingual Practice: Literature, National Culture, and Translated Modernity, China 1900–1937*. Stanford, CA: Stanford University Press.

Luo Zhitian 羅志田 2000. "Zouxiang guoxue yu shixue de 'Sai xiansheng'" 走向國學 與史學的賽先生, *Jindai shi yanjiu* 近代史研究 (3): 59–94.

MacPherson, Kerrie 2002. *A Wilderness of Marshes: The Origins of Public Health in Shanghai 1843–1893*. Oxford: Lexington Books.

Masini, Federico 1993. *The Formation of Modern Chinese Lexicon and Its Evolution toward a National Language: The Period from 1840 to 1898*. Berkeley, CA: Journal of Chinese Linguistics.

Meng, Yue 1999. "Hybrid Science versus Modernity: The Practice of the *Jiangnan Arsenal* 1864–1897," *East Asian Science, Technology, and Medicine* 16: 13–52.

Nathan, Carl F. 1974. "The Acceptance of Western Medicine in Early 20th Century China: The Story of the North Manchurian Plague Prevention Service," in John Bowers and Elizabeth Purcell, eds., *Medicine and Society in China*. New York: Josiah Macy, Jr., Foundation 55–75.

Needham, Joseph, et al. 1957–. *Science and Civilisation in China*. 7 vols. Cambridge: Cambridge University Press.

Numata, Jirō 1992. *Western Learning: A Short History of the Study of Western Science in Early Modern Japan*, translated by R. C. J. Bachofner. Tokyo: Japan-Netherlands Institute 60–95.

Pong, David 1994. *Shen Pao-chen and China's Modernization in the Nineteenth Century*. Cambridge: Cambridge University Press.

Pyenson, Lewis, and Susan Sheets-Pyenson 2000. *Servants of Nature: A History of Scientific Institutions, Enterprises, and Sensibilities*. New York: Norton.

Reardon-Anderson, James 1991. *The Study of Change: Chemistry in China 1840–1919*. Cambridge: Cambridge University Press.

Reynolds, Douglas 1993. *China 1898–1912: The Xinzheng Revolution and Japan*. Cambridge, MA: Council on East Asian Studies, Harvard University Press.

Rogaski, Ruth 2004. *Hygienic Modernity: Meanings of Health and Disease in Treaty-Port China*. Berkeley: University of California Press.

Schmalzer, Sigrid 2007. "On the Appropriate Use of Rose-Colored Glasses: Reflections on Science in Socialist China," *Isis* 98, 3: 571–583.

Schneider, Laurence 1988. "Genetics in Republican China," in J. Bowers, J. Hess, and N. Sivin, eds., *Science and Medicine in Twentieth-Century China: Research and Education*. Ann Arbor: Center for Chinese Studies, University of Michigan 3–29.

Shen, Grace 2007. "Murky Waters: Thoughts on Desire, Utility, and the 'Sea of Modern Science,'" *Isis* 98, 3: 584–596.

Sheng, Jia 1995. "The Origins of the Science Society of China 1914–1937." Ph.D. dissertation, Cornell University.

Smith, Crosbie 1978. "The Transmission of Physics from France to Britain 1800–1840," *Historical Studies in the Physical Sciences* 9: 1–62.

Tan Sitong 譚嗣同 1981. *Tan Sitong quanji* 譚嗣同全集. 2 vols. Beijing: Zhonghua shuju.

Tian, Miao 2003. "The Westernization of Chinese Mathematics: A Case Study of the *duoji* Method and Its Development," *East Asian Science, Technology, and Medicine* 20: 63–70.

Wang Bing 王冰 1994. "Shijiu shiji zhongqi zhi ershi shiji chuqi Zhongguo he Riben de wulixue" 十九世紀中期至二十世紀初期中國和日本的物理學, *Ziran kexueshi yanjiu* 自然科學史研究 13, 4: 326–329.

____ 1996. "Jindai zaoqi Zhongguo he Riben zhi jian de wulixue jiaoliu" 近代早期中國和日本之間的物理學交流, *Ziran kexueshi yanjiu* 自然科學史研究 15, 3: 227–233.

____ 1999. "On the Physics Terminology in Chinese and Japanese during Early Modern Times," in Yung Sik Kim and Francesca Bray, eds., *Current Perspectives in the History of Science*. Seoul: Seoul National University Press 517–521.

Wang, Hui 1995. "The Fate of 'Mr. Science' in China: The Concept of Science and Its Application in Modern Chinese Thought," *positions: east asia cultures critique* 3, 1 (Spring): 14–29.

Wang, Y. C. 1966. *Chinese Intellectuals and the West 1872–1949*. Chapel Hill: University of North Carolina Press.

Wang Yangzong 王楊宗 2000. "1850 niandai zhi 1910 nian Zhongguo yu Riben zhi jian kexue shuji de jiaoliu shulue" 一八五零年代至一九一零年中國與日本之間科學書籍的交流述略, *Tōzai gakujutsu kenkyū kiyō* 東西學術研究紀要 (Kansai University) 33 (March): 139–152.

Wang, Zuoyue 2007. "Science and the State in Modern China," *Isis* 98, 3: 558–570.

Weston, Timothy 2004. *The Position of Power: Beijing University, Intellectuals, and Chinese Culture 1898–1929*. Berkeley: University of California Press.

Wright, David 1998. "The Translation of Modern Western Science in Nineteenth-Century China 1840–1895," *Isis* 89: 653–673.

____ 2000. *Translating Science: The Transmission of Western Chemistry into Late Imperial China 1840–1900*. Leiden: E. J. Brill.

Wu, Yi-li 1998. "Transmitted Secrets: The Doctors of the Lower Yangzi Region and Popular Gynecology in Late Imperial China." Ph.D. dissertation, Yale University.

Xu, Xiaoqun 1997. "'National Essence' vs. 'Science': Chinese Native Physicians' Fight for Legitimacy 1912–37," *Modern Asian Studies* 31, 4: 847–877.

Yu Yue 俞樾 1897. "Xu 序," in Wang Renjun 王仁俊, *Gezhi guwei* 格致古微. Wuchang: Zhixue hui edition 1a–2b.

Zhao Yuanyi 趙元益, ed. 1897. *Gezhi shuyuan jiawu keyi* 格致書院甲午課藝. Shanghai lithograph.

Historiography of Science and Technology in China

The First Phase

Iwo Amelung

Abstract

The history of science and technology is almost ubiquitous in present-day China. It is a strongly institutionalized field, taught and researched at many universities and specialized institutes. In this essay I examine how the history of science and technology entered Chinese discourse during the beginning of the twentieth century. While related to earlier developments, especially to the theory of the Chinese origins of the Western sciences (Xixue zhongyuan shuo) during the late nineteenth century and the epistemic change in China during the first years of the twentieth century, I argue that the history of science and technology in China became a seriously researched subject only when it was integrated into a twofold discourse on identity. This discourse dealt, on the one hand, with national identity. China, aspiring to become a modern state, needed a past that could be linked with a Chinese modernity, and the history of Chinese science and technology was one means to achieve this. On the other hand, the professional identity of the newly formed group of scientists and engineers was also at stake. As in the West, Chinese scientists and engineers wanted to legitimize their professions and enhance their social positions by drawing attention to the importance of science and technology in the glorious Chinese past. I also show that Japan and the West were of considerable importance for the early phase of historiography of science and technology in China. They offered not only a blueprint and inspiration for the discourse but also research results, which in due time were naturalized and became an indispensable part of the Chinese discourse.

In this essay, I will make a first attempt to analyze the development of historiography of science and technology in China during the first half of the twentieth century. This might be considered a contribution toward filling a glaring lacuna in Chinese studies, for despite the large amount of research on the development of historiography in China, there is almost no research on the historiography of science and technology. In the view of many observers, the attention accorded to history of science and technology is mainly due to the work of Joseph Needham (1900–1995). There indeed is no reason to question the immense importance of his work; it is necessary, however, to understand that Needham's preoccupation with the topic was closely linked to the emergence of a discourse on the history of Chinese science and technology and that to quite some degree his work could profit from the results of the research of Western and Chinese historians and scientists.

Historiography of science and technology can be understood as an academic discipline. My goal in this essay, however, is not to provide a disciplinary history—although there is no doubt that such a history would be quite useful as well. Rather, I will look into the discourse on the history of Chinese science and technology as it has developed in China since the beginning of the twentieth century. It is my contention that the emergence of the historiography of Chinese science and technology needs to be understood as part of a number of larger developments:

1   The reception and appropriation of Western knowledge since the late imperial era.
2   The acceptance of the idea that science constitutes the "great denominator of nationhood."[1]
3   The desire to create a Chinese identity that, while tapping into the riches of the Chinese cultural tradition, embodied modernity—or as Schneider has put it in another context, "to be modern but to remain Chinese."[2]

The development of the historiography of Chinese science and technology touches upon other topics as well, particularly attempts to establish a relationship between China's natural environment and modern science—a practice called "tropicalization of science" by Prakash[3]—and the issue of the "indigenization" of science, which was related to the question of the professional identity of Chinese scientists. It should be noted here that in this essay I will refrain from differentiating between history of science and history of technology. In China since the beginning of the twentieth century, both fields are, in fact, quite often intermingled or at least treated together. More often than not, the generic designation *kexueshi* 科學史 (history of science) was, and is, employed for both.

The Notion of History of Science in Early Twentieth-Century China

The modern word for "history of science," *kexueshi*, entered China—as many other modern terms did—from Japan, during the first years of the twentieth century. While it is still unclear when the word was first used, we can be certain that early usage of the term and its predecessors referred, not to the Chinese technological and scientific tradition, but entirely to the West's historical experience in relationship to science and technology. One early example can be

1   Chakraborty 2000, 186.
2   Schneider 1971, 95.
3   Prakash 1999, 6.

found in the "Statutes of the Imperial University" (Jingshi daxuetang zhangcheng 京師大學堂章程) of 1902, which mandated that students specializing in the Classics take a course in "history of science of foreign countries."[4] It is not clear, however, what the contents of the course were. Still, the requirement is rather illustrative for the "history of science" in China at this time: the history of science and technology was important, but this history certainly was understood as the history of the sciences in the West. This can be clearly seen from journals and magazines that were dedicated to the sciences and their popularization and that were published in China during the first ten years of the twentieth century. Quite a number of these periodicals regularly dealt with issues of history of science, more often than not by giving brief biographies of important scientists, engineers, or technicians. All these biographies were devoted to Western (or at times Japanese) figures. In many cases such biographies were intended to serve as an example—as the biographer of Watt stated in 1904: "I drafted this biography of Watt in order to let my compatriots know that industry is an armoured warship and a highly explosive bomb in the competition over production, and 'science' is the endowment on which industry rests."[5] This author also believed that in China there was no science to speak of, and this was the opinion of many other scholars and intellectuals as well. Liang Qichao 梁啟超 (1873–1929), for example, in 1902 published a brief article entitled "Brief History of the Development of Science" (Gezhixue yange kaolüe 格致學沿革考略) in which he intended to demonstrate the prowess of Western sciences. Liang even states that his motivation for composing the article was to show that "China does not lack anything more than science."[6] The well-known writer Lu Xun 魯迅 (1881–1936), who in 1907 was one of the first to employ the word *kexueshi* in the title of an article, did not see any reason to include Chinese developments in his exposition. In fact, his only reference to China in that article is a biting critique of those conservatives—he calls them "those who embrace the national essence to their deaths"—who claim that all discoveries and inventions of the West could be found in Chinese history as well, or who were willing to accept Western inventions and discoveries only if these were supposedly of Chinese origin.[7] Yu Heyin's 虞和寅 (1879–1944) "A Short History of Botany" (Zhiwuxue lüeshi 植物學略史) from 1903 also constitutes an interesting example. Yu here mentions Li Shizhen 李時珍 (1518–1593) and his *Compendium of Materia Medica* (*Bencao gangmu* 本草綱目), but

---

4 "Qinding Jingshi daxuetang zhangcheng" 2000, 98.

5 Wang Benxiang 1904.

6 Liang 1902.

7 Lu 1980.

he stresses that Li focused on pharmaceutical application and that this is "absolutely different from our present-day understanding of botany." He thus refrains from treating it in his history.[8] A similar but somewhat fuller explanation was provided by Zhang Fu in "A History of Zoology" (Dongwuxue lishi 動物學歷史) in the *Xuebao* 學報 of 1907:

> In our country the sciences have developed earliest, but only the names of the things can be found in some canonical scripts and miscellaneous writings. There have never been scholars who did specialized research and published specialized books ... There is no [book] that in detail describes the facts of biology or that could be called "zoology." Even the saints like Confucius did not know more than the names of the birds and animals and the plants and grasses, and that was it. This was a result of the situation of their times and not because the men of antiquity were not as good as those today. Since Han and Tang times, however, there have been Zhang Hua's 張華 [232–300] *Bowuzhi* 博物志 and Li Shizhen's *Bencao gangmu*, but their language was not detailed, and if it was detailed, it was not correct. (Our country's science was developed best during the Ming dynasty. Shizhen needed 26 years to complete this book. It consists of 52 *juan* and it deals with 1,871 different species. Although his main emphasis was on pharmaceutical application, [his results] do not tally with experimental practice. The mistakes are numerous. Nevertheless, the richness of the collected information extends to botany, zoology, and mineralogy; it thus in tendency covers the whole field of "natural history." In fact, there never has been a work as gigantic as this in the Eastern Ocean.) What is said in it, however, is not sufficiently detailed, and what is detailed in it is not correct. It thus is sufficient to constitute the shining light of our country's scientific circle but it is of no help for those of us who study the history of biology.[9]

This is a striking example of a phenomenon that Levenson aptly has called "patriotic schizophrenia." Denouncing the Chinese scientific tradition here obviously was seen as a means to convince compatriots of the necessity of introducing and adopting Western science as a means for national survival.[10] As "saving China through science" (*kexue jiuguo* 科學救國) this idea has a

---

8   Yu 1903.
9   Zhang Fu 1907.
10   Levenson 1959, 136.

positive connotation even today.[11] On the other hand, however, the exposition in the *Xuebao* also could serve nationalist and patriotic feelings, which need to have some roots in a not completely disastrous past.

The idea that China had no history of science and technology to speak of remained very dominant for a long time. This can be quite clearly seen from Zhang Zigao's 張子高 (1886–1976) *Brief History of the Development of Science* (*Kexue fada shilüe* 科學發達史略), which was based on a lecture that the American-educated Zhang Zigao had given at Dongnan University (Dongnan daxue 東南大學) in 1920. It was, to my knowledge, the first monograph on the history of science and technology published in China. While the original lecture was a rather conventional, concise description of the development of science in the West, in the publication of 1923 a section on the "past and future of science in China" was added. Zhang makes the caustic remark: "Science in China, positively spoken (*changyan zhi* 長言之), has a history of several thousand years and infinite hope. Negatively spoken, its past achievements are minimal and its future is hard to predict." Zhang then briefly described the early development of Chinese mathematics, astronomy, and medicine. He pointedly noted that Western scholars often accord the discovery of magnetism and gunpowder to China (thus corroborating that this notion was not commonly known in China at this time) but that unfortunately there was no theoretical treatment of these discoveries in China. In fact, he was greatly distressed by the lack of the physical sciences in China, which in his eyes was due to the lack of interest in the natural world by Chinese scholars, the lack of fundamental concepts (*jiben guannian* 基本觀念), and the practice of explaining but not conducting experiments.[12] It should be noted here that after the founding of the People's Republic, Zhang became an eminent historian of Chinese chemistry and published an important book on that topic as well as many articles.

Even more striking is another book on history of science, written by Sha Yuyan 沙玉彥 (1903–1961) and simply called *History of Science* (*Kexueshi* 科学史).[13] Published in 1931, this book is completely devoted to history of science and technology in the West and even states that printing had been invented by no one other than Gutenberg. For someone like Wang Zhixin 王治心 (1881–1968), who in 1930 published an article entitled "The Discoveries of Traditional Chinese Science" (*Zhongguo gudai kexue shang de faming* 中國古代科學上

---

11   See Wang Zuoyue 2002.

12   Zhang Zigao 1923, 241–249.

13   Sha 1931.

的發明), this resulted in the necessity to explain to his readers why he would write on such an apparently strange topic:

> When I propose such a title, my readers certainly are extremely surprised. This is because at the present time what China lacks most is exactly science. Everybody admits that China is backward in the realm of science, so how can there be any scientific discoveries to speak about? Saying that China is backward in science is possible, but to claim that China has not made any scientific discoveries is incorrect. And it is not only this. In fact, Chinese scientific discoveries are earlier than those of all other countries of the world ... What I want to propose here is by no means intended to exaggerate our own [achievements] or to suggest that we should abandon Western science and return to our own ancient inventions and use them again. No, what I intend to stress is that we should make extraordinary efforts and study Western sciences since our own ancestors have made such efforts in this realm![14]

The "Chinese Origins of the Western Sciences"

While a certain or even outright skepticism toward the Chinese scientific and technological tradition constituted the mainstream of the discourse during the late imperial and early Republican era, it is possible to find examples that suggest a completely different approach. One is the "The Chinese Physicist Mozi" (Zhongguo wulixuejia Mozi 中國物理學家墨子) published in the *Journal of the Natural and Applied Sciences* (*Lixue zazhi* 理學雜誌) in 1906 and which, according to the author, should serve to spur the "scientific spirit of the Chinese."[15] In the same way, the biography of the "scientist" (*lixuejia* 理學家) Fang Yizhi 方以智 (1611–1671) aimed to prove the equivalence of "Chinese wisdom" and the "pride of the West" (i.e., the sciences).[16] This strain of reasoning harks back to the so-called "theory of the Chinese origins of the Western sciences" (*Xixue Zhongyuan shuo* 西學中源說), which was very popular in the late nineteenth century. As is well known, this theory goes back to the assumption, first proposed by Chinese Daoists, that Buddhism, which was introduced

---

14   Wang Zhixin 1930, 167–168.
15   Jue Chen 1906.
16   Gong 1906.

to China since in the first century CE, was nothing more than a debased form of Daoism, which supposedly had been introduced to India by Laozi.[17]

During the Jesuit mission of the seventeenth and eighteenth centuries, this theory was widely employed in astronomy and mathematics. The attempt to establish correspondences between the Chinese and the Western mathematical praxis resulted in a revival of traditional Chinese mathematics.[18] During the late nineteenth century the idea of the "Chinese origins of the Western sciences" was particular widespread and was now extended to all areas of "Western knowledge," from Christianity to military knowledge expertise.[19] This development coincided with a renewed interest in the non-orthodox pre-Qin text of the *Mozi* 墨子—especially the *Mohist Canon (Mojing* 墨經), which had been transmitted in mutilated form. Against the background of the incoming Western knowledge, attention especially centered on the passages on mechanics and optics in the *Mohist Canon.*[20]

The theory of the Chinese origins of the Western sciences had two functions: (1) it served to confirm Chinese superiority since the incoming knowledge supposedly was of Chinese origin and (2) it made the application of new discoveries and technologies digestible to conservatives since the "new" knowledge supposedly was autochthonous.[21] To make the Chinese and Western traditions comparable (or, more exactly, to relate pieces of the Chinese tradition to modern Western knowledge), a common denominator was necessary. This in all cases was the Western knowledge, which meant that the Chinese sources related to science and technology had to be reclassified along the lines of the Western sciences, as they became known in China.

It was on this basis that Mozi suddenly could become a physicist. It was precisely the search for parallels between Western scientific discoveries and technological inventions and references to science and technology in traditional Chinese literature that made a completely new view of China's own tradition possible. Instead of making use of canonical epistemological approaches developed over hundreds of years, one now could employ theorems discovered in the Western sciences or descriptions of inventions, which in many cases gave a completely new meaning to traditional Chinese literature. Propagators

---

17   Liu Dun 2002.
18   Wang Yangzong 1997, 83.
19   Quan 1935.
20   Amelung 2001, 2004.
21   At this time, when the theory of the Chinese origins of the Western sciences was at its peak in China, Western missionaries who were active in China and many Western scholars, of course, were claiming that all culture derived from the West. If anything, this claim was even more sweeping than the Chinese one; see Huters 2005, 28–29.

of the theory of the Chinese origins of the Western sciences used certain aspects of Western sciences in a highly selective way, if this knowledge could—apparently or actually—be found in traditional Chinese texts. Chinese historiography on science and technology in this way acquired a strangely ahistorical nature, which can be observed in quite a number of related publications even today.[22] The Chinese-origins theory, however, can be interpreted in a different way as well, namely as a classical approach of history of science, as in the West, as a process of "translation"—translation into a modern language, or the language of modernity.[23]

Appropriating the Chinese Scientific Tradition by Studying the West

Many Chinese historians of science have criticized Western historians of science for neglecting the Chinese side of the story. While this, to a certain extent, may be true for general histories of science, it certainly cannot be said that Western and Japanese Sinologists and historians of science were not interested in China's history of science. To the contrary, during most of the nineteenth and early twentieth centuries Western and Japanese scholars were probably more interested in questions of the history of science and technology in China than Chinese scholars were themselves. This situation had far-reaching consequences for the development of Chinese historiography of science. Time and again Chinese ventures into the subject were influenced by Western research results or by Western (and Japanese) approaches. The eminent historian of mathematics Li Yan 李儼 (1892–1963), for example, claimed that he began to do research on the Chinese mathematical tradition because he had observed that Westerners and Japanese had become interested in the topic[24] and because it was imperative for Chinese to do research on the Chinese mathe-

---

22   One example is whether the refractive index was known in ancient China. The question is contested and probably impossible to answer since it involves obscure passages in the *Mohist Canon*. Even if the *Canon* indeed refers to the refractive index, the question of whether or how it was employed or developed by later Chinese "scientists" has not been treated at all.

23   For the Western case, see Kragh 1987, 89.

24   This quite clearly relates to the work of Mikami Yoshio 三上義夫 (1875–1950); see Mikami 1913. One Chinese mathematician who was directly influenced by Mikami after having met him in Japan was Zhou Da 周達 (1879–1949); see Feng 2002.

matical tradition, since otherwise "national learning (*guoxue* 國學) would be lost."[25]

The most spectacular case of appropriating the results of Western research revolves around the coining of the phrase "four great inventions" (*si da faming* 四大發明). As is well known, the phrase is ubiquitous today, widely used not only in textbooks for primary education and in museums but in the popular media as well. It is clear that the idea goes back to Francis Bacon, who in his *Novum Organum* (bk. 1, aphorism 129) assigned a world-shaking importance to the "three Great Inventions"—compass, gunpowder, and printing—without, however, being aware of their Chinese origins. This idea was taken up by others, among them Karl Marx, who in a similar fashion stressed their importance for the development of world history but equally failed to mention their Chinese origins.[26] Already during the nineteenth century some Western scholars were aware that these inventions appeared first in China. Missionary Sinologists such as W. A. P. Martin (1827–1916) stressed, however, that despite these important inventions, the arts and the sciences in China had not developed in the same way as in the West.

In 1902 Liang Qichao published an article titled "The Relationship between Physical Geography and Civilisation" (Dili yu wenming zhi guanxi 地理與文明之關係). Liang pointed out here that European and American civilization is very much indebted to Asia. He states that Christianity originally came from Asia, and except for Roman law, which was an original European contribution, Greek philosophy and the Chinese civilization of the Sui and Tang had been transmitted to Europe by way of Arab civilization.

> The biggest contributions to the progress of modern European and American civilization are the compass, which is used for nautical travel and finding places; firearms, which are used to strengthen the army and to protect the country; and printing, which is used to circulate thoughts. And the Europeans were not able to invent these three things. They have studied them from the Arabs, and the Arabs have studied them from our China! Now the Europeans continue to produce [all sorts of] techniques and confer them to the Eastern countries. This in fact is no more than paying a debt of gratitude and returning past nurture with some added interest in order to make up past debts! How is it then that the Europeans

---

25   Li 1917.
26   Marx 1982, pt. 6, 1928.

can disdain us in such a way? Whether or not we accept their recompense, however, is our own choice![27]

This statement contradicts Liang's skeptical attitude concerning Chinese science, mentioned above. The ideas, however, did not come from Liang himself but from the Japanese source of his article, namely Ukita Kazutami's 浮田和民 (1859–1946) book *Discussion of the Origins of Historiography* (*Shigako genron* 史學原論): Liang's article was an annotated translation of this book's fifth chapter.[28] The source of Ukita's information is not clear, however, and it is hard to say whether Liang's article had any influence on the situation in China at that time. In any case, the catchphrase "four great inventions" entered Chinese discourse only when in 1925 the American scholar Thomas Francis Carter published his *Invention of Printing in China and Its Spread Westward*, which coined the slogan, adding the art of papermaking to the compass, gunpowder, and printing.[29] The influence of this book on the development of Chinese historiography of science and technology can hardly be overestimated. By 1926 it had already been partly translated into Chinese,[30] and it was this translator, the well-known historian Xiang Da 向達 (1900–1966), who in 1930 for the first time used the phrase *si da faming* in the title of an article, which he concludes with the following sentences:

> During the Middle Ages, Chinese paper, the art of printing, the compass, and gunpowder, these four great inventions of China, were introduced to Europe and then started Europe's modern civilization. During the last one hundred years, Western civilization has been introduced to China, and this has resulted in large-scale transformations of the Chinese social system and economic organization. The scope of these transformations is unprecedented in history. Printing, which is closely related to opening up the intelligence of the people, spread to Europe from the East one thousand years ago. There its brilliance became even stronger. Now it again enters our motherland, and this will be closely related to the future renaissance of China.[31]

---

27   Liang 1999, 947.
28   Ukita 1898.
29   Carter 1925.
30   Carter 1926.
31   Jue Ming (i.e., Xiang Da) 1930, 18.

Carter's book was instrumental too for sparking intensive Chinese research into the history of printing. In this way it also was of considerable importance in establishing Shen Kuo 沈括 as a polymath, which he is commonly considered to be today. Hu Daojing 胡道靜 (1913–2003), the most important commentator and editor of Shen Kuo's *Brush Notes from the Dream Creek* (*Mengxi bitan* 夢溪筆談) became interested in this book because it contains a description of printing with movable type, to which his attention had been drawn by Carter's book.[32]

A somewhat more complex example of the use of Western research and ideas can be observed in the realm of chemistry. The notion, commonly held even today, that alchemy originates with Daoist techniques for achieving immortality probably shows up for the first time in an article by Joseph Edkins (1823–1905) published in 1859.[33] The idea was taken up by W. A. P. Martin, who in 1879 published an article "Alchemy in China."[34] Interestingly enough, Martin also included this article in his *Lore of Cathay,* published in 1901, where he changed the title to "Alchemy in China, the Source of Chemistry."[35] It seems as if this contention fed into both the Chinese and the Western discourses. It most likely was Martin's *A Brief Examination of Western Learning* (*Xixue kaolüe* 西學考略*)*, where Martin also elaborates on his view, that was instrumental for its incorporation into the *Xixue zhongyuan* discourse.[36] One prominent example is an article by Tang Caichang 唐才常 (1867–1900) on the Chinese sources of Western knowledge found in the works of Zhu Xi 朱熹 (1130–1200). Zhu Xi, of course, is well known for having written a commentary to the *Cantongqi* 參同契. Tang mentions this work as well and makes quite clear that he considered the origins of alchemy to be with the Daoists.[37] The first modern Chinese scientist who took up the idea was Wang Jin 王璡 (1888–1966) in a series of articles in *Kexue* beginning in 1920. Wang Jin stresses that the sciences in China had not developed in the past, but that in two realms in China there "was no lack of men doing research" and these two realms were chemistry and mathematics. Wang reaches two interesting conclusions. On the achievements of Chinese alchemy he states:

32   Zhou 1993.

33   Cf. Edkins 1859.

34   Martin 1879.

35   Martin 1901, 44–72.

36   Martin 1883, 62b.

37   Tang 1903.

From the description given above it becomes clear that prior to the seventeenth century Chinese researches in chemistry were not inferior to Western researches in chemistry at the same time. Although the theoretical approaches of Chinese scholars are not more than bluster, the technique of the artisans of that time was very refined; it actually was even better than that of the different countries of the West. However, since the eighteenth century, the chemists of the West have grown as ferociously as spring bamboo; they have advanced the merits of academics like one thousand *li* on one day, so that we could only look on them from behind.[38]

On the function of doing research on the history of Chinese alchemy, Wang states:

The rise of modern chemistry only started after the discovery of the atomic theory, and the formation of the atomic theory required the idea that a compound is formed from fixed proportions. But the Daoists and the physicians of our country did not have this concept. For these reasons, if there were no European theories coming in from the outside, the progress of chemistry in China without any doubt would be very slow. When doing research on traditional Chinese chemistry, the point, which may ease my mind, is that the development of science in the East and the West follows the same rules. For a very long time the alchemical and the iatrochemical phases of our country were comparable with the situation in Europe. Moreover, the achievements also were comparable; thus, it is not necessarily so that the East is inferior to the West. If it is like this, why then should China not rise and in this scientific age catch up quickly so that within some ten years it could compare favorably to the Europeans?[39]

It is not clear what inspired Wang Jin to do this research. Certainly, Western influences can be observed, as Wang Jin notes that he had read Alexander Smith's *Introduction to Inorganic Chemistry*, which claims that the Chinese knew about chemical reactions with oxygen due to their yin-yang theory.[40]

Other Chinese chemists apparently had less interest in the Chinese tradition. Ding Xuxian 丁緒賢 (1885–1978), for example, who in 1925 published *A History of Chemistry (Huaxue tongkao* 化學史通考), merely briefly hints in his foreword that he has interest in the Chinese history of chemistry, but except

---

38   Wang Jin 1920a   564.
39   Wang Jin 1920b   684.
40   Smith 1917, 79.

for this delivers a completely Western history of the subject, without even changing his narrative regarding the invention of gunpowder.[41] It is striking as well that Zhang Zigao, who after 1949 would become one of the most important historians of Chinese chemistry, in his 1923 book mentioned above does not point out any relationship between Chinese alchemy and the development of modern chemistry. Interestingly enough, at the same time in the West, the interest in Chinese alchemy was steadily on the rise. This was true not only for historians of science but for general chemists as well. In an article for the *Scientific Monthly* in 1922, William Henry Adolph recalled his first course in industrial chemistry and how the lecturer introduced each subject with the statement "This substance was first discovered by the Chinese."[42] Apparently, it required the publication of a work by another foreigner before Chinese historians of chemistry were finally convinced that chemistry indeed may have been of Chinese origin. This book was O. S. Johnson's *The Study of Chinese Alchemy*, published in English in 1928.[43] Huang Sufeng 黃素封 (1904–1960), a historian of chemistry who in 1935 published a *History of the Development of Chemistry* (*Huaxue fada shi* 化學發達史),[44] was so impressed by Johnson's book and especially the idea of chemistry's Chinese origins that he published a full translation of the book in 1937.[45]

It should be noted that using Western research results remained a very important feature of Chinese history of science and technology for a rather long time, on occasion with curious results. The story of Wan Hu, for example, who today in China is widely celebrated as the founding father of manned space flight, since he supposedly fixed forty-seven rockets to a chair, hoping that this would carry him to the moon, is entirely based on a Western account: namely on the book of a writer of popular books on science and technology in the United States, who probably meant it as a joke.[46] Liu Xianzhou 劉仙洲 (1890–1975), however, the pioneering historian of technology and vice-president of Qinghua University, took up the story. He went to considerable lengths to explain why the name Wan Hu cannot be found in Chinese sources (claiming that it was either a wrong transliteration or an official title rather than a name).[47] However, even Liu failed to find firsthand evidence. The fact that he

---

41   Ding 1925.

42   Adolph 1922, 441.

43   Johnson 1928.

44   Huang 1935.

45   Johnson 1937.

46   Zim 1945, 31–32.

47   Liu Xianzhou 1962.

considered the story seriously, however, quite clearly contributed greatly to its popularity, and today Wan Hu is an indispensable part of the canon of the triumphs of Chinese science.

Science as "National Studies" (*Guoxue*)

In recent years it has been stressed that nature and a nature-related discourse played an important role in the National Essence Movement of the early twentieth century.[48] While in the reasoning of early National Essence (Guocui) scholars, quite clearly "natural history" (*bowu* 博物) was most important,[49] already in 1907 first attempts to integrate science and technology into the National Studies (Guoxue 國學) discourse were undertaken. The famous proposal to establish a school for "national essence" published in the first issue of the third volume of the journal of the National Essence Movement, the *Guocui xuebao*, in 1907 states at the beginning: "Since antiquity China several times has experienced the destruction of the country. But every time when the country was lost, the learning could be preserved. But today the country is not lost but the learning is lost first. For this reason the loss of national learning is even crueler than the disaster brought about by the Qin dynasty and the Mongols." The National Essence to be studied in this school, however, was to extend to topics relevant to history of science as well. In the field of "natural history" (*bowuxue* 博物學), for example, during the first term students should be required to take a class on "history of Chinese natural and applied sciences" (*Zhongguo like xueshi* 中國理科學史), before moving on to Chinese botany, zoology, and mineralogy. Considered most important, however, was the course in "calendrical and mathematical sciences" (*lishuxue* 曆數學), which was supposed to stretch over three years. During the first term students were expected to study the different mathematical schools and the *Nine Chapters of Mathematical Art* (*Jiuzhang suanshu* 九章算術). The second term would be devoted to Chinese algebra, the third one to the study of the "four elements," the fourth to similarities and differences between Chinese and Western mathematics, and the fifth and sixth terms would discuss calendrical sciences and "the meaning of Chinese mathematics."[50]

The assumption that Chinese mathematics was an indispensable part of "National Essence" and "National Studies" was quite widespread, as we have

---

48   Fan 2004.

49   Cheng 2006.

50   "Ni she guocui xuetang qi" 1907.

already seen in the case of Li Yan. Other historians of mathematics forwarded similar arguments. Ye Qisun 葉企孫 (1898–1977), for example, later to become one of China's foremost physicists, in his youth published several articles on the history of mathematics. In one he pointed out that one of the chapters of the *Nine Chapters* has to be considered as National Essence (*Shenzhou guocui* 神州國粹) and thus "has to be known."[51] Qian Baocong 錢寶琮 (1892–1974), one of the most important historians of Chinese mathematics, claims that he himself originally had the idea of preserving the National Essence, which, however, he abandoned in the course of the May Fourth Movement. Influenced by the writings of Hu Shi 胡適 (1891–1962) and Qian Xuantong 錢玄同 (1897–1939), he became convinced of "the need to reorganize the national past" and "to develop national studies."[52] Liang Qichao tried to include history of natural sciences into a framework of National Studies when he was working at the National Studies Institute (Guoxue yuan 國學院) at Qinghua University.[53] The fact that the most important bibliography on National Studies, which was published between 1929 and 1936 in four volumes, includes work dedicated to the history of science and technology shows that, for scholars dealing with National Studies, science and technology could not be neglected.[54]

History of Science and Patriotism

While many proponents of the National Studies Movement, especially those influenced by Hu Shi's idea of "reorganizing the national past" (*zhengli guogu* 整理國故), maintained a rather critical attitude toward the Chinese tradition, the line distinguishing this critical perception from an approach that celebrated Chinese history as a source for patriotic feelings was quite thin. When the conflict with Japan grew more acute during the 1930s, Chinese intellectuals—even the most critical historians—turned to the values of the Chinese tradition.[55] This, of course, included historians of science and technology as well, who began to functionalize the history of science and technology for patriotic propaganda. Yan Dunjie 嚴敦杰 (1917–1988), for example, who after 1949 would become one of the most important Chinese historians of mathematics, wrote in 1936 (when he was only nineteen years old):

---

51   Ye 1916, 59.
52   He 1998, 553.
53   Yao and Gao 2002, 98.
54   Cf. Beiping Beihai tushuguan bianmuke 1929–.
55   Schwarcz 1985, 231–235; Q. Wang 2001, 149–151.

When in recent years normal people, especially high school students, have read books on Western mathematics, they have been as surprised and excited as astronomers during Qing times who had read the "nine arts" (*jiu shu*) of Jartoux, so that they exclaimed: "Oh how intelligent are these foreigners. How did they manage to invent a discipline of this degree of refinement which is so wonderful?" Do not overstress the achievements of others and cancel out our own ones! Our China! Our China with its 5,000 years of civilization! Is it really possible that there was nobody who knew about mathematics? Is it really possible that there was nobody who was able to think in such a refined way? But, alas, there was not only one but, in fact, a number of mathematicians and astronomers (*chouren*) who have discovered important mathematical theorems and methods ... Zu Chongzhi discovered the formula for calculating the circumference of a circle.[56]

In due time issues related to the history of science were also taken up by the political establishment, including Guomindang 國民黨 chairman Jiang Jieshi 蔣介石 (1887–1975)[57] and the leader of the Communists Mao Zedong 毛澤東 (1893–1976), who referred to the Chinese inventions of the past, which in his opinion served as a proof for the outstanding history of civilization in China.[58] This patriotic and at times even nationalist appropriation of the scientific and technological tradition would become much more pronounced after 1949, when it was influenced by the Soviet experience.

Professional Identity and Localization of Science

In late nineteenth- and early twentieth-century Europe and the West, the history of science and technology, as well as its representation in museums, had a number of different functions. Scientists and engineers stressed the identity of scientific disciplines[59] and—probably even more important for them—highlighted the importance of scientists and especially engineers within the society.[60] Since the 1910s in China a considerable number of scientists and engineers studied the supposed history of their respective disciplines in China. The most

---

56   Yan 1936, 37.
57   Jiang Jieshi 1984a   14; 1984b.
58   Mao 1976.
59   Christie 1990, 12.
60   See, e.g., Menzel 2001, 86–100.

well-known example is the geologist Ding Wenjiang 丁文江 (1887–1936), who not only intensively researched Xu Xiake 徐霞客 (1587–1641), in whom he saw a sort of a protogeographer,[61] but also did extensive editing work related to the *The Exploitation of the Works of Nature* (*Tiangong kaiwu* 天工開物), which he—rightly—considered close to his professional identity.[62]

For Chinese scientists and engineers, researching the Chinese history of their respective traditions would serve another purpose as well. It would help to more closely connect the modern sciences with China and thus help to attain the goal of (as it was termed then) "Sinification" (Zhongguohua 中國化) of the natural sciences. In 1935 Liu Xianzhou called this "taking the special situation of our country into account" (*zhuzhong ben guo qingxing* 注重本國情形), which in his opinion meant that, for example, one should do research on traditional Chinese irrigation techniques and apply them in the present day, albeit in an improved form.[63] Already in 1926 the first issue of the journal *Natural World* (*Ziran jie* 自然界) quite programmatically argued for the indigenization of the sciences:

> We should start to investigate the sciences and techniques that originally existed in our country, such as in the lacquer industry, the dyeing industry, the soy industry, and the brewing industry. The methods they use are in accordance with the scientific method. It seems as if our ancestors already have done reliable research in the realm of organic chemistry and bacteriology. The plans drawn up for arched bridges are related to research in mechanics, the development of salt-and fire-wells in Sichuan is related to geology, in the books of ancient alchemy there are indeed records of how to produce sulfuric acid, and in the works on music there are exact calculations about musical scales. We think that the scientifc contributions of our people are by no means limited to the invention of the compass and gunpowder. It is absolutely necessary to examine the scriptures that our ancestors have left to us; this work is of greatest importance. About the things happening in our own family we know less than people from other families do. What has been written in our own books is unknown to us; this, indeed, is utterly embarassing.[64]

---

61   Ding wrote that Xu's "searching spirit is just what has distinguished Europeans and Americans during the last one hundred years" (quoted from Tan 1942, 42).

62   Furth 1970, 89–90.

63   Liu Xianzhou 1935, 89.

64   *Ziran jie* "Fakan zhiqu" 1926.

In the mid-1930s, many of the ingredients that after 1949 would form the discourse on Chinese history of science and technology had fallen into their places. These included the first reconstructions of traditional Chinese scientific apparatus (such as the seismograph of Zhang Heng 張衡 [78–139]—in this case on the basis of models first done in Japan) that later on would be of great importance for the process of popularization.[65] When in 1936 Zhu Qianzhi 朱謙之 (1899–1972), dean of the history department of Sun Yat-sen University（Zhongshan daxue 中山大學) in Guangzhou, proposed the establishment of a "Chinese Society for the History of Science" (Zhongguo kexueshi she 中國科學史社), the research proposal outlined in the draft statute of the society quite neatly summarized its long-term goals: the new society should serve the development of "China's indigenous scientific culture," "propagate new forms of history," and promote the cooperation between historians and natural scientists. The society had four main research areas:

1   China's disciplinary histories, such as the histories of astronomy, mathematics, physics and chemistry, biology, medicine, psychology, earth sciences, and engineering.
2   Special topics from the Chinese history of science, such as gunpowder, the compass, printing, traditional seismographs, Zu Chongzhi's method of determining pi, traditional explanations of music, etc.
3   Theoretical issues of Chinese history of science, such as the characteristics of China's science, Chinese scientific thought, Chinese contributions to science, as well as the question of China's lack of development in the sciences.
4   How China's science spread abroad and how foreign science was received in China.[66]

The Needham Factor

Joseph Needham's multivolume *Science and Civilisation in China* began publication in the 1950s. Not all Chinese historians of science and technology are convinced of the quality and the accuracy of Needham's descriptions and assessments.[67] Even they, however, are quite aware that Needham's work was instrumental for putting the history of Chinese science and technology on the

---

65   Milne 1886, 14; Zhang Yinlin 1956 [1928]; Wang Zhenduo 1989.
66   "Fulu er: Zhongguo kexueshi she changcheng" 1936, 3.
67   Guo 2007.

world stage. It is important not to forget that *Science and Civilisation* is based on Needham's interactions with China and with Chinese scholars, which had already begun in the 1930s, when Needham fell in love with one of his Chinese collaborators in Cambridge. When he was the head of the Sino-British Scientific Cooperation Office in China between 1943 and 1946, he used this position to collect material for *Science and Civilisation*, which at the beginning was not planned as a work of the giant scale it finally would become. Needham maintained close contact with Chinese scientists, and during these years he met many of the best Chinese scientists of the time. Needham normally acknowledged the influence of those scientists on the cover pages of his work. In fact, however, the influence of Chinese scientists on Needham was much more profound than commonly perceived. Needham's interpretation of the *Mohist Canon* (*Mojing* 墨經), for example—key to his treatment of Chinese physics—was directly influenced by the physicist Qian Linzhao 錢臨照 (1906–1999). Needham met Qian in Kunming in 1943, when Qian had just finished working on his seminal paper on Chinese mechanics and optics.[68] Qian reports that Needham was greatly excited by his paper.[69]

While there are many passages in Needham's work in which a Chinese influence is visible, I will limit myself here to a brief discussion of the question of why China had not developed modern science. This question was intensively discussed in China, and it was—in a somewhat modified form—central to Needham's work. Today in China it is known as the "Needham riddle" (*Li Yuese nanti* 李約瑟難題).[70] When asking this question, Needham was highly influenced by Karl August Wittfogel's (1896–1988) ideas developed in *Wirtschaft und Gesellschaft Chinas* and a number of articles.[71] We know that Needham

---

68   Qian 1942.

69   Qian 2001, 12–13.

70   Needham's own description of the Needham riddle runs as follows: "When I first formed the idea, about 1938, of writing a systematic objective and authoritative treatise on the history of science, scientific thought and technology in the Chinese culture-area, I regarded the essential problem as that of why modern science had not developed in the Chinese civilization (or Indian) but only in Europe? As years went by, and as I began to find out something at last about Chinese science and society, I came to realize that there is a second question at least equally important, namely, why between the first century BC and the fifteenth century AD, Chinese civilization was much more efficient than occidental in applying human natural knowledge to practical human needs?" (Needham 1964, 385).

71   On the covers of two offprints of Wittfogel's articles that Wittfogel gave Needham during their meeting in New York, Needham noted "precious" and "A unique paper! This and Hudsons' book are the only two accounts known to me which really go right to the root of

used to discuss his ideas on the nonemergence of modern science in China with a large number of Chinese scientists. By October 1944, however, these scientists had a new basis for discussing this question, namely the translation of a chapter of Wittfogel's work: "Why Did China Not Develop Modern Science?" (Zhongguo weishenmo meiyou chansheng ziran kexue 中國為甚麼沒有產生自然科學). [72] The journal *Kexue shibao* 科學時報, in which the article was published, was one of the most important leftist journals dedicated to the popularization of science, and we can be quite sure that it was read by many scientists.

Just a few weeks later, Needham, in Meitan 湄潭 on the occasion of the thirtieth anniversary of the foundation of the Chinese Science Society (Zhongguo kexueshe 中國科學社), gave a talk on the history of science in China. From the diary of Zhu Kezhen we know that on this occasion a common chord was struck between Needham and Chinese scientists interested in the history of science in China, [73] certainly partly because both sides had been exposed to Wittfogel's ideas on the development (and nondevelopment) of science in China. The relationship between Needham and his Chinese colleagues was to last. It is quite indicative that when Needham gave assistance to scholars from the People's Republic of China to participate in the International Conference on the History of Science held in Italy in 1956, the three scholars he supported were Zhu Kezhen, Li Yan, and Liu Xianzhou. Zhu was vice-president of the Academy of Sciences and the "gray eminence" behind the establishment of the history of science and technology in China as an academic discipline. Li had been interested in the history of mathematics since the 1910s because he wanted to prevent the loss of Chinese learning. Liu was a historian of technology who went to great lengths to prove that certain important inventions and machines that had shaped the development of the world derived from China.

Concluding Remarks

In this paper, I have attempted to sketch some of the more important factors leading not only to the formation of a discourse on the Chinese scientific and technological tradition but also to the emergence of a new academic disci-

---

the matter!" These offprints are stored at the Needham Research Institute in Cambridge, UK. I would like to thank John Moffet of the Needham Research Institute for helping me obtain access to this material.

72   Wittfogel 1944.
73   Zhu 2006, vol. 9, 207–208.

pline: the history of Chinese science and technology. While this did not take place until the 1950s, and was at least partially inspired by the Soviet model, it is important to note that the foundation for this development had been laid in the beginning of the twentieth century. Almost all the scientists and scholars mentioned in this paper continued to play important roles during the first twenty to thirty years of the People's Republic. The period analyzed in this paper thus certainly can be viewed as crucial for the development of the field. One could argue that it was the tension between self-assertion and self-doubt, combined with the need of Chinese scientists and engineers to legitimatize their positions in a quickly changing society, which paved the way for the emergence of the field of history of science and technology in China. Equally important, however, was the constant influence of the West. Certain aspects of this influence were indirect—Chinese scholars and scientists wanted to prove that China had its own tradition of those factors that had made the West rich and strong. Chinese researchers thus adopted a whole bundle of values and ideas related to nationhood and an understanding of modernity as being mainly defined by scientific prowess; all these views had been prevalent in the West since the end of the eighteenth century. At the same time, direct influences were also at play, as Chinese researchers integrated the results of Western research into their own efforts.

When Chinese scholars and scientists did research on the history of science and engineering in China, they could not but employ a Western paradigm of the classification of the sciences. This development already can be seen during the late Qing, in the last phase of the application of the theory of the Chinese origins of the Western sciences. However, it became ever more important during the first half of the twentieth century. Although during the early phase there were scholars who consciously excluded the Chinese tradition from their treatment of history of science, this massive reclassification, which of course applied to more than just the realm of science and technology, continues to dominate modern Chinese research on the history of science and technology in China.

Bibliography

Adolph, William Henry 1922. "The History of Chemistry in China." *Scientific Monthly* 14, 5: 441–446.

Amelung, Iwo 2001. "Weights and Forces: The Reception of Western Mechanics in Late Imperial China," in Michael Lackner, Iwo Amelung, and Joachim Kurtz, eds., *New*

Terms for New Ideas: Western Knowledge and Lexical Change in Late Imperial China. Leiden: E. J. Brill   197–232.

——— 2004. "Naming Physics: The Strife to Delineate a Field of Modern Science in Late Imperial China," in Michael Lackner and Natascha Vittinghoff, eds., *Mapping Meanings: Translating Western Knowledge into Late Imperial China.* Leiden: E. J. Brill   381–422.

Beiping Beihai tushuguan bianmuke 北平北海圖書館編目科, comp. 1929–. *Guoxue lunwen suoyin* 國學論文索引 [Index to essays on National Studies]. Beijing: Zhonghua tushuguan. [Subsequent volumes were published under the names *xubian* (sequel), *sanbian* (third collection), and *sibian* (fourth collection); all Beijing: Zhonghua tushuguan xiehui.]

Carter, Thomas F. 1925. *The Invention of Printing in China and Its Spread Westward.* New York: Columbia University Press.

Carter, Thomas Francis [Kate 卡特] 1926. "Zhi zi Zhongguo chuanru Ouzhou kaolüe" 紙自中國傳入歐洲考略 [Did paper come to Europe from China?], translated by Xiang Da 向達, *Kexue* 科學 11, 6: 735–743.

Chakraborty, Pratik 2000. "Science, Nationalism, and Colonial Contestations: P. C. Ray and His Hindu Chemistry," *Indian Economic and Social History Review* 37, 2: 185–213.

Cheng Meibao 程美寶 2006. "Wan Qing guoxue dachao zhong de bowuxue zhishi—lun Guocui xuebao zhong de bowu tuhua" 晚清國學大潮中的博物學知識－論國粹學報中的博物圖畫 [Natural history knowledge in late Qing dynasty National Learning tide], *Shehui kexue* 2006, 8: 18–31.

Christie, John R. R. 1990. "The Development of the Historiography of Science," in Robert C. Olby et al., eds., *Companion to the History of Modern Science.* London: Routledge   5–22.

Ding Xuxian 丁緒賢 1925. *Huaxueshi tongkao* 化學史通考 [A history of chemistry], Guoli Beijing daxue congshu 11. Beiping: Guoli Beijing daxue chubanshe.

Edkins, Joseph 1859. "Phases in the Development of Tauism," *Transactions of the China Branch of the Royal Asiatic Society*   1st series   5: 83–99.

Fan, Fa-ti 2004. "Nature and Nation in Chinese Political Thought: The National Essence Circle in Early-Twentieth-Century China," in Lorraine Daston and Fernando Vidal, eds., *The Moral Authority of Nature.* Chicago: University of Chicago Press   409–437.

Feng Lisheng 馮立升 2002. "Zhou Da yu Zhong Ri shuxue jiaowang" 周達與中日數學交往 [Zhou Da and the exchange between Chinese and Japanese mathematics], *Ziran bianzhengfa tongxun* 1: 68–71.

"Fulu (er): Zhongguo kexueshi she zhangcheng (caoan)" 附錄(二):中國科學史社章程(草案) [Appendix two: Statute for the Chinese Society for the History of Science] 1936. *Xiandai shixue* 3, 2: 3–4.

Furth, Charlotte 1970. *Ting Wen-chiang: Science and China's New Culture*, Harvard East Asian Series 42. Cambridge, MA: Harvard University Press.

Gong Xia 公俠 1906. "Erbai liushi nian qian de Lixue dajia Fang Yizhi zhuan" 二百六十年前的理學大家方一智傳 [Biography of the great scientist Fang Yizhi, who lived 260 years ago], *Lixue zazhi* 2: 1–12.

Guo Jinhai 郭金海 2007. "Li Yuese Zhongguo kexue jishushi yu Zhongguo ziran kexueshi yanjiushi de chengli" 李約瑟中國科學技術史與中國自然科學史研究室的成立 [Needham's *Science and Civilisation in China* and the founding of the Institute for the History of Chinese Natural Sciences], *Ziran kexueshi yanjiu* 26, 3: 273–291.

He Shaogeng 何紹庚 1998. "Qian Baocong xiansheng zhuan" 錢寶琮先生傳 [A biography of Mr. Qian Baocong], in *Li Yan, Qian Baocong kexueshi quanji* 李儼錢寶琮科學史全集. Shenyang: Liaoning jiaoyu chubanshe, vol. 10, 551–557.

Huang Sufeng 黃素封 1935. *Huaxue fada shi* 化學發達史 [A history of the development of chemistry], Wanyou wenku, dier ji 247. Shanghai: Shangwu yinshuguan.

Huters, Theodore 2005. *Bringing the World Home: Appropriating the West in Late Qing and Early Republican China*. Honolulu: University of Hawai'i Press.

Jiang Jieshi 將介石 1984a. "Kexue de daoli" 科學的道理 [The foundations of science], in Qin Xiaoyi 秦孝儀, ed., *Xian zongtong Jiang gong sixiang yanlun zongji* 先總統蔣公思想言論總集. Taibei: Zhongguo guomindang zhongyang weiyuanhui dangshi weiyuanhui, vol. 13, 13–31.

—— 1984b. "Zhongguo zhi mingyun" 中國之命運 [The fate of China], in Qin Xiaoyi 秦孝儀, ed., *Xian zongtong Jiang gong sixiang yanlun zongji* 先總統蔣公思想言論總集. Taibei: Zhongguo guomindang zhongyang weiyuanhui dangshi weiyuanhui, vol. 4, 1–166.

Johnson, O. S. 1928. *The Study of Chinese Alchemy*. Shanghai: Commercial Press.

—— 1937. *Zhongguo liandanshu kao* 中國練丹術考 [The study of Chinese alchemy], translated by Huang Sufeng 黃素封, Baike xiao congshu. Shanghai: Shangwu yinshuguan.

Jue Chen 覺晨 1906. "Zhongguo wulixuejia Mozi" 中國物理學家墨子 [The Chinese physicist Mozi], *Lixue zazhi* 4: 63–70; 6: 75–87.

Jue Ming 覺明 [Xiang Da] 1930. "Zhongguo si da faming kao zhi yi: Zhongguo yinshuashu de qiyuan" 中國四大發明考之一。中國印刷術的起源 [Researches in one of the four great inventions of China: The Chinese origins of printing], *Zhongxuesheng* 5: 1–18.

Kragh, Helge 1987. *An Introduction to the Historiography of Science*. Cambridge: Cambridge University Press.

Levenson, Joseph R. 1959. *Liang Ch'i-chao and the Mind of Modern China*. Cambridge, MA: Harvard University Press.

Li Yan 李儼 1917. "Zhongguo shuxueshi yulu" 中國數學史餘錄 [Records on the history of Chinese mathematics], *Kexue* 3, 2: 238–241.

Liang Qichao [Zhongguo zhi xinmin] 梁啟超 [中國之新民] 1902. "Gezhixue yange kaolüe" 格致學沿革考略 [Brief history of the development of science], *Xinmin congbao* 10: 14.

Liang Qichao 梁啟超 1999. "Dili yu wenming zhi guanxi" 地理與文明之關係 [The relationship between physical geography and civilization], in Liang Qichao, *Liang Qichao quanji* 梁啟超全集. Beijing: Beijing chubanshe   943–947. [Originally published 1902.]

Liu Dun 劉頓 2002. "Cong 'Laozi huahu' dao 'Xixue zhongyuan'" 從 "老子化胡" 到 "西學中源" [From the idea that Laozi changed the Barbars to the theory of the Chinese origins of Western learning], *Faguo hanxue* 法國漢學 6 (Qinghua daxue chubanshe).

Liu Xianzhou 劉仙洲 1935. "Xuexi jixie gongcheng ying zhuyide de ji dian" 學習機械工程應注意的的幾點 [Some points worth considering when studying mechanical engineering], *Qinghua zhoukan*   87–89.

____ 1962. *Zhongguo jixie gongcheng faming shi: Diyi bian* 中國機械工程發明史。第一編 [History of discoveries about machinery in China: First collection]. Beijing: Kexue chubanshe.

Lu Xun 魯迅 1980. "Kexueshi jiaopian" 科學史教篇 [Essay on the instruction in history of science], in Lu Xun, *Fen* 墳 [The grave]. Beijing: Renmin wenxue chubanshe   18–36.

Mao Zedong 毛澤東 1976. "Zhongguo geming yu Zhongguo gongchandang" 中國革命與中國共產黨 [The Chinese Revolution and the Chinese Communist Party], in *Mao Zedong ji 7 Yan'an qi III (1939, 9–1941, 6)* 毛澤東集 7 延安期 III (1939, 9–1941, 6). Hong Kong: Yishan tushu   97–125.

Martin, W. A. P. 1879. "Alchemy in China." *China Review, or Notes & Queries on the Far East* 7, 4: 242–255.

____ 丁韙良 1883. *Xixue kaolüe* 西學考略 [A brief examination of Western learning]. Beijing: Tongwenguan.

____ 1901. *The Lore of Cathay, or the Mind of China*. New York: F. H. Revell.

Marx, Karl 1982. *Zur Kritik der politischen Ökonomie (Manuskript 1861–1863)*. In *Karl Marx, Friedrich Engels Gesamtausgabe, Zweite Abteilung: "Das Kapital" und Vorarbeiten*, vol. 3, pts. 1–6. Berlin (Ost): Dietz-Verlag.

Menzel, Ulrich 2001. "Die Musealisierung des Technischen: Die Gründung des 'Deutschen Museums von Meisterwerken der Naturwissenschaft und Technik' in München." Doctoral dissertation, Technical University of Braunschweig.

Mikami Yoshio 1913. *The Development of Mathematics in China and Japan*, Abhandlungen zur Geschichte der mathematischen Wissenschaften mit Einschluß ihrer Anwendungen 30. Leipzig: Teubner.

Milne, John 1886. *Earthquakes and Other Earth Movements*, International Scientific Series 55. New York: Appleton.

Needham, Joseph 1964. "Science and Society in East and West," *Science and Society* 28, 4: 385–408.

"Ni she guocui xuetang qi" 擬設國粹學堂啟 [Proposal to establish a school for National Essence] 1907. *Guocui xuebao* 3, 1: n.p.

Prakash, Gyan 1999. *Another Reason: Science and the Imagination of Modern India.* Princeton, NJ: Princeton University Press.

Qian Linzhao 錢臨照 1942. "Shi Mojing zhong guangxue lixue zhu tiao" 釋墨經中光學力學諸條 [Explanations on the passages on optics and mechanics in the *Mojing*], in *Li Shizeng xiansheng liushi sui jinian wenji.* Kunming: Guoli Beiping yanjiuyuan 135–162.

_____ 2001. "Qian Linzhao zizhuan" 錢臨照自傳 [Autobiography of Qian Linzhao], in Zhu Qingshi, ed., *Qian Linzhao wenji.* Hefei: Anhui jiaoyu chubanshe 2001, 3–13.

"Qinding Jingshi daxuetang zhangcheng" 欽定京師大學堂章程 (GX 28/11 [1902/3]). [Statutes of the Imperial University compiled by imperial order] 2000. In Wang Xuezhen and Guo Jianrong, eds., *Beijing daxue shiliao*, diyi juan 1898–1911. Beijing: Beijing daxue chubanshe 87–130.

Quan Hansheng 全漢昇 1935. "Qingmo de Xixue yuanchu Zhongguo shuo" 清末的西學源出中國說 [The theory of the Chinese origins of Western learning during the late Qing], *Lingnan xuebao* 4, 2: 57–102.

Schneider, Laurence A. 1971. *Ku Chieh-kang and China's New History: Nationalism and the Quest for Alternative Traditions.* Berkeley: University of California Press.

Schwarcz, Vera 1985. *The Chinese Enlightenment: Intellectuals and the Legacy of the May Fourth Movement of 1919.* Berkeley: University of California Press.

Sha Yuyan 沙玉彥 1931. *Kexueshi* 科學史 [History of science]. Shanghai: Shijie shuju.

Smith, Alexander 1917. *Introduction to Inorganic Chemistry.* New York: Century.

Tan Qixiang 譚其驤 1942. "Lun Ding Wenjiang suo wei Xu Xiake dili shang zhi zhongyao faxian" 論丁文江所謂徐霞客地理上之重要發現 [On the important geographical discoveries of Xu Xiake as pointed out by Ding Wenjiang], in *Jinian Xu Xiake shishi sanbai nian jinian kan* 紀念徐霞客逝世三百年紀念刊. N.p.: Zhejiang daxue 42–50.

Tang Caichang 唐才常 1903. "Zhuzi yulei yi you Xiren gezhi zhi li tiaozheng" 朱子語類已有西人格致之理條證 [The proof for the existence of the ideas of Western science in the classified conversations of Zhu Xi], in *Xinxue da congshu* 新學大叢書. Shanghai: Jishan qiaoji shuju, J. 65, 1a–4b.

Ukita Kazutami 浮田和民 1898. *Shigako genron* 史學原論 [Discussion of the origins of historiography]. Tokyo: Tokyo Senmon gakko.

Wang Benxiang 王本祥 1904. "Qiji da famingjia Wate zhuan" 汽機大發明家瓦特傳 [The great inventor of the steam engine Watt], *Kexue shijie* 9.

Wang Jin 王璡 1920a. "Zhongguo gudai jinshu yuanzhi zhi huaxue" 中國古代金屬原質之化學 [The chemistry of metallic elements of ancient China], *Kexue* 5, 6: 555–564.

——— 1920b. "Zhongguo gudai jinshu huahewu zhi huaxue" 中國古代金屬化合物之化學 [The chemistry of metallic compounds in ancient China], *Kexue* 5, 7: 672–684.

Wang, Q. Edward 2001. *Inventing China through History: The May Fourth Approach to Historiography*, SUNY Series in Chinese Philosophy and Culture. Albany: State University of New York Press.

Wang Yangzong 王揚宗 1997. "Mingmo Qingchu 'Xixue zhongyuan' shuo xinkao" 明末清初西學中源說新考 [A new investigation into the question of the theory of "the Chinese origins of Western learning" during the late Ming and the early Qing], in Liu Dun and Han Qi, eds., *Keshi xinkao: Qingzhu Du Shiran xiansheng congshi kexueshi yanjiu 40 zhounian xueshu lunwenji.* Shenyang: Liaoning jiaoyu chubanshe 71–83.

Wang Zhenduo 王振鐸 1989. "Zhang Heng houfeng didongyi de fuyuan yanjiu" 張恆候風地動儀的復原研究 [Research on the reconstruction of Zhang Heng's seismograph], in Wang Zhenduo, *Keji kaogu luncong* 科技考古論叢. Beijing: Wenwu chubanshe 287–344.

Wang Zhixin 王治心 1930. "Zhongguo gudai kexue shang de faming" 中國古代科學上的發明 [The discoveries of traditional Chinese science], *Xieda xueshu* 1: 167–179.

Wang, Zuoyue 2002. "Saving China through Science: The Science Society, Scientific Nationalism, and Civil Society in Republican China," *Osiris* 17: 291–322.

Wittfogel, Karl August, 魏特夫 1944. "Zhongguo weishenme mei you chansheng ziran kexue" 中國為甚麼沒有產生自然科學 [Why did China not develop modern science?], translated by Wu Zaoxi 吳藻溪, *Kexue shibao*, October 1.

Yan Dunjie 嚴敦杰 1936. "Zhongguo suanxuejia Zu Chongzhi ji qi yuanzhoulü zhi yanjiu" 中國算學家祖沖之及其圓周率之研究 [The Chinese mathematician Zu Chongzhi and his researches on the number π], *Xueyi* 學藝 (Wissen und Wissenschaft) 15, 5: 37–50.

Yao Yaxin 姚雅欣 and Gao Ce 高策 2002. "Qinghua guoxueyuan shiqi Liang Qichao yu 'Zhongguo ziran kexueshi' yanjiu kuangjia de fuchu" 清華國學院時期梁啓超與中國自然科學史研究框架的浮出 [Liang Qichao at the time when he was working at the Institute for National Studies and the emergence of a framework for researching Chinese natural science], *Kexue jishu yu bianzhengfa* 22, 5: 96–100.

Ye Qisun 葉企孫 1916. "Kaozheng Shang Gong" 考正商功 [On Shang Gong], *Qinghua xuebao* 2, 2: 59–87.

Yu Heyin 虞和寅 1903. "Zhiwuxue lüeshi" 植物學略史 [A short history of botany], *Kexue shijie.* [I was unable to determine the issue of *Kexue shijie* in which this article appeared since the issue(s) I could consult had been unbound and rebound according to topics.]

Zhang Fu 彰孚 1907. "Dongwuxue lishi" 動物學歷史 [A history of zoology], *Xuebao* 1: 6.

Zhang Yinlin 張蔭麟 1956 [1928]. "Zhongguo lishi shang zhi 'qiqi' ji qi zuozhe" 中國歷史上之奇器及其作者 [Wonderful machines and their producers in Chinese history], in Weiliang Lun 倫偉良, ed., *Zhang Yinlin wenji* 張蔭麟文集. Taibei: Jicheng tushu gongsi 64–85.

Zhang Zigao 張子高 1923. *Kexue fada shilüe* 科學發達史略 [A brief history of the development of science]. Shanghai: Zhonghua shuju.

Zhou Zhaoji 周肇基 1993. "Zhuming kejishi xuejia Hu Daojing jiaoshou" 著名科技史學家胡道靜教授 [The eminent historian of science Hu Daojing], *Zhongguo keji shiliao* 中國科技史料 14, 1: 43–49.

Zhu Kezhen 竺可楨 2006. *Zhu Kezhen quanji* 竺可楨全集 [The complete works of Zhu Kezhen]. Shanghai: Shanghai keji jiaoyu chubanshe.

Zim, Herbert Spencer 1945. *Rockets and Jets*. New York: Harcourt, Brace.

Ziran jie "Fakan zhiqu" 自然界發刊旨趣 [*Natural World*, "Announcement of Publication"] 1926.

Disciplining the National Essence

Liu Shipei and the Reinvention of Ancient China's Intellectual History

Joachim Kurtz

Abstract

Around the year 1900, the taxonomies in which knowledge had been organized in China for centuries were unsettled and gradually superseded by a Western-inspired disciplinary matrix. The introduction of new curricula of higher education and the abolition of the civil examination system unmistakably marked the demise of the old regime of learning. In the natural sciences the transition led to a rapid denigration of Chinese knowledge. In the realms of the humanities and social sciences, the transformation proved to be more complex. To defend the validity of values enshrined in canonical and some noncanonical texts, late Qing scholars suggested various ways to preserve China's moral heritage in the incipient world of global knowledge. One of the earliest and boldest attempts to secure a sustainable place for China's embattled moral sciences was formulated by Liu Shipei in the context of the "National Essence" (Guocui) Movement. This essay aims to reconstruct the key elements of Liu's reinvention of ancient China's intellectual history in Euro-American terms and to analyze the violent conceptual transformations that this effort required and that foreshadowed many reformulations of the histories of Chinese science and thought published throughout the Republican period.

In the decades around the year 1900, the taxonomies in which knowledge had been organized in China for centuries were unsettled and gradually superseded by a Western-inspired disciplinary matrix. The introduction of new curricula of higher education from 1898 onward[1] and the abolition of the civil examination system in 1905[2] unmistakably marked the demise of the old regime of learning. In the field of the natural sciences the transition quickly led to an almost complete denigration of the "old" Chinese knowledge.[3] In the realms of the humanities and social sciences, however, the transformation proved to be more complex. To defend the universal validity of values enshrined in canonical and some noncanonical writings, late Qing scholars suggested various ways to preserve China's moral heritage in a new world of knowledge defined in globalized terms of Euro-American origin. Their efforts contributed to the emergence of new genres of writing aimed at ensuring the survival of substantial portions of traditional thought through translations into a new lan-

---

1 Weston 2002.

2 Elman 2000, 585–618.

3 Reynolds 1991; Elman 2005, 396–421.

guage of scholarship and a Europeanized disciplinary framework that continue to shape our understanding of "Chinese philosophy," "Chinese logic," "Chinese religion," and many other areas of learning.

One of the earliest and boldest attempts to secure a place for China's embattled moral sciences in the incipient world of global knowledge was formulated in 1905 by the intellectual prodigy and self-styled "radical no. 1"[4] Liu Shipei 劉師培 (1884–1919) in the context of the movement for the preservation of the "National Essence" (Guocui 國粹). Arguing against efforts to save Chinese learning through the creation of a protected reservoir of "National Studies" (Guoxue 國學), Liu drafted a master plan for a new grand narrative of ancient China's intellectual history that was structured by disciplinary boundaries and, as such, raised questions that none of the now all-too-familiar histories of "Chinese philosophy," and so on, could afford to ignore. This essay aims to identify the key elements of Liu's radical effort at translating early Chinese knowledge into disciplinary terms and to illustrate the violent conceptual transformations necessary for its completion. I will focus on a close reading of Liu's deliberations on what he considered to be the core disciplines of the modern humanities—psychology, ethics, and sociology—and concentrate in particular on the three most significant aspects of his reinvention of ancient China's intellectual history: first, his transformation of argumentative strategies derived from the popular theories of the "Chinese origins of Western knowledge" (Xixue Zhongyuan shuo 西學中源說); second, his adaptation of philological methods borrowed from the reinvigorated studies in noncanonical masters (zhuzixue 諸子學); and, third, his use of an untested scholarly vocabulary imported from Japan in interpretations of ancient Chinese texts.

In recent years, the formation of modern academic disciplines in late Qing and early Republican China, not only in the humanities and social sciences, has attracted growing interest. Pioneering Chinese studies by Chen Pingyuan,[5] Sang Bing,[6] Luo Zhitian,[7] Zuo Yuhe,[8] inter alia, have greatly enhanced our understanding of the "intellectual resources" (sixiang ziyuan 思想資源) and "conceptual tools" (gainian gongju 概念工具)[9] applied in the reordering of late Qing China's discursive terrain. More recently, a series of edited volumes emerging from a project initiated by John Makeham has traced the formation

---

4   Wang Xiaoling 1998, 163.
5   Chen 1998.
6   Sang 2001.
7   Luo 2003.
8   Zuo 2004.
9   Wang Fansen 2001.

of individual disciplines.[10] Other works have reconstructed the institutional transformations accompanying, and more often than not fueling, this complex process;[11] studied the emergence of the new media that facilitated the public negotiation of domestic and imported knowledge;[12] examined the new terms in which novel and old ideas alike came to be circulated;[13] and investigated the subtexts of changing gender and ethnic relations underlying these multilayered transformations.[14] What I hope to add to this sizable body of scholarship is a detailed, and deliberately myopic, analysis of the complexities of conceptual change on the microhistorical level of a single text that anticipated many of the problems, albeit hardly any of the solutions, besetting the rewriting of ancient China's intellectual history in disciplinary terms throughout the Republican period and beyond.

I   Early Inspirations

The role of Liu Shipei in the late-Qing reinvention of Chinese intellectual history is not very well understood. The precocious offspring of a distinguished scholarly lineage from Yangzhou,[15] a center of Han Learning since the seventeenth century, is best known for his chameleonlike political radicalism that led him from a vicious breed of anti-Manchu nationalism, via a romanticized anarchism with feminist inclinations, to a thoroughly anachronistic monarchism after the demise of the imperial order.[16] In the early years of the twentieth century, however, Liu was seen above all as a rising star on China's intellectual scene due to his unusual philological skills, his elegant style, and his versatility in absorbing foreign ideas and blending them with Chinese concepts.[17] In his preanarchist phase, between 1903 and 1907, Liu applied all these abilities in a series of books and articles devoted entirely or in part to a reinterpretation of ancient China's intellectual heritage. In combination with his increasingly outspoken political opinions, these reflections almost instantly

---

10   To date, three volumes on the formation of history, sociology, and anthropology, as well as philosophy, as modern academic disciplines have appeared in this series: Moloughney and Zarrow 2011; Dirlik, Li, and Yen 2012; and Makeham 2012.

11   Guan 2000.

12   Mittler 2004; Janku 2003.

13   Lackner, Amelung, and Kurtz 2001.

14   Fong, Qian, and Zurndorfer 2004.

15   For an overview of this lineage, see Wan 2003, 328.

16   Zarrow 1990, 32–45; Fang-yen Yang 1999, chaps. 2–3.

17   Chang 1987, 146–179.

earned him nationwide acclaim.[18] Although they did not exert long-lasting influence, Liu's symptomatic reflections are of special interest because they illustrate the violent transformations necessary to discipline traditional Chinese learning in globalized terms much more vividly than the more sophisticated attempts that soon rendered his tentative deliberations obsolete.

A crucial element of these transformations was a shift in the languages and taxonomical schemes in which ancient Chinese knowledge was discussed. Liu Shipei played a significant role in initiating this transition. His first book, *The Quintessence of China's Social Contract* (*Zhongguo minyue jingyi* 中國民約精義), published in 1903,[19] followed the formal example of mainstream "Chinese-origins" anthologies, most notably Wang Renjun's 王仁俊 (1866–1914) monumental *Gezhi guwei* 格致古微 (Ancient subtleties of science   1896).[20] Like Wang Renjun, Liu subsumed the ideas he culled from a partial Chinese adaptation of a Japanese translation of Rousseau's *Contrat social*[21] under a roughly chronologically ordered selection of supposedly equivalent, or at least related, quotations from classical Chinese sources.[22] In his second work, the *Book of Expulsion* (*Rangshu* 攘書, 1903), a radical anti-Manchu treatise that again contained some passages with a scholarly focus, Liu was less strict in his insistence on the formal priority of Chinese learning and the divisions in which it was customarily presented. Thus, in a chapter "On the Correct Use of Names" (Zhengming pian 正名篇), Liu not only borrowed terms from Yan Fu's 嚴復 (1859–1921) rendition of John Stuart Mill's *Logic* to corroborate Xunzi's 荀子 (ca. 313–238 BCE) insights into the properties of "names" (*ming* 名) but, wherever it suited his argumentative purposes, also used Xunzi to support points argued by Yan Fu's Mill.[23] By freely exchanging the order of *explanans* and *explanandum*, Liu deliberately transgressed the interpretive conventions of the Chinese-origins genre. His uninhibited wandering between two conceptual frames whose priority had to be determined in each particular instance was a new element in discourses dedicated to the amalgamation of Chinese and Western knowledge. Chinese-origins authors had invariably insisted on the precedence of Chinese categories until 1897, when the order was abruptly reversed. Beginning with Jiang Biao's 江標 (1860–1899) *Records on the Essence*

---

18   Li Fan 2003, 131–148.

19   Liu 1903c.

20   Wang Renjun 1993. The most comprehensive study of the Chinese-origins genre remains
      Quan 1935; for more recent discussions, see Lackner 2003; and Huters 2005, 23–42.

21   Yang Tingdong 1902. The translation was based on Harada 1883.

22   Wang Xiaoling 1998; Zarrow 1990, 36–38; Angle 1998.

23   Liu 1903b   15–17. See Yan 1902–1905.

of Science (*Gezhi jinghua lu* 格致精華錄, 1897), ancient Chinese knowledge that resonated with concerns of modern natural science began to be redefined to fit into a Westernized disciplinary matrix.[24] Still, the taxonomies of Chinese and Euro-American knowledge had remained strictly separated, not least to protect China's traditional moral and political order. What Liu seemed to have in mind in his *Book of Expulsion* and a series of roughly simultaneous essays, however, was an eclectic fusion of just these taxonomies, in both the natural and moral sciences, in the dual hopes of opening up new, ideologically consequential paths in the domestication of Western and the selective preservation of Chinese learning.

Yet, as in so many other instances, Liu did not immediately pursue his daring idea any further. It resurfaced only when he became the editor of the *National Essence Journal* (*Guocui xuebao* 國粹學報) in 1905. By then, Liu's main inspiration, borrowed from Japanese precedents,[25] was an idealized European renaissance that he translated into the hope of preserving traditional Chinese scholarship and values through a reinterpretation in Western-derived terms.[26] This hope, which was shared by many of his contemporaries on both sides of the intellectual divide between the self-styled "reformers" and "revolutionaries,"[27] was the driving force behind his most ambitious project to date: the compilation of a comprehensive intellectual history of the late Zhou that was to be written along the lines of modern academic disciplines.

II   Redefining a Genre

There is strong evidence indicating that Liu never intended to complete this project.[28] All he accomplished was a series of seventeen "Prefaces" (*xu* 序; in our context more appropriately translated as "Prolegomena") that appeared in the first five issues of the *National Essence Journal*.[29] Although totaling no more than fifty double-faced pages, Liu's "Prolegomena to an Intellectual History of the Late Zhou" (*Zhoumo xueshushi xu* 周末學術史序) aroused considerable excitement in intellectual circles, even envy, due to their glamor-

---

24   Jiang 1897. See Amelung 2004.

25   Fan 2004, 413–416.

26   Wang Dongjie 2000; Zheng 1997, 132–139; Bernal 1976, 106. On late Qing and early Republican views of the European renaissance more generally, see Eber 1975.

27   Luo 2003, 90–106.

28   Li Jinxi 1936.

29   Liu 1905. In the following, Liu's "Prolegomena" will be cited directly in the text (enclosed by parenthesis) by page number in the annotated reedition in Liu 1998, 211–288.

ous novelty.[30] Liu's was indeed the first Chinese attempt to translate China's moral sciences into a Western-inspired taxonomy of knowledge. Moreover, the "Prolegomena," as I shall refer to them henceforth, could claim several other "firsts" that excited or disconcerted many of his contemporaries. First, the "Prolegomena" were, to my knowledge at least, the first Chinese text advocating the possibility, or even necessity, of compiling separate histories of "Chinese philosophy," "Chinese logic," "Chinese religion," and so on—projects that are customarily credited to the self-declared masters of intellectual innovation of the May Fourth generation or the ubiquitous Liang Qichao 梁啓超 (1873–1929).[31] Second, Liu was the first author to draw radical consequences from Confucius's imminent demotion from most venerated saint to the modest head of "one among the nine schools of late Zhou thought" (215). While many earlier authors, most notably Zhang Xuecheng 章學誠 (1738–1801), whose thinking had inspired generations of scholars in Liu's native Yangzhou, had declared that the "Six Classics are all histories" (liu jing jie shi 六經皆史),[32] none had as yet dared to reject all methodological distinctions between canonical (jing 經) and noncanonical (zi 子) writings, or to historicize Confucius and his teachings as unforgivingly as Liu set out to do in the "Prolegomena."[33] Finally, and perhaps most consequentially, Liu established a new genre of Chinese intellectual historiography: the xueshushi 學術史, or the history of what, for want of a better English term, is usually rendered as "scholarship."[34] Liu, however, had a narrower meaning in mind when choosing this designation for his historiographical project. Xueshu 學術, he wrote, drawing on definitions in his favorite reference tool, Xu Shen's 許慎 second-century Explanation of Writing through the Analysis of Characters (Shuowen jiezi 説文解字), must be understood as a compound in which xue 學 denotes "understanding," or any kind of specific knowledge, while shu 術 refers to the "path," that is, the ability to put this knowledge into practice (212). Xueshu was thus intended to refer to a unity of specialized knowledge and a methodology for its successful application. According to the precepts of the new historiography emerging in the first years of the twentieth century, as Liu Shipei understood them,[35] writing a history of this ideal unity meant tracing its origins and development up until the time when knowledge was divorced from practice and

---

30   Wu 1995, 172–173; Li Xiaoqian 2001b.

31   Cua 2000.

32   Nivison 1966, 201–204; Wang Hui 2004, 458–486.

33   Zheng 1994.

34   Li Fan 2000, 2004.

35   Li Hongyan and Zhong Weimin 1994.

Chinese thought had lost its vitality. (As Liu reiterated in the closing para-graphs of almost all his "Prolegomena," this moment had arrived embarrass-ingly early: it coincided with the canonization of the Confucian Classics and the simultaneous suppression of all other schools of thought in 136 BCE.) By his own account, Liu's attempt to compile such a history required a break with the previously dominant model of Chinese intellectual history: the "case studies of learning" (*xuean* 學案) as exemplified by Huang Zongxi's 黃宗羲 (1610–1695) *Case Studies of the Learning of Ming Scholars* (*Mingru xuean* 明儒學案).[36] Where the "case studies," according to Liu, focused on "people" (*ren* 人), that is to say, individual thinkers and their scholarly lineages, his "history of scholar-ship" aimed at "collecting insights of all different schools and arranging them by discipline (*xue* 學)" (212).

Although Liu's "Prolegomena" were uncharacteristically nonbelligerent in tone, the aim of his new disciplinary histories was unmistakably subversive. In Europe, Liu proclaimed in his introductory "General Preface," the Catholic Church had controlled learning for centuries. Only with the decline of the church's power after 1500 had this prerogative begun to shift into the hands of "the multitude," thus fostering Europe's unprecedented rise. Similarly, in ancient China, control of learning had initially rested with the hereditary offices of court astrologers and historians (*shiguan* 史官).[37] Only after central power collapsed in the Eastern Zhou could "talented and educated common-ers," as well as unemployed bureaucrats, realize their dormant potential and establish independent schools of thought (211). While these schools failed to agree on most issues—according to Liu mainly because they lacked a "science of disputation" (*bianxue* 辯學) akin to European logic (245)—the transition from "official learning" (*guanxue* 官學) to "private learning" (*sixue* 私學) had led to the productive dissemination of knowledge among the lower echelons of society (212). Contrary to traditional assessments, the resulting "diversity" of opinions was not a threat but a potential asset if channeled into the secure boundaries of scholarly disciplines—a strategy, Liu claimed, already imple-mented by Confucius and his disciples, who had preserved the wisdom of the early sage-kings by teaching it in four separate "courses of study" (*ke* 科) (240). The demand addressed to the Qing educational regime that Liu implicitly raised in his historiographical ruminations was hence to loosen control over scholarship and allow private learning to flourish again without fear of suppression,[38] for the first time since the ideological stratification under the

36   Zhu 2001.
37   Li Xiaoqian 2001a.
38   Hon 2003, 249–251.

Han emperor Wudi 武帝 (156–87 BCE), as long as it moved within the conceptual confines of the emerging disciplinary conventions.

III   Delineating Boundaries

How these conventions should be defined, and more specifically where Chinese knowledge could be situated within this new disciplinary matrix, were fiercely contested issues when Liu started to publish his "Prolegomena." In accordance with the worn-out *ti/yong* formula that asserted that Chinese knowledge was the "substance" (*ti* 體) and appropriate Western knowledge was exclusively for purposes of "application" (*yong* 用), the educational curricula promulgated between 1898 and 1904 had all reserved a protected space for canonical studies and several other traditional subjects alongside the newly introduced sciences from the West and Japan.[39] Liu and some of his associates feared that this artificial ghettoization could not be sustained. If Chinese learning was to survive, they argued, it had to compete with Western knowledge in the same arena, and the National Essence circle strove to demonstrate that it could do so when fitted creatively into new global categories and terms.

Liu left no hint as to how he selected the disciplines treated in his "Prolegomena," or whether he intended his taxonomy as a complete and coherent disciplinary system or simply a cumulative list of basic fields. At first sight, the sixteen chapters to be explored in his new intellectual history certainly seem rather eclectic:

[1] Psychology (*xinlixue* 心理學)   [9] Education (*jiaoyuxue* 教育學)
[2] Ethics (*lunlixue* 倫理學)   [10] Natural Sciences (*likexue* 理科學)
[3] Logic (*lunlixue* 論理學)   [11] Philosophy (*zhelixue* 哲理學)
[4] Sociology (*shehuixue* 社會學)   [12] Calculating Skills (*shushuxue* 數術學)
[5] Religion (*zongjiaoxue* 宗教學)   [13] Philology (*wenzixue* 文字學)
[6] Political Science (*zhengfaxue* 政法學) [14] Technology (*gongyixue* 工藝學)
[7] Economics (*jixue* 計學)   [15] Penal Law (*falüxue* 法律學)
[8] Military Science (*bingxue* 兵學)   [16] Composition (*wenzhangxue* 文章學)

Contrary to the claims of most recent commentators, Liu clearly did not suggest "a purely Western taxonomy."[40] Although Euro-American disciplines formed the backbone of his system, he did not hesitate to make adapta-

---

39   Beijing daxue xiaoshi yanjiushi 1993, 81–130.
40   See, e.g., Wu 1995; Li Xiaoqian 2001b; Luo 2003.

tions dictated either by his own scholarly predilections or by the need to find a place for all kinds of late Zhou knowledge included in traditional bibliographical catalogs. The omission of a separate category for "history," for instance, may have been due to Liu's already-mentioned conviction that the Classics were "all histories." More difficult to explain was his slighting of mathematics, a prominent subject in late Zhou thought, which he relegated to the status of an auxiliary to astronomy and touched upon only in the very brief chapter on the "natural sciences" (242–243). Conversely, the addition of a section on "calculating skills," a generic label for techniques of divination, astrology, soothsaying, and numerological speculation introduced in early bibliographical treatises such as *The Seven Categories* (*Qilüe* 七略) and the "Bibliographical Record" in the *Book of Han* (*Hanshu yiwenzhi* 漢書藝文志),[41] was probably intended to acknowledge the historical importance of these arts, even if Liu remained skeptical as to their scholarly value. This chapter was also the place where he discussed medicine, another striking omission from his list of chapters, on the grounds that in the late Zhou physicians and diviners, as well as their respective skills, were regarded as closely related (256–257). Finally, the call to compile separate histories of "philology" (or, more precisely, the morphological, phonological, and semantic study of Chinese characters) and "composition" (i.e., the art of writing elegant texts in diverse literary and nonliterary genres)[42] may have reflected Liu's belief that traditional Chinese philology, or "lesser learning" (*xiaoxue* 小學), was the foundation of all true knowledge, ancient or modern, and thus an indispensable part of any comprehensive disciplinary matrix.[43]

The sequence in which Liu arranged his projected chapters was also far from being unequivocally "modern" or Western. From his sketchy explanations, we may infer that chapters 1 through 4—"Psychology," "Ethics," "Logic," and "Sociology"—were to be considered as "moral sciences," elucidating, respectively, "human nature," personal and family morality, the individual's place in society (which, according to Liu, was the domain of "the science of names," logic),[44] and the rules and patterns of human association. Chapter 5, "Religion," bridged the gap between society and state by emphasizing a cosmological foundation for both and linking behavior beneficial to the community, such as the demands of "filial piety" (*xiao* 孝), to the sphere of the spiritual or divine (223–225). Chapters 6 through 9—"Politics," "Economics," "Military

---

41   Zuo 2004, 44–50.

42   Zhang Huien and Zhong Humei 1994.

43   Liu 1903a.

44   For discussion, see Kurtz 2011, 289–301.

Science," and "Education"—were concerned with classic domains of state activity. Drawing on his political writings, Liu used these sections to highlight sprouts of a vaguely democratic vision of government that treated the state and its resources as the common property of all citizens instead of the private domain of one ruling clan. Chapters 10 and 11, "Natural Sciences" and "Philosophy" (or rather what was left of the latter since ethics and logic had already been discussed), addressed quintessential areas of modern scholarship that had no obvious equivalents in traditional classifications. Reverting to an unmitigated Chinese-origins mode of argumentation, Liu here listed bits and pieces of ancient knowledge that presumably anticipated theories whose adoption was propagated as most urgent for the survival of the Chinese state and nation. In the "Philosophy" section, he thus identified fragmentary precedents for the fashionable doctrines of "evolutionism" (*tianyan xuepai* 天演學派), "utilitarianism" (*leli xuepai* 樂利學派), and "cosmopolitanism" (*datong xuepai* 大同學派) (247–254). The linkages between the remaining disciplines, if any were intended, were more tenuous. With the exception of chapter 14, "Technology," in which Liu presented a teleological history of human civilization through quotations describing the progressive development of "tools" (*qi* 器) in Chinese antiquity, the most obvious characteristic shared among these sections was the fact that they dealt with subjects that had been seen as more significant in the history of Chinese thought than in the West. Neither "Calculating Skills" nor "Philology," "Penal Law," or "Composition" would have occupied comparably prominent places in nineteenth-century histories of European ideas. Yet, even if the modified Western taxonomy in which Liu proposed to cast his enterprise did not achieve ultimate systematic unity, it certainly provided a useful structure for his projected reinvention of China's intellectual history in disciplinary terms.

IV   Filling the Frame

Unfortunately, however, the way in which Liu filled this structure in the "Prolegomena" was for the most part far from convincing. Hardly any of his concrete suggestions of parallels or equivalences between ancient Chinese and modern Western theories and concepts stood the test of time. The main reason for this failure was Liu's sketchy knowledge of the European disciplines into which he intended to translate China's intellectual heritage. Liu did not read any Western language, and one cannot help the impression that he tackled the many Japanese titles quoted in his early writings through Liang Qichao's infamous "method of reading Japanese texts as Chinese" (*Hewen Handu fa* 和

文漢讀法).[45] In his search for new ideas Liu depended much more heavily than he was willing to concede on a limited number of the most ornate Chinese renditions of Western and Japanese texts—in particular, works translated by Yan Fu and Zhang Binglin 章炳麟 (1869–1936).[46] Moreover, he amplified the problem by approaching these translations with the same philological methods that he had been trained to apply to classical writings in preparing for the civil service examinations. Many of his most audacious comparisons were based on interpretations of terms that had only recently been invented—in more or less arbitrary fashion—to translate foreign notions. Liu nonetheless expounded them as if they were the sanctified words in classical scriptures, which had been glossed and revised by generations of devoted scholarly minds, and he based the most far-reaching conclusions on such pseudo-etymological analyses.

Results could be disastrous, especially in cases where he built his arguments exclusively on fanciful but ultimately misguided glosses of the names or definitions of entire disciplines. One such example, that I have discussed in more detail elsewhere,[47] is his gloss and subsequent discussion of *lunlixue* 論理學, logic, or literally "the science of reasoning," as *mingxue* 名學, "the science of names." Although *mingxue* had indeed been proposed by Yan Fu as a rendition of the European term "logic," this fact alone did not warrant Liu's conclusion that logic as a whole should be treated as a forgotten sub-branch of Chinese philology expounding the properties of "names" in the sense of "designations of rank," and thus as a discipline prescribing and perpetuating individual positions in social hierarchies (217–219).

But even if Liu based some of his ideas on misguided interpretations of Western notions, most of his deliberations were far from absurd. On the contrary, more often than not Liu's reflections anticipated argumentative strategies that continue to reverberate in influential works on the history of "Chinese philosophy" and cognate fields. A case in point is his opening of the "Prolegomenon to a History of Psychology" (Xinlixue shi xu 心理學史序). Liu started this chapter with a review of the discipline's history in Europe:

> Some time ago, I looked into the history of scholarship in the Far West. In antiquity there were ten states in the Far West, of which the most important was Greece. The first school to gain prominence in Greece was that of the Ionians, who based their teachings on the "patterns of things" (*wuli*

---

45   Xia 2006, 277–286.
46   Li Fan 2003, 83–87.
47   Kurtz 2011, 294–301.

物理, physics); that is, they focused on that which is below form. Later, the Italians [Eleatics] emerged, whose theories were founded on the "patterns of the heart" (*xinli* 心理, psychology); that is, they concentrated on that which is above form. This is the sequence in which scholarship develops.

In a second step, Liu attributed universal validity to the occidental experience by claiming that developments there had unfolded with obvious and indisputable necessity:

In high antiquity, humans lived in jungles and were haunted by beasts; they were not yet enlightened, and therefore, the idea of looking into the heart (*guanxin* 觀心) had not yet come into existence.[48] When humans are born, their hearts are in a state of rest; only when affected by things are [their hearts] stirred. The things thus successively give rise to self-understanding and prevent vain pondering. In consequence, the first sprouts of the idea of looking into things (*guanwu* 觀物) germinated at this time. The people of mid-antiquity slowly developed new knowledge. They understood that things are derived from consciousness and that consciousness depends on the heart. Hence, they examined things in the distance as well as, near at hand, their own selves, and thus, the notion of looking into one's own heart spontaneously emerged.

Finally, in the third step of his argument, Liu confirmed the universality of the European example by pointing out parallels in China's intellectual history:

If we look back to the times of the Flame Emperor and the Yellow Emperor, we see that scholarship successively became more and more complete. But there was a tendency to focus on the factual world and venerate practical activity. In this, [the scholars of that time] were close to the Ionians. Only under the Xia and Shang did they start to talk about the patterns of the heart, that is, psychology. (213)

After thus establishing the parallels in the formation of scholarship in China and the West, Liu moved on to discuss the specific contents of the discipline. In the case of psychology, he traced early Chinese theories of "human nature"

---

48   Note: *For the term* guanxin, *see the Buddhist canon.* (Here and in the following, footnotes starting with "Note:" and set in *italics* reproduce Liu Shipei's own interlinear annotations and comments.)

(*xing* 性) from the first attested mention of the term *xing* in the *Classic of Documents* (*Shujing* 書經) to the well-known debate between Mengzi 孟子 (ca. 370–290 BCE) and Xunzi, which, according to Liu, marked the early climax and premature end of the development of "psychology" in ancient China.

Liu apparently aimed at presenting structurally similar arguments in all his chapters but was not always able to construct equally smooth parallels. Among the more felicitous examples is his outline of the "Prolegomenon to a History of Ethics" (Lunlixue shi xu 倫理學史序). In this instance, he even managed to infer the universal standard of the discipline from traditional Chinese beliefs:

> The Chinese Way revolves around the art of humaneness (*renshu* 仁術). [The character for] "humaneness" consists of the components "two" (*er* 二) and "humans" (*ren* 人); therefore, the great Way of humaneness becomes manifest only when humans are united ... When humans get together, patterns of relation (*lunli* 倫理) emerge.[49] Theories of the patterns of human relations began to sprout in the age of [the prehistoric emperors] Tang 唐 and Yu 虞. In the statement "He made the able and virtuous distinguished so that he would be close to the nine classes of his kindred," the "Canon of Yao" (Yaodian 堯典) [in the *Documents*] infers [the art of] regulating the household from the way of self-cultivation.[50] When Xie 契 was charged with education [under the prehistoric emperor Shun 舜], he reverently spread the Five Teachings ... ; these too derive the Way of state and society from the regulation of the household.[51] This is the sequence in which ethics is discovered. (215)

While reasserting the early Chinese emphasis on personal and family ethics as the beginning of all morality, Liu Shipei ended even this chapter, after a critical review of the ethical views of various pre-Qin schools of thought, with the somber note that ideological restrictions after the canonization of the

---

49 Note: *The two terms* lun 倫 *and* li 理 *both refer to a definite order. In his* Annotations on the Meaning of Characters in the Mengzi (Mengzi ziyi shuzheng 孟子字義疏證), *Dai Dongyuan* 戴東原 [Dai Zhen 戴震 (1724–1777)] *explains* li 理, "pattern," with "proper order" (tiaoli 條理). *I should add that the Five Human Relations* (wulun 五倫) *are all relative terms* (duidai zhi mingci 對待之名辭).

50 Note: *"Self-cultivation" is what today's Westerners call "personal ethics"; "regulating the household" is what they call "family ethics." For this [art of] "regulating the household" is derived from "self-cultivation."*

51 Note: *When Westerners talk about ethics, they move from family ethics to societal ethics and from there on to state ethics in order to highlight the individual's duties toward society and state.*

Confucian Classics had led to a one-sided and ultimately untenable focus on "private morality" to the detriment of "public spirit"—an aberration that Liu strove to correct through his roughly simultaneous *Textbook on Ethics* (*Lunli jiaokeshu* 倫理教科書, 1905), which he designed in hopes of having it adopted by the state for high school education[52] (217).

Liu's introduction to the "Prolegomenon to a History of Sociology" (Shehuixue shi xu 社會學史序) is an example of a much less smooth reconciliation between classical Chinese scholarship and a modern European discipline. I cite this chapter in some detail, on the one hand, to illustrate the awkward results that Liu's philological method could produce when applied to freshly minted terms and, on the other, to underline that Liu's intuitions may nonetheless have seemed quite plausible in the discursive context of the time. Liu's basic claim was that the scope and methods of contemporary sociology, as he understood the discipline from his reading of Zhang Binglin's *Sociology* (*Shehuixue* 社會學, 1902),[53] had been anticipated in two of China's most venerated Classics: the *Classic of Changes* (*Yijing* 易經) and the *Spring and Autumn Annals* (*Chunqiu* 春秋). His presentation of affinities between sociology and the *Changes*, especially, makes for rather curious reading today:

> Examining the hexagrams in the *Zhou Changes*, we find that [the text] first lists "judgments" (*tuan* 彖) and "images" (*xiang* 象), then "line commentaries" (*yaoci* 爻辭). *Tuan*, "judgments," refers to "material" (*cai* 財),[54] that is, "facts and objects" (*shiwu* 事物);[55] *xiang*, "images," means "resemblances," that is, "phenomena" (*xianxiang* 現像); and *yao*, "line commentaries," alludes to "imitations" or "models" (*xiao* 效), that is to say, to the "proper order" (*tiaoli* 條理).[56] The sociology of contemporary Western scholars aims to gather phenomena of the human world and discover the patterns of human association in order to fathom the foundations of facts and objects. The American [Franklin H.] Giddings once said: "The origin of society is the consciousness of similarity in kind (*tong-*

---

52   Angle 1998, 624–633.

53   Zhang Binglin 1902. Zhang's translation was based on Kishimoto 1900.

54   Note: *The "Attached Verbalizations"* [Xici 繫辭] *says: "The judgments are the material." "Material" here means "raw material."*

55   Note: *Mr. Ruan Yuntai* 阮雲臺 [Ruan Yuan 阮元 (1764–1849)] *glosses the ancient character* ter tuan, *"judgments," with* li 蠡, *"worms," "seashells," or "fragments"; and* li *in turn can be glossed as* fen 分, *"part" or "lot." Thus,* tuan *and* li *refer to things and events having a definite order.*

56   Note: *The six lines are arranged in due sequence; this confirms the meaning of "proper order."*

lei yishi 同類意識). Initial disturbances arise from differences (*chabie* 差別); if we wish to see [these differences] subdued, we must imitate (*moxiao* 模效) human nature." Now, *tuan*, "judgments," refers to one's "part" or "lot" (*fen* 分), which is equal to "differences" (*chabie*); and *yao*, "line commentary," can be glossed as "to copy" (*fang* 仿), which is the same as "to imitate" (*moxiao*). Therefore, the words of the sociologists are close to the teachings of our *Changes*. (220)

After this already preposterous beginning, Liu went on to outline more concrete parallels that seem hardly less far-fetched:

The Way of the *Zhou Changes* lies in no more than storing what is past and scrutinizing what is to come (*cangwang chalai* 藏往察來)[57] and in exploring the mysterious and probing subtleties (*tanze suoyin* 探賾索隱).[58] "Storing what is past" is based upon exploring the mysterious; it focuses on facts and objects. The Westerners call this "social dynamics" [or, in more literal retranslation, "dynamic sociology"] (*dong shehuixue* 動社會學). "Scrutinizing what is to come" is based upon probing subtleties; it focuses on patterns. The Westerners call this "social statics" (*jing shehuixue* 靜社會學) ["static sociology"]. The usefulness of "storing what is past" lies in bringing together kinds and differentiating groups (*julei fenqun* 舉類分群).[59] [The sages] assisted the beginnings and asked

---

57 Note: *The "Attached Verbalizations" says: "[The sages] stored what was past and scrutinized what was to come"; "Unscrutinized, past and present are called continuity"* [quotation inaccurate, JK]; *and also "Divinity lies in scrutinizing what is to come; wisdom lies in storing what is past." Mr. Jiao Litang* 焦理堂 [Jiao Xun 焦循 (1763–1820)] *wrote in his* Discussions of the Changes (Yi hua 易話): *"When studying the* Changes, *we must first of all know that, prior to Fu Xi's* 伏羲 *creation of the eight trigrams, it was impossible to demarcate the ages." This shows that the* Changes *provides sociological evidence.*

58 Note: *The "Attached Verbalizations" says: "[The sages] went to utmost depths and ground the sprouts" and "[They] expounded depths and reached far"; this relates to "probing subtleties."*

59 Note: *The "Attached Verbalizations" says: "Squares* [i.e., clearly bounded sets of individual things] *are brought together according to kind; [individual] objects are differentiated according to group." In his* Discussions of the Changes, *Mr. Jiao Litang extended the meaning of this statement. From the principle that the people give birth to and nourish one another and the division of labor* (yishi tonggong 易事通功) *among them, he inferred that the best public administration is born entirely from the people's group* [collective] *nature* (qunxing 群性). *His interpretation is extremely refined. Similarly, the Westerners call "sociology"* (shehuixue 社會學) *the "science of groups"* (qunxue 群學 [a term introduced by Yan Fu, JK]); *this means the same as "differentiating objects according to group."*

for the end; [they] planned the appearance [of facts and objects] and designed the desired outcome[60] in order to push on and record the changes and alternations of the past and present.[61] This is the science of "exploring the mysterious." The usefulness of "knowing what is to come" lies in being without desire and taking no purposive action.[62] [The sages] purified their hearts and preserved the secrets;[63] [they] verified the patterns of the fluctuations of full and empty in order to reverse the numbers and know what is to come.[64] In this, the teachings of the *Changes* resonate with sociology. (220–221)

Cutting through the dense veils of philological smoke in which Liu Shipei shrouded his case, we may grant that his basic intuition was not entirely outlandish: in view of the limited knowledge available to him, his discovery that both the *Changes* and late nineteenth-century sociology were designed as tools for social engineering operating on the basis of numbers that served as "material" for numerological speculation or quantitative analysis appeared certainly as original or idiosyncratic but not necessarily unreasonable. No less plausible was his presentation of the *Spring and Autumn Annals* as a storehouse of historical evidence bolstering the claim of a steady progress of humankind as put forward by many contemporary sociologists, historians, and archeologists. Drawing on Kang Youwei's utopian reading of the *Gongyang* 公羊 commentary to the *Spring and Autumn Annals*,[65] Liu found in the latter traces of the same ideology of progress that characterized the positivistic sociology of Herbert Spencer and his followers in Japan and China:

---

60   Note: *The* Changes *says: "[The sages] planned all appearances" and "[They] designed the desired outcome."*

61   Note: *The* Changes *says: "For the sake of the Way, change frequently" and "One opening and one closing, this is an alternation." The* Zhou Changes *says to this: "It is difficult to know what is above [form]" but "easy to know what is below [form]."*

62   Note: *The* Changes *says: "Change is without desire and without purposive action." "[Change] is still and unmoving; when stimulated, it gradually comprehends the causes of all-under-heaven."*

63   Note: *The* Changes *says: "The Superior Man, by relying on them [i.e., the virtues of the stalks, trigrams, and lines], purifies his heart; he retreats and preserves them in secrecy."*

64   Note: *The* Changes *says: "[The sages] gradually knew the things to come" and "Those who know study the words of the judgments and wish to overcome [adversity] step by step." This is called "holding on to definite numbers in order to reverse what has not yet come to be."*

65   Hsiao 1975, 72–78.

For the general meaning of the *Spring and Autumn Annals*, we need to look into the *Gongyang* tradition. There are three [two?] courses of study in the *Gongyang*: the first is called "preserving the three unities" (*cun santong* 存三統); the second, "comprehending the three ages" (*tong sanshi* 通三世). The three ages—the age of disorder; the age of ascending peace; and the age of perfect harmony—serve to confirm the traces of progress (*jinhua* 進化) in human societies.[66] Now, although human groups experience changes and transformations, their achievements follow a definite order; they must move through natural stages. When [Confucius] established the text of the three ages in the *Spring and Autumn*, he venerated the trajectory of the past and knew the route of things to come;[67] this almost means the same. Modern scholars have used the *Gongyang* tradition to corroborate the "Rites Cycle" (Liyun 禮運) chapter [in the *Rites Records* (*Liji* 禮記)]. In my view, by enumerating the details of food and drink, palaces and dwellings, by tracing them back from the times after the sage-kings had produced them to the times before the sage-kings' creation, the "Rites Cycle" proves that, while facts and objects are numerous, each undergoes progressive changes, and this corresponds indeed to the general meaning of the *Spring and Autumn*. In this, the *Spring and Autumn Annals* resonates with sociology. (221)

Finally, Liu borrowed the authority of Sima Qian 司馬遷 (ca. 145–85 BCE) to underline his claim of affinities between the *Changes*, *Spring and Autumn Annals*, and Western sociology, curiously by citing a passage that Yan Fu had used some years earlier to prove the Chinese origins of Baconian logic:[68]

> The Grand Historian once said: "The *Changes* proceeds from the hidden to the manifest, while the *Spring and Autumn* infers the visible from the hidden." For the *Changes* mainly speaks about patterns, whereas the *Spring and Autumn* mainly speaks about facts. Who could deny that they are similar [to sociology] in this?

---

66   Note: *Modern scholars have verified the meaning of "comprehending the three ages" in several canonical works, most often in the* Odes *and* Documents.

67   Note: *Thus, when Confucius instructed Zizhang by saying, "The ten thousand ages can be known," he also referred to holding on to definite numbers in order to reverse what had not yet come to be.*

68   Yan 1897, 1320. See Kurtz 2011, 280–281.

Notwithstanding his peculiar methodology, Liu Shipei presented a few valuable insights in the remainder of his "Prolegomena," particularly in his sketches of future histories of religion, politics, economics, education, and penal law in ancient China. In these parts of his text, Liu's enumerations of classical quotations added up to useful and original overviews of the similarities and differences among the various pre-Qin schools as seen through the prism of Western-derived notions. To be sure, these sections, too, confirmed how fuzzy his understanding of some new concepts remained, and they also mirrored his political concerns and ideological preferences. Yet, the creativity and thoughtfulness with which he regrouped traditional insights in these outlines earned him high praise for his effort, not only among his National Essence peers.[69]

One final aspect of the "Prolegomena" should be highlighted. In his selection of passages that seemed to fit into the framework of a particular discipline, Liu intentionally parted with the Chinese-origins practice of piling up as many Chinese "equivalents" of Western ideas as could possibly be found. Instead of trying to preserve as much as possible of China's intellectual heritage, Liu used his translation as an opportunity for a radical, merit-based selection. Since he entertained characteristically strong opinions on most subjects he introduced, Liu carefully chose what he considered valuable in late Zhou sources while purging what he thought was shallow or morally dubious. Nowhere was this more obvious than in his "Prolegomenon to a History of Logic." According to Liu, such a history should exclusively focus on Xunzi's "On the Correct Use of Names," supplemented by a few legible fragments from the notoriously opaque Mohist Canon (Mojing 墨經). Other schools and texts, including those that are customarily raised in discussions of Chinese logic today—notably the thinkers associated with the School of Names (Mingjia 名家), the Yinwenzi 尹文子, and parts of the Zhuangzi 莊子—were explicitly excluded from his envisioned disciplinary history on the grounds that their authors only "split hairs and messed with words in order to elevate themselves" and, hence, that they were "sophists" (guibianjia 詭辯家) and not "logicians" in the sense of true "scientists of names" (219). In this and several other chapters of the "Prolegomena," Liu anticipated a strategy of selective appropriation (which I call the "shrink-to-fit" approach to Chinese thought) that has become particularly common in modern accounts of "Chinese philosophy": the tendency to ignore or suppress everything that cannot be understood in the key terms defined by the European gestalt of a certain discipline. At the same time, he proved immune to the temptation of pretending that each of the "nine schools" had something to contribute to the subject matter of all the sixteen

---

69   Li Jinxi 1936.

fields of learning whose histories he aimed to reconstruct. In hindsight at least, Liu appears as a much less defensive interpreter of China's intellectual traditions than most later historians of "Chinese philosophy," "Chinese logic," "Chinese science," and so on, who continue to outdo one another by amassing more and more, albeit not necessarily more convincing, evidence for the claim that "China has always had it, too." A good illustration of this difference is Liu Shipei's concise two-page sketch of late Zhou psychology, which cannot stand in more pronounced contrast to the multivolume compilations of the discipline's history that were produced in the second half of the twentieth century. In marked distinction to the large and unperturbed camp of modern and contemporary precedent-hunters, Liu retained enough faith in China's traditions that he was able to forgo such vain exercises in feigned cultural pride.

V   Concluding Remarks

It would be futile to try and gauge the precise impact of Liu's "Prolegomena" on the rewriting of ancient China's intellectual history. As even the few examples quoted in the preceding pages should have made abundantly clear, hardly any of the disciplinary parallels he suggested would appear convincing from today's comfortable vantage point. Most had already been forgotten or rejected by the time the first actual histories of late Zhou thought, as opposed to mere outlines or sketches of such, started to be compiled. But influence is hardly the only criterion by which the significance of a work of such dazzling novelty and sweeping ambition as Liu's "Prolegomena" can be measured. Even if, or perhaps precisely because, Liu failed to secure a safe haven for China's endangered moral heritage, his radical effort to translate ancient Chinese knowledge into the language and categories of globalized academic disciplines can serve as a vivid illustration of the violent epistemic changes that reshaped Chinese scholarly discourses in the early years of the twentieth century and paved the way for the naturalization of modern academic disciplines in the Republican period.

What can we learn from this symptomatic text? First of all, the "Prolegomena" confirmed that conceptual "firsts" alone hardly guarantee success in an enterprise as daring as Liu's new intellectual history. On the contrary, Liu's futile attempt to ensure a measure of continuity with the scholarly tradition he aimed to defend could just as well be read as a "last." By revealing the ineptness of his philological methods in coming to adequate terms with the new knowledge he so eagerly wanted to embrace, Liu's often misguided deliberations underlined that evidential scholarship, irrespective of how expertly it was handled, had irrevocably outlived its utility. Probing the "original meanings" of characters used to render imported words and notions by means of classical tools like the second-century *Explanation of Writing* could do little to facilitate appropriate understanding.

In the context of contemporary discursive practices, it is nonetheless difficult not to be impressed by the boldness and creativity of Liu Shipei's efforts. Even if they failed to bring about the idealized renaissance he envisioned, Liu's labors opened up a new path in the late Qing search for alternative traditions. His idea of a *xueshushi* established a precedent for a new genre of writing Chinese intellectual history. By disciplining China's national essence, Liu's "Prolegomena" offered a way out of the ghetto of national studies narrowly understood that many of his contemporaries, and an even greater number of later scholars down to the present,[70] saw as the only hope for the survival of traditional learning and values. At the same time, his interpretive enterprise underscored the price that such a translation inevitably had to pay. Most ancient ideas could be fitted into the new categories only after being thoroughly trimmed and decontextualized—a phenomenon that continues to plague many histories of Chinese thought, be they concerned with philosophy, ethics, logic, sociology, religion, or any science. Finally, Liu's more often than not surprising choices of disciplinary abodes for ancient theorems serve as a reminder that even the seemingly most obvious and innocent interpretations of old Chinese texts in modern terms have not always been as self-evident as they may appear today. If nothing else, his ruminations thus call for a "denaturalizing" genealogy of the ways in which ancient China's intellectual history has been disciplined.[71]

The need to rethink our understanding of the formation of modern academic disciplines in China, which a close reading of Liu Shipei's "Prolegomena" and similar symptomatic texts suggests, applies more generally to the historiography of science and thought in the late Qing and Republican periods. Needhamian reveries of a more or less seamless integration of traditional Chinese knowledge into modern disciplinary histories have, as several contributions to this volume confirm, proven incapable of doing justice to the dynamic epistemic values underlying studies of nature, man, and society in imperial times. Driven more by the desire to claim civilizational parity than the search for adequate understanding, such histories focus on unearthing similarities and differences in a predetermined set of characteristics regarded as essential to modern knowledge. By highlighting only decontextualized frag-

---

70   Dirlik 2011.
71   Bevir 2008.

ments that can be taken to resonate with modern concerns, they obscure the plurality of styles of inquiry and reasoning that competed in nineteenth- and early twentieth-century China. To recover these styles in their complexity, more open-ended studies that interpret these consequential conceptual transformations in their local and global contexts, such as those assembled in this book, are indispensable.

Bibliography

Amelung, Iwo 2004. "Naming Physics: The Strife to Delineate a Field of Modern Science in Late Imperial China," in Michael Lackner and Natascha Vittinghoff, eds., *Mapping Meanings: The Field of New Learning in Late Qing China*. Leiden: Brill   381–422.

Angle, Steven C. 1998. "Did Someone Say 'Rights'? Liu Shipei's Concept of *Quanli*," *Philosophy East and West* 48, 4: 623–651.

Beijing daxue xiaoshi yanjiushi 北京大學校史研究室, ed. 1993. *Beijing daxue shiliao: Diyi juan 1898–1911* 北京大學史料：第一卷 1898–1911 [Historical materials related to Peking University: Volume 1, 1898–1911]. Beijing: Beijing daxue chubanshe.

Bernal, Martin 1976. "Liu Shih-p'ei and National Essence," in Charlotte Furth, ed., *The Limits of Change: Essays on Conservative Alternatives in Republican China*. Cambridge, MA: Harvard University Press   90–112.

Bevir, Mark 2008. "What Is Genealogy?" *Journal of the Philosophy of History* 2: 263–275.

Chang, Hao 1987. *Chinese Intellectuals in Crisis: Search for Order and Meaning   1890–1911*. Berkeley: University of California Press.

Chen Pingyuan 陳平原 1998. *Zhongguo xiandai xueshu zhi jianli: Yi Zhang Taiyan, Hu Shizhi wei zhongxin* 中國現代學術之建立：以章太炎、胡適之為中心 [The formation of modern Chinese scholarship, with special regard to Zhang Binglin and Hu Shi]. Beijing: Beijing daxue chubanshe.

Cua, Anthony S. 2000. "Emergence of the History of Chinese Philosophy," *International Philosophical Quarterly* 40, 4: 441–464.

Dirlik, Arif 2011. "*Guoxue* / National Learning in the Age of Global Modernity." *China Perspectives* 1, 4–13.

Dirlik, Arif, Guannan Li, and Hsiao-pei Yen, eds. 2012. *Sociology and Anthropology in Twentieth-Century China: Between Universalism and Indigenism*. Hong Kong: Chinese University Press.

Eber, Irene 1975. "Thoughts on Renaissance in Modern China: Problems of Definition," in Lawrence G. Thompson, ed., *Studia Asiatica: Essays in Asian Studies in Felicitation of the Seventy-Fifth Anniversary of Professor Ch'en Shou-yi.* San Francisco, CA: Chinese Materials Center 189–218.

Elman, Benjamin A. 2000. *A Cultural History of Civil Examinations in Late Imperial China.* Berkeley: University of California Press.

——— 2005. *On Their Own Terms: Science in China 1550–1900.* Cambridge, MA: Harvard University Press.

Fan, Fa-ti 2004. "Nature and Nation in Chinese Political Thought: The National Essence Circle in Early-Twentieth-Century China," in Lorraine Daston and Fernando Vidal, eds., *The Moral Authority of Nature.* Chicago: University of Chicago Press 409–437.

Fong, Grace, Nanxiu Qian, and Harriet Zurndorfer, eds. 2004. *Beyond Tradition and Modernity: Gender, Genre, and Cosmopolitanism in Late Qing China.* Leiden: E. J. Brill.

Guan Xiaohong 關曉紅 2000. *Wan Qing xuebu yanjiu* 晚清學部研究 [A study of the late Qing Department of Education]. Guangzhou: Guangdong jiaoyu chubanshe.

Harada Sen 原田潛 1883. *Min'yakuron fukugi* 民約論覆義 [A rendering of the *Social Contract*]. Tokyo: Shunyōdō.

Hon, Tze-ki 2003. "National Essence, National Learning, and Culture: Historical Writings in *Guocui xuebao, Xueheng,* and *Guoxue jikan," Historiography East and West* 1, 2: 242–286.

Hsiao, Kung-chuan 1975. *A Modern China and a New World: K'ang Yu-wei, Reformer and Utopian 1858–1927.* Seattle: University of Washington Press.

Huters, Theodore 2005. *Bringing the World Home: Appropriating the West in Late Qing and Early Republican China.* Honolulu: University of Hawai'i Press.

Janku, Andrea 2003. *Nur leere Reden: Politischer Diskurs und die Shanghaier Presse im China des späten 19. Jahrhunderts.* Wiesbaden: Harrassowitz.

Jiang Biao 江標, ed. 1897. *Gezhi jinghua lu* 格致精華錄 [Records on the essence of science]. Shanghai.

Kishimoto Nobuta 岸本能武太 1900. *Shakaigaku* 社会学 [Sociology]. Tokyo: Tai Nippon tosho.

Kurtz, Joachim 2011. *The Discovery of Chinese Logic.* Leiden: E. J. Brill.

Lackner, Michael [Lang Mixie 朗宓榭] 2003. "Yuan zi dongfang de kexue? Zhongguoshi ziduan de xingshi" 源自東方的科學？—中國式自斷的形式 [*Ex oriente scientiae?* A Chinese genre of self-assertion], *Ershiyi shiji* 76: 85–95.

Lackner, Michael, Iwo Amelung, and Joachim Kurtz, eds. 2001. *New Terms for New Ideas: Western Knowledge and Lexical Change in Late Imperial China.* Leiden: E. J. Brill.

Li Fan 李帆 2000. "Qingmo minchu xueshu shi boxing chaoliu shulun" 清末民初學術史勃興潮流述論 [An account of the flourishing of the trend of intellectual history during the late Qing and early Republic], *Jilin daxue shehui kexue xuebao* 5, 9: 63–67.

___ 2003. *Liu Shipei yu Zhong Xi xueshu* 劉師培與中西學術 [Liu Shipei and Chinese and Western learning]. Beijing: Beijing shifan daxue chubanshe.

___ 2004. "Xueshushi: Qingmo minchu de xianxue" 學術史：清末民初的顯學 [Intellectual history: A prominent genre in the late Qing and early Republican periods], *Guangming ribao*, April 27.

Li Hongyan 李洪岩 and Zhong Weimin 仲衛民 1994. "Liu Shipei shixue sixiang zonglun" 劉師培史學思想總論 [A general discussion of Liu Shipei's historical thinking], *Jindaishi yanjiu* 3: 253–272.

Li Jinxi 黎錦熙 1936. "Xu" 序 [Preface]. Reprinted in Liu Shipei 1997, vol. 1, 26.

Li Xiaoqian 李孝遷 2001a. "Liu Shipei 'guxue chu yu shiguan lun' tanxi" 劉師培 "古學出於史官論" 探析 [An analysis of Liu Shipei's theory that "traditional learning originated with the official scribes"], *Shehui kexue jikan* 5: 121–125.

___ 2001b. "Liu Shipei yu jindai zhuzixue yanjiu" 劉師培與近代諸子學研究 [Liu Shipei and modern Chinese studies in noncanonical philosophies], *Fujian luntan* 4: 90–95.

Liu Shipei 劉師培 [Liu Guanghan 劉光漢] 1903a. "Guowen zaji" 國文雜記 [Miscellaneous notes on the national language]. Reprinted in Liu Shipei 1997, vol. 3, 463–466.

___ [Liu Guanghan 劉光漢] 1903b. *Rangshu* 攘書 [Book of expulsion]. Reprinted in Liu Shipei 1997, vol. 2, 1–17.

___ [Liu Guanghan 劉光漢] 1903c. *Zhongguo minyue jingyi* 中國民約精義 [The quintessence of China's social contract]. Reprinted in Liu Shipei 1997, vol. 1, 560–597.

___ [Liu Guanghan 劉光漢] 1905. "Zhoumo xueshushi xu" 周末學術史序 [Prolegomena to an intellectual history of the late Zhou]. Reprinted in Liu Shipei 1997, vol. 1, 500–525.

___ 1997. *Liu Shipei quanji* 劉師培全集 [The complete works of Liu Shipei]. 4 vols. Beijing: Zhonggong zhongyang dangxiao chubanshe.

___ 1998. *Liu Shipei xinhai qian wenxuan* 劉師培辛亥前文選 [Selected writings by Liu Shipei prior to 1911], edited by Zhu Weizheng 朱維錚. Beijing: Sanlian shudian.

Luo Zhitian 羅志田 2003. *Guojia yu xueshu: Qingji Minchu guanyu "guoxue" de sixiang lunzhan* 國家與學術：清季民初關於 "國學" 的思想論爭 [Scholarship and the state: Ideological debates about "National Studies" in the late Qing and early Republic]. Beijing: Sanlian chubanshe.

Makeham, John, ed. 2012. *Learning to Emulate the Wise: The Genesis of Chinese Philosophy as an Academic Discipline in Twentieth-Century China*. Hong Kong: Chinese University Press.

Mittler, Barbara 2004. *A Newspaper for China? Power, Identity, and Change in Shanghai's News Media 1872–1912*. Cambridge, MA: Harvard University Press.

Moloughney, Brian, and Peter G. Zarrow 2011. *Transforming History: The Making of a Modern Academic Discipline in Twentieth-Century China*. Hong Kong: Chinese University Press.

Nivison, David S. 1966. *The Life and Thought of Chang Hsüeh-ch'eng (1738–1801)*. Stanford, CA: Stanford University Press.

Quan Hansheng 全漢升 1935. "Qingmo de Xixue Zhongyuan shuo" 清末的西學中源說 [The late Qing theory of the Chinese origin of Western knowledge], *Lingnan xuebao* 4, 2: 57–102.

Reynolds, David C. 1991. "Redrawing China's Intellectual Map: Images of Science in Nineteenth-Century China," *Late Imperial China* 12, 1: 27–61.

Sang Bing 桑兵 2001. *Wan Qing Minguo de guoxue yanjiu* 晚清民國的國學研究 [National Studies in the late Qing and Republic]. Shanghai: Shanghai guji chubanshe.

Wan Shiguo 萬仕國 2003. *Liu Shipei nianpu* 劉師培年譜 [Annalistic biography of Liu Shipei]. Yangzhou: Guangxia shushe.

Wang Dongjie 王東杰 2000. "*Guocui xuebao* yu 'guxue fuxing'" 《國粹學報》與 "古學復興" [The *National Essence Journal* and the revival of ancient learning], *Sichuan daxue xuebao* 5: 102–112.

Wang Fansen 王汎森 2001. "'Sixiang ziyuan' yu 'gainian gongju'—Wuxu qianhou de jizhong Riben yinsu" 思想資源與概念工具—戊戌前後的幾種日本因素 ["Intellectual resources" and "conceptual tools"—Some Japanese factors before and after 1898], in Wang Fansen, *Zhongguo jindai sixiang yu xueshu de xipu* 中國近代思想與學術的系譜 [Modern Chinese thought and the genealogy of scholarship]. Shijiazhuang: Hebei jiaoyu chubanshe 149–164.

Wang Hui 汪暉 2004. *Xiandai Zhongguo sixiang de xingqi* 現代中國思想的興起 [The rise of modern Chinese thought]. 4 vols. Beijing: Sanlian shudian.

Wang Renjun 王仁俊 1993. *Gezhi guwei* 格致古微 [Ancient subtleties of science], in Ren Jiyu 任繼愈, ed., *Zhongguo kexue jishu dianji tonghui: jishu juan* 中國科學技術典籍通彙：技術卷 [Anthology of classical works of Chinese science and technology: Technology]. Zhengzhou: Henan jiaoyu chubanshe 791–886. [Originally published 1896.]

Wang Xiaoling 1998. "Liu Shipei et son concept de *contrat social chinois*," *Études chinoises* 17, 1–2: 155–190.

Weston, Timothy B. 2002. "The Founding of the Imperial University and the Emergence of Chinese Modernity," in Rebecca E. Karl and Peter Zarrow, eds., *Rethinking the 1898 Reform Period: Political and Cultural Change in Late Qing China*. Cambridge, MA: Harvard University Press 99–123.

Wu Guangxing 吳光興 1995. "Liu Shipei dui Zhongguo xueshushi de yanjiu" 劉師培
對中國學術史的研究 [Liu Shipei's researches in Chinese intellectual history],
Xueren 7: 163–186.

Xia Xiaohong 夏曉紅 2006. *Yuedu Liang Qichao* 閱讀梁啓超 [Reading Liang
Qichao]. Beijing: Sanlian shudian.

Yan Fu 嚴復 1897. "Yi 'Tianyanlun' zixu" 譯天演論自序 [Translator's preface to the
Tianyanlun (On evolution)]. Reprinted in Yan Fu 1986, 1319–1321.

———, trans. 1902–1905. *Mule mingxue* 穆勒名學 [Mill's *Logic*]. Shanghai:
Shangwu yinshuguan.

___ 1986. *Yan Fu ji* 嚴復集 [The works of Yan Fu], edited by Wang Shi 王栻. 5 vols.
Beijing: Zhonghua shuju.

Yang, Fang-yen 1999. "Nation, People, Anarchy: Liu Shih-p'ei and the Crisis of Order in
Modern China." Ph.D. dissertation, University of Wisconsin–Madison.

Yang Tingdong 楊廷棟 1902. *Lusuo Minyuelun* 路索民約論 [Rousseau's *Social Contract*]. Shanghai: Wenming shuju.

Zarrow, Peter 1990. *Anarchism and Chinese Political Culture*. New York: Columbia University Press.

Zhang Binglin 章炳麟, trans. 1902. *Shehuixue* 社會學 [Sociology]. Shanghai:
Guangzhi shuju.

Zhang Huien 張會恩 and Zhong Humei 鐘虎妹 1994. "Liu Shipei wenzhangxue sixiang chutan" 劉師培文章學思想初探 [A first exploration of Liu Shipei's ideas
on composition], *Zhongguo wenxue yanjiu* 2: 72–77.

Zheng Shiqu 鄭師渠 1994. "Wan Qing guocuipai lun Kongzi" 晚清國粹派論孔子
[The late Qing National Essence circle on Confucius], *Luodi shifan xuebao* 3: 75–81.

___ 1997. *Wan Qing guocuipai: Wenhua sixiang yanjiu* 晚清國粹派：文化思想研
究 [The National Essence circle in the late Qing: Studies in cultural thought].
Beijing: Beijing Shifan daxue chubanshe.

Zhu Honglin 朱鴻林 2001. "Weixue fang'an—xuean zhuzuo de xingzhi yu yiyi" 為學
方案－學案著作的性質與意義 [Plans for learning: The nature and meaning
of the *xuean* genre], in Xiong Bingzhen 熊秉真, ed., *Rang zhengju shuohua:
Zhongguo pian* 讓證據説話：中國篇 [Let evidence speak: China]. Taibei:
Maitian chuban 287–318.

Zuo Yuhe 左玉河 2004. *Cong sibu zhi xue dao qike zhi xue—xueshu fenke yu jindai
Zhongguo zhishi tixi zhi chuangjian* 從四部之學到七科之學－學術分科與
近代中國知識體系之創建 [From the learning of the Four Treasuries to the
learning of the Seven Faculties: The disciplinary division of scholarship and the formation of the modern Chinese knowledge system]. Shanghai: Shanghai guji
chubanshe.

Science in Translation

Yan Fu's Role

Shen Guowei

Abstract

Chinese characters are considered an adaptable system, open to expansion and revision. Throughout history, the creation of new characters was one of the most important solutions to enlargements of the conceptual repertoire. Both scholars of "Dutch Learning" in Japan and missionaries active in nineteenth-century China used Chinese characters in their translations of Western concepts. From a methodological point of view, Japanese scholars mostly coined compound words rendering the literal meanings of the Western terms, while translators in China, invigorated by the success of the new characters devised for chemical elements, believed that drafting new characters was more in line with the characteristics of the Chinese language. However, notwithstanding the painstaking efforts with which they were created, the new characters proposed by missionaries were eventually replaced by compound terms first used in Japanese adaptations. This essay examines the different practices and attitudes of Yan Fu toward the Japanese creation of new characters for kexue 科学 to translate "science."

I   Introduction: From Concepts to Words

Yan Fu 严复 (1854–1921), the famous Chinese translator, spent considerable energy inventing new words to translate foreign concepts. For everyday concepts, Yan Fu believed that it was sufficient if a translation could be understood by society;[1] for the key words of the era, however, Yan Fu held that it was necessary to search the Chinese Classics in order to ensure accuracy.[2] For words in the latter category, such as "liberty" (自由), "authority" (权利), or "economics" (经济), Yan Fu carefully considered the deeper meanings. When translating another of the era's most important buzzwords, "science," however, Yan Fu appears to have made only a cursory effort. "Science" (*kexue* 科学) is one of the core concepts of the modern age. How, then, did Yan Fu, living as he did at the close of the nineteenth century and the dawning of the twentieth, understand SCIENCE, and what did he intend when he chose 科学 to represent

---

1   Yan 1986, vol. 3, 518.
2   Yan 1986, vol. 3, 519.

© KONINKLIJKE BRILL NV, LEIDEN, 2014 | DOI 10.1163/9789004268784_006

it?[3] I believe this is a fascinating question, and in this essay I will examine *The Collected Writings of Yan Fu* (严复集) and some of his early translations, in particular his thought process when analyzing "science" and related concepts.[4] This kind of analysis should shed some new light on Yan Fu's conception of science.

II   Origin of *Kexue* 科学 as a Word: Its Creation and Spread

The first knowledge from Europe to be incorporated into the body of Japanese scholarship was Dutch Learning (Rangaku 蘭学), which was introduced to Japan in the middle of the eighteenth century. In translating the new concepts from Europe, the Rangaku scholars used traditional Chinese vocabulary, such as *qiongli* 穷理 (searching for governing principles), whenever possible. They were deeply aware, however, of the chasm between the Chinese and European systems of knowledge and methodology.[5] In the Meiji era (1868–1912), Japan began its campaign of full-scale absorption of the new learning from Europe. At the close of 1870, Nishi Amane 西周 (1829–1897) noted in the conclusion of his encyclopedia, the *Hyakugaku renkan* 百学連環, that Western learning encompassed a myriad of disciplines, each with its own array of fields, and that detailed academic inquiry was the pursuit only of scholars, who were each experts in their own fields, and who never strayed into other disciplines. Although traditional Chinese classical studies (*kangaku* 漢学) also distinguished between the Five Classics and the official histories of each era, there was no equivalent in Japan of the Western concept of academic fields. Nishi also published material from the introduction to his encyclopedia in an article entitled "Theories of Knowledge" (Chisetsu 知説) in the *Meiroku zasshi* (*Journal of the Meiji Six Society*). In his treatise Nishi expounded upon the word *gakujutsu* 学術 (academics): the first character, 学, represented the seed germ

---

3   English words written in capital letters denote concepts, while words in lowercase letters denote the lexemes themselves.

4   Yan 1986. The most famous of Yan Fu's translations are Huxley 1981; Smith 1981; and Mill 1981.

5   See, e.g., Utagawa 1980. It is also important to keep in mind that at the beginning of the Meiji period, the lexeme *kagaku* also meant "studying for the [Chinese] civil service examination" (科挙之学). For example, in the April 1869 issue of *Kōgisho nisshi* 公議所日誌, there is the following passage: "Even so, since studying for the civil service examination (*kagaku*) has become a useless exercise in empty rhetoric, proctors must scrutinize the candidates for their virtue or wickedness and their actions" (然レドモ科学ハ空文無益ニ成行モノ故試官ヨク其人ノ正邪ト実行トニ注意スベシ。). See Sōgō 1986.

of knowledge, the observation of objective truth, and consideration of its meaning; the second character, 術, described adherence to known methodologies and experimentation. The proper order for conducting research was first to gather facts and ponder their meaning (学), and then to conduct experiments (術). Nishi further pointed out that the basis for 学 was investigation, of which there were several valid methods. The three methods current in Europe at the time were observation, experimentation, and proof. None of these steps could be omitted.[6] Nishi Amane impressed upon his readers that the most important research methods were deduction and induction. Nishi laid out his argument as follows:

> "Science" consists of the induction of a general truth from the observed facts, stating this truth as a hypothesis, expounding the conclusions that follow, and recording the result as a monograph, to serve as an example. Once the facts have become clear from investigation, the art of science lies in making this truth useful to mankind. Thus, the aim of the inquiry must not ignore the benefit or harm the discovered truth might have in a narrow quest for truth for its own sake. The art of science consists of making use of discovered truth in order to eliminate evils and loss, and promote benefit and gain for mankind, etc. In this way, inquiry serves to open up new frontiers of learning, while the art of science aids in the progress of technology.
>
> Though the aim of inquiry and the art of science differ, when united as the study of science, they are difficult to separate. Chemistry is a good example. Although analytical chemistry may be classed as inquiry and synthetic [organic] chemistry as the art of science, the two are all but inseparable.[7]

Nishi added that, while European scholarship was enjoying an unprecedented flowering, it had never had an overarching, unifying principle. Isidore Auguste Marie François Xavier Comte (1798–1857) ordered the five disciplines from simplest to most complex: astronomy, physics, chemistry, biology, and sociology; thus, the various disciplines and fields discussed in Nishi Amane's encyclope-

---

6 Yamamuro and Nakanome 2009, middle vol., 202. I have translated all citations from this work from the Japanese. English words within the citations are the editors' (Yamamuro and Nakanome) reconstruction from transliterations in the original work.

7 *Meiroku zasshi* 明六雑誌 22 (December 19, 1874), in Yamamuro and Nakanome 2009, middle vol., 236.

dia, *Hyakugaku renkan*, were much more complicated.[8] It was in this context that the compound *kagaku/kexue* 科学 first appeared. This use of the word *kagaku/kexue* has been touted as the first known instance of a translation for "science," making Nishi the first person to coin the Japanese term. More recently, however, the consensus within the Japanese academic community has been that *kagaku* did not mean "science" but rather "subject" or "discipline."[9] In other words, the lexeme *kagaku* may well be a mistake for *gakka* 学科, which appears frequently in the *Hyakugaku renkan*. There is another basis for this conclusion besides the context for each usage of *kagaku*. That is, Nishi Amane did not use this term in his other original works until a much later point in his career. Another plausible explanation is that Nishi originally intended *gaku* 学 as a translation for "science," *jutsu* 術 for "art," and *kagaku* as a variant of *gakka*, which would mean "subject" or "discipline." Nishi explained the latter as the study of specialized subjects.

During the same period, Nakamura Masanao 中村正直 (1832–1891) was also using the term *gakujutsu* 学術 to render the term "science."[10] Nakamura pointed out that, according to Western theories, Western academics could be divided into two categories: metaphysics and physics. The former category included literature, logic, theology, ethics, law, and politics; the latter category contained physics, craftsmanship, chemistry, medicine, and agriculture. Nishi Amane's list of disciplines and their hierarchy were adapted by Nakamura as the two categories "metaphysics" and "physics." These terms also contain the implication of hierarchy.[11]

In 1877, three years after publication of "Theories of Knowledge," Nishi gave a lecture entitled "Science Lies in Deepening Understanding of the Source" (学問ハ淵源ヲ深クスルニ在ルノ論) at Tokyo University, in which he state the following:

> First, when deepening understanding of science (as stated in the title), it is necessary to meet the urgent needs of the era, and although one should not neglect to choose the quickest, most efficient approaches, nevertheless when studying science, effort should be made to grasp the deeper workings of each discipline, even when it does not seem to have immediate application; though such pains may seem profitless, being able to expound on the theories of science requires a thorough understanding.

---

8   Yamamuro and Nakanome 2009, middle vol., 237. Also see Fan 1988.
9   Hida 2002, 205.
10   Nakamura 1874, in Yamamuro and Nakanome 2009, first vol., 341.
11   *Meiroku zasshi* 6 (September 22, 1874), in Yamamuro and Nakanome 2009, middle vol., 87.

As with the proverbial bounty of enriching water flowing from river to sea, so it is with gathering all the varied truths and assembling them into one, unified truth, where left meets right on common ground.[12]

The word *gakumon* in the title is clearly meant as "science," and since the term *kagaku* is modified by the adjective "each," it refers to the various disciplines that make up "science." This lecture reflects the thought of Nishi Amane on the subject of the relationship between the "myriad subjects" (百科諸学) and philosophy. Two years later, Nakamura Masanao used *kagaku* and *gakumon* to render "science" in one of his translations. This is the first undoubtedly deliberate use of *kagaku* as a translation[13] for "science," and there are even sample sentences demonstrating how to use the word, some of them with a phonetic rendering of "science" printed in small characters above them.[14] In fact, the word *kagaku* was never again limited to academic disciplines but was extended to mean the scientific method as well.[15] The 1881 *Dictionary of Philosophy* (哲学字彙) firmly established *kagaku* as the standard translation of "science," and this became the accepted term throughout Japanese society.

From the second decade of the Meiji era (1887–), *kagaku* became a buzzword in Japan.[16] Judging from the definitions in Japanese reference works, the word *kagaku* was more strongly associated with the natural sciences, as it is in the 1893 *Japan Great Dictionary* (日本大辞書), "*kagaku*—another name for physical science" (*rigaku* 理学); in the 1896 *Great Imperial Dictionary* (帝国大辞典), "There are set principles that govern everything, and academic inquiry into these principles is called 'science' (*kagaku*). Science is physics, as opposed to philosophy, which is metaphysics"; and in the 1897 *Japan New Dictionary* (日本新辞林), "Science and philosophy complement each other." These dictionary definitions point to a tendency in Japan to see science and philosophy as a pair, in a dualistic paradigm. Regarding the relationship between science and philosophy in nineteenth-century Japan, Tsuji Tetsuo commented:

---

12   Nishi 1960, 572. Also see Tsuji 1973, 178.

13   Nakamura 1879. The original work was *The Science of History*, by G. G. Zerffi, which was written in response to Japanese demand. This work had an important role in the establishment of the positivist school of history in Japan. See Katō 1991, 260. The passage quoted is from Hida 2002 (my translation).

14   The Japanese lexeme *kagakuteki* 科学的 corresponds to "scientific."

15   "Scientific" (*kagakuteki* 科学的) refers not only to an empirical methodology but also to logical reasoning.

16   Hida 2002, 206–210.

In Japan, when modern science was adopted, the academic methodologies of science and the nature of logical inquiry were not understood as being an integral part of modern science. Science was imported to provide immediate practical benefits through specialized expertise; this was because the methodologies and system of knowledge of science could not be readily understood, and the situation at the time was too desperate to prioritize attempts to grasp the deeper significance of Western scientific thought. The uniquely Japanese adaptation of modern philosophy came to (unexpectedly) fill this gap.[17]

Next, let us turn our attention to China. As Fan Hongye 樊洪业 pointed out, the traditional method for Chinese gentlemen to cultivate cultured manners was to "examine the phenomena and ponder the truth" (*gewuzhizhi* 格物致知).[18] Jesuit missionaries arrived in China at the close of the Ming dynasty (1368–1644). Regarding the scholarship they brought with them, Xu Guangqi 徐光启 (1562–1633) wrote: "Roughly speaking, there are three kinds: the greatest is to develop morality and serve heaven; the least is to examine the phenomena and discover the truth."[19] He treated the latter as an appendix to theology.[20]

In the nineteenth century, the latest European science once again flowed into China, this time brought by Protestant missionaries. The traditional terms *gewu* 格物 or *gezhi* 格致 came to denote the natural sciences, although these terms were also used for physics and chemistry or simply physics.[21] The modern Chinese word *kexue* 科学 was borrowed from Japanese. Who adopted this word, and when? How was it used?[22] The answers to these questions are the subject of many studies on the history of modern academics.[23] In the article quoted above, Fan Hongye conjectures that the honor belongs to Kang Youwei

---

17   Tsuji 1973, 179–180.

18   Fan 1988, 40.

19   Xu 1965a.

20   Xu 1965b.

21   Fan 1988, 44–45.

22   The definition of *kexue* in the *Xin erya* 新尔雅, a 1903 collection of technical terms compiled by Chinese students in Japan, is "In research of the phenomena of the world, a systematic ordering of knowledge is called *kexue*."

23   Early research includes Yuan 1985; and Fan 1988. The greater part of Fan's work discusses in depth the shift from *gezhi* to *kexue*. More recent works include Jin and Liu 2008. Chapter 12 of this book utilizes new methodologies such as statistical analysis to examine the prevalence of *gezhi* and *kexue* in the Chinese-speaking world during the early modern era, as well as other issues in the history of thought. Both works have served as inspiration

康有为 (1858–1927), but Zhu Fajian 朱发建 disagrees, stating that Taiwanese scholarship has found evidence that the memorial in which *kexue* appears is a later forgery and thus cannot be taken as conclusive. Kang Youwei's *Annotated Bibliography of Japanese Books (Riben shumu zhi)* (日本书目志) does contain the lexeme *kexue*, but only as part of a book title. This can hardly be said to be the first known Chinese usage of the word.[24] Zhu Fajian believes that the first person to adopt *kagaku/kexue* was Wang Guowei 王国维 (1877–1927).[25] In his introduction to the *Essentials of Oriental History* (东洋史要), published in December 1899 as the Chinese translation of Kuwabara Jitsuzō's 桑原隲蔵 (1871–1931) *Intermediate History of the Orient* (中等東洋史), he wrote:

> My schoolmate Fan Bingqing (樊炳清), of Shanying [present-day Zhejiang Province], translated Kuwabara Jitsuzō's *Essentials of Oriental History*. It has just been published. My teacher Professor Fujita then discussed the main points of the book and had Wang Guowei write in the foreword that modern history is a science. Therefore, it is not permissible for there to be no order among facts; no matter what the discipline, if there is any scholar who lacks order in his inquiries, his work cannot be considered science.[26]

Although there is no way to know today what kinds of discussions took place between Fujita Toyohachi 藤田豊八 (1869/70–1929) and Wang Guowei, it is very clear that Fujita taught his students that the essence of science was the relationship between the myriad phenomena and our knowledge, and that history was no exception to this principle. This assertion coincides with the tenets of Japanese positivist history as pioneered by Nakamura Masanao. Even so, the following excerpt, written by Liang Qichao 梁启超 (1873–1929), is older:

> This then is the future of the Pacific. As all the races of men advance in politics, commerce, religion, and academics, there are disputes and wars

---

for my own work. Besides these, academic works that explore *kexue* include Elman 2000; Zhu Jianfa 2005; Zhou 2009; Zhang 2009.

24   Also see Shen 2003.

25   See Zhu Fajian 1899. Zhou Cheng (2009) believes that Tang Tingshu 唐廷枢 "was the first person in China to use *kexue*," but the examples cited are all compound words such as *jiaokexue* 教科学 (theory of education) or *jiaokeshu* 教科书 (textbook), which should be analyzed as *jiaoke+xue*.

26   Zhu Fajian 1899. For an analysis of the circumstances behind publication of this book and the translation challenges involved, see Sanetō 1970, 216; Shen 1994, 222–268 (223–272 in the 2008 ed.).

of aggression. The relationship is significant, and there is no doubt that progress leads to world war. When this happens, there must be a direct cause. In fact, there are two such phenomena: the advance of science and the balance of power among the great powers.[27]

We must remember, however, that this piece by Liang Qichao is merely a translation of a Japanese publication. It would have been extremely difficult for Liang to have as deep an understanding of *kexue/kagaku* as did Wang Guowei, since he did not study under Fujita Toyohachi.

After the turn of the twentieth century, large numbers of translated Japanese books and magazines flooded into China, and educational reforms within China contributed to the surge in the number of instances of *kexue* appearing in Chinese publications, including government documents. For example, Zhang Zhidong 张之洞 (1837–1909) wrote in his *Guidelines for Educational Affairs* (*Xuewu gangyao* 学务纲要): "All teachers should lecture scientifically, and students respond scientifically, and the language used must not veer into the vulgar and crude."[28] At this time, *kexue* still meant "academics divided into disciplines." As a proponent of "Chinese content, Western practicality" (中体西用), however, Zhang Zhidong failed to understand that introducing Western academic structures and systems of knowledge, as exemplified by SCIENCE, would also require the adoption of the underlying methodologies that were completely foreign to China's academic tradition, not to mention the specialized terminology that he had characterized as "vulgar and crude." It was in such an awkward milieu that Yan Fu found himself confronted with the problem of translating "science."

III   Yan Fu's Relationship with SCIENCE

On February 4, 1895, Yan Fu published "On the Speed of World Change" (Lun shibian zhiji 论世变之亟) in the *Zhibao* newspaper (直报), outlining his beliefs regarding the reasons for the strength of the Western nations.[29] A month later, he published "On the Origin of Strength" (Yuanqiang 原强), in which he introduced Chinese readers to sociology for the first time.[30]

---

27   Liang 1899.

28   Zhang Zhidong, *Xuewu gangyao* 学务纲要, September 1903. Cited from *Jindai Zhongguo jiaoyu shiliao* 1928, 8–30.

29   Yan 1986, vol. 1, 1–5.

30   Yan 1986, vol. 1, 5–15.

Describing it as a field that "greatly explains matters of ethics," he also stated that the study of mathematics, logic, physics, and chemistry was a prerequisite for sociology. In this way, Yan considered the "physics" subjects, such as mathematics and the practical sciences, to be the foundation for metaphysical disciplines that "greatly explain matters of ethics." He proceeded to divide knowledge into the three categories of heaven, earth, and man, commenting that the study of man was the most urgent because it included physiology and psychology, which constituted the foundations of sociology. These remarks illustrate Yan's view of SCIENCE at this time, which was clearly influenced by the theories of Comte.[31]

From May 1 to May 8, 1895, Yan Fu serially published his "On Our Salvation" (Jiuwang juelun 救亡决论) in Zhibao.[32] He asserted that China must reform by "losing no time in abolishing the eight-legged essay [the writing portion of the civil service examination]." Regarding reforms in the academic establishment, Yan believed it imperative to "excise the eight-legged essay and to lecture on Western learning." After painstakingly outlining the various grievous ills caused to man and nation by eight-legged essays, Yan discussed how Western physics and chemistry, or gezhi 格致, were completely different from Chinese learning, because with their proviso that "the proof of every theorem or law, every phenomenon, must be tested and only then accepted cautiously," all theories had to be tested through experimentation. "Western gezhi" in this context refers especially to the Western natural sciences. Yan was careful to mention that "Western scholars say, however, that all study seeks not only to know the unknown but also to probe what is possible and impossible. Astronomers do not spend their lives merely examining the heavens; chemists are not limited to experimenting with substances; as for botanists, they do not farm; zoologists do not need to practice animal husbandry. The great, exquisite effect lies in training the mind and practicing the manipulation of the heart, so that those who learn to sink and float with the times or those who learn sincerity cannot indulge in preposterousness." Emphasizing the nonutilitarian nature of science and its character-building effects on people, Yan pointed out that science was equivalent to the Chinese concept of the study of governing principles (lixue 理学) and that it would make an adequate substitute for the traditional method that Chinese gentlemen used to cultivate cultured manners, namely, to "examine phenomena and ponder the truth" (gewuzhizhi 格物致知). In this way Yan distinguished his stance from that of Zhang Zhidong, who simply wanted to borrow the trappings of Western science.

---

31   Fan 1988, 45–46.
32   Yan 1986, vol. 1, 40–54.

Regarding how the Western concept of scientific investigation, which Yan Fu translated as *xue* 学, qualified as scholarship in the traditional Chinese sense, which was also denoted by the lexeme *xue* 学, Yan stated that in order for knowledge of any kind to be elevated to the status of scholarship, it must have organization, be systematic and provable, and have strictly defined terminology. This kind of scholarship (学) could then be investigated according to logic and be applied to human society.

In his preface to *Tianyanlun*, his 1898 translation of Thomas Huxley's *Evolution and Ethics*, Yan Fu wrote that, while the ancients had divided learning into physics and metaphysics, the two were now united into one as a result of the acknowledgment that metaphysics also follows the three principles of physics. The three principles were measurement, broad applicability of universal truths, and experimentation. All three were necessary elements in science, but "experimentation in particular is vital."[33] The following passage is likely the original English source for these statements: "And the business of the moral and political philosopher appears to me to be the ascertainment, by the same method of observation, experiment, and ratiocination, as is practised in other kinds of scientific work, of the course of conduct which will best conduce to that end."[34]

In his "The Effects of Western Learning" (Xixuemenjing gongyong 西学門径功用), which he published in the periodical *Guowenbao* 国闻报 on September 22 and 23, 1898, Yan Fu remarked that, in order to probe truth through investigation (学), it was first necessary to observe the objective facts and then to organize the findings according to the properties of each fact.[35] Only then would the proper foundations be laid for an analysis of the phenomena, which was always conducted in a strictly methodical, logical manner. The scholars of ancient Europe and China, no matter how great their accomplishments, had many errors in their conclusions because they had followed only two of the three necessary procedures. This was why modern science came about, with its emphasis on experimentation.

It is possible to observe the influence of Nishi Amane's thought (as quoted above) in Yan Fu's statements regarding science; in fact, the two are largely in agreement. In this way these two men had a great impact on their countrymen, although there was nearly a thirty-year gap between when they introduced science to their respective societies. Both men were writing under circumstances in which there was as yet no set translation for the new science vocabulary. As

---

33   Huxley 1981, 44.

34   Huxley 1902, 43.

35   Yan 1986, vol. 1, 92–95.

explained in the next section, Yan's understanding of science was based on *An Inquiry into the Nature and Causes of the Wealth of Nations* by Adam Smith (1776) and *A System of Logic* by John Stuart Mill (1843).

IV   Yan Fu and "Science"

As outlined above, Yan Fu used such expressions as "examine the phenomena and search for governing principles" (格物穷理), "investigation" (学), "academics" (学问), "scholarship" (学术), and "physics and chemistry" (格致) in his early works in order to convey the concepts contained in "science." In other words, although he had a deep understanding of the meaning of science itself, he had not yet adopted a single term to encompass all of the above aspects. Yan first used the word *kexue* 科学 in his translations of Adam Smith's *An Inquiry into the Nature and Causes of the Wealth of Nations* and John Stuart Mill's *A System of Logic*, which were published after 1895. There are about a dozen instances of this word in the latter work. Around 1900, perhaps influenced by his experiences translating, Yan began to use the term in his own writings. There are 143 instances of *kexue* in *The Collected Writings of Yan Fu* (严复集).

Yan Fu used the word mostly in the following three groups of works:

1. His translations of *An Inquiry into the Nature and Causes of the Wealth of Nations* and *A System of Logic* contain eighteen instances of *kexue*. These examples reflect Yan Fu's understanding of SCIENCE and his choice of Chinese words during the translation process.

2. His "Letter to the Editor of the *Waijiaobao* on Education" (与外交报主人书) and "On the Burning Issue of Physical Science and Education" (論今日教育应以物理科学為当務之急) together contain forty instances of *kexue*. Both were published around 1903.

3. There are thirty-seven instances of *kexue* in "Lectures on Politics" (政治讲义).[36] The period around 1906 was also marked by frequent use of this word.

The use of *kexue* in these works will be analyzed in the next sections.

---

36   Yan 1986, vol. 5, 1241–1316.

"Science" in Yan's Translations of An Inquiry into the Nature and Causes of the Wealth of Nations and A System of Logic

An Inquiry into the Nature and Causes of the Wealth of Nations is a book about economics written in 1776 by Adam Smith. In book 5, chapter 1, part 3, article 2, "Of the Expense of the Institution for the Education of Youth," Smith discusses the origins of the cost of primary, middle, and higher education, the organization of the faculties, and issues regarding the educational environment from the viewpoints of both teachers and students.[37] In this passage, which is more or less distant from the topic of economics, there are twenty-seven instances of the word "science."[38] There are also other expressions, some of them contrasting "science" with "art." For example:

> (1) In its nature, it is arbitrary and discretionary; and the persons who exercise it, neither attending upon the lectures of the teacher themselves, nor perhaps understanding the sciences which it is his business to teach, are seldom capable of exercising it with judgment.[39]

Yan Fu simply rendered the general meaning and did not provide a translation for the individual lexeme "science" (621).

> (2) If in each college, the tutor or teacher, who was to instruct each student in all arts and sciences, should not be voluntarily chosen by the student, but appointed by the head of the college …[40]

Yan's translation of the above-mentioned passage contains the first use of *kexue*. Here, however, it is used to denote "arts and sciences" (622).

> (3) In the universities, the youth neither are taught, nor always can find any proper means of being taught the sciences, which it is the business of those incorporated bodies to teach.[41]

The "sciences" in the above-mentioned passage appear as "specialized sciences" in Yan Fu's translation (624).

---

37   For Smith's original, I used Smith 1995. Pages numbers for citations from this work are given in parentheses in the text. I also used Smith 1981, 2008.
38   The work as a whole features forty-three instances of "science."
39   Smith 1904, vol. 2, 251.
40   Smith 1904, vol. 2, 252.
41   Smith 1904, vol. 2, 254.

(4) The parts of education which are commonly taught in universities, it may perhaps be said, are not very well taught. But had it not been for those institutions, they would not have been commonly taught at all; and both the individual and the public would have suffered a good deal from the want of those important parts of education.[42]

In this example Yan Fu has used *kexue* for "education" (624).

In the 1770s, when Adam Smith was writing, the core meaning of "science" was "a particular branch of knowledge or study; a recognized department of learning.[43] While Yan Fu's *kexue* cannot be considered a perfect match, it does convey the implication of there being many disciplines, and this was the standard usage in Chinese at the time.

Thus, it seems plausible that Yan Fu's notion of "science" is derived from Adam Smith—or perhaps even more plausibly from Mill. Yan Fu started translating *An Inquiry into the Nature and Causes of the Wealth of Nations* and *A System of Logic* at around the same time, and the latter had a large influence on Yan.[44] In Mill's original work, "art" (*shu* 术) and "science" (*xue* 学) seem to form a contrast, with science superior to art; science is the basis of art; and without science art is shallow. In that case, what are the factors that distinguish science from art? Can art be elevated to science and, if so, how? The process of drawing up a systematic theory based on observed phenomena is indispensable. Several of the "arts" cannot be called "sciences" because they consist of the combination of multiple sciences. Logic is considered a unifying discipline that provides the methodology for all other disciplines, which is why Yan saw fit to introduce it to his Chinese readers. Yan believed that "science" included "examining the phenomena and pondering the truth, leading to medicine, which is the sum of logic, mathematics, chemistry, and physics." He also pointed out that logic governs its own relationship with the other disciplines: "Though logic is the unifying principle underlying all subjects, it is itself an independent discipline." Yan ended his remarks with a discussion on the problems of specialized science terminology: "The ideas of science are sparkling jewels and the terminology is accurate, which is why its rules are the strictest of all."[45]

Where did Yan Fu get the word *kexue*? In Chinese-language works written in China around the year 1900, there are almost no examples of *kexue*. In Japanese-

---

42   Smith 1904, vol. 2, 254.

43   *Oxford English Dictionary*   2nd ed. (1989).

44   For Mill's original, I used Mill 1848. I also used Mill 1981; and a Japanese translation, Mill 1949.

45   Mill 1981, 3.

influenced journals that published translations from Japanese, such as *Dissenting News* (*Qingyibao* 清议报) or *A Collection of Translated Books* (*Yishuhuibian* 译书汇编), however, there are already instances of *kexue*. It is possible that Yan Fu had encountered new Japanese words in publications such as these, including "philosophy" (*zhexue* 哲学), which he used numerous times in his translations of *An Inquiry into the Nature and Causes of the Wealth of Nations* and *A System of Logic*.[46] On the other hand, although Japanese undeniably had an influence on the form of the new words, Yan's *kexue* does not draw its meaning from Japanese. "Science" may be either singular or plural in English, but Chinese nouns do not decline according to gender, number, or case. Yan tried to distinguish between science as a whole and the various sciences by using *xue* 学 and *kexue*, respectively. In other words, for Yan, "science" was not a collective noun.[47]

Kexue *in "Letter to the Editor of the* Waijiaobao *on Education"*

The "Letter to the Editor of the *Waijiaobao* on Education" (与外交报主人书) and the March 4, 1902, issue of the *Waijiaobao* that prompted it belong to the second group of writings. There are sixteen uses of *kexue*, the meaning and usage of which are in the same vein as in *An Inquiry into the Nature and Causes of the Wealth of Nations* and *A System of Logic*.

During this period, there was a profusion of arguments regarding reform of the educational system. Some of the most representative were: (1) retain Chinese learning as the foundation, with pragmatic borrowings from Western technology, and (2) adopt Western politics as the foundation and embellish with Western arts. The stance in the *Waijiaobao* was that education should be conducted in Chinese and not in foreign languages.[48] Yan Fu's rebuttal included

---

46   Yan Fu was not satisfied with *zhexue* (哲学) because "the Western name for governing principles (*lixue* 理学) reveals its origins in the study of temperaments, with observation of phenomena as its opposing concept. Japanese have rendered theology, anthroposophy (*zhixue* 智学), and philosophy (*aizhixue* 爱智学) all alike, as *zhexue*. I hope that the most recent studies on this subject will all be called *aizhixue*, with all subjects pertaining to the spirit classed as *xinxue* (心学), since the term *zhexue* has not yet become standard" (理学其西文本名谓之出形气学，与格物诸形气学之对，故亦翻神学，智学，爱智学，日本人谓之哲学。顾晚近科学独有爱智以名其全，而一切性灵之学归于心学，哲学之名似尚未安也。) (Mill 1981, 12).

47   The usage "one science, two sciences" (一科学，二科学) in Smith 1981 and Mill 1981 reflects this circumstance. The elements that make up true compound nouns cannot be modified by external adjectives or other modifiers. Thus, "very big sea" (很大海) and "very long residence" (很旧居) are incorrect expressions in contemporary Chinese.

48   I will discuss Yan Fu's ideas on a national language (*guoyu* 国语) in depth elsewhere.

a discussion of *kexue*, especially in his refutation of the idea of adopting Western politics and a little art. He says of this approach that it has "everything backward." What, after all, is meant by "art"? Isn't it in fact "science"? Logic, mathematics, chemistry, and physics are all "sciences." All these sciences are based on principles and laws, which are also the foundation of the best aspects of Western politics. As Huxley pointed out, since Western politics did not yet completely conform to the principles of "science," it would not remain at its current level. Chinese politics would be left further and further behind, rendering China unable to take its place among the great nations, because Chinese governance was not in accordance with the principles of science and was, in fact, in violation of them. In Yan's eyes, the "Western arts" as commonly perceived actually embodied the modern scientific spirit, with its emphasis on "observe, generalize, experiment," so that, if "science" and "Western arts" are equivalent, then the notion that "the basis of Western politics is Western arts" would be true but not the reverse. Some may argue that "Western arts" and "science" are not synonymous, but even if that is the case, surely science is the foundation of both Western arts and Western politics—in other words, they would be component concepts included in "science"—and are thus inseparable, like the left and right hands of the same person. In the context of the letter, Yan uses *kexue* in both its narrow and its wide senses: the former as the "physics" corollary to logic; and the latter as a system of knowledge encompassing both "physics" and metaphysics. It is clear that Yan was using the word here primarily in its narrow sense.

Included in the same group of manuscripts is "On the Burning Issue of Physical Science and Education" (論今日教育应以物理科学為当務之急), the script for a lecture (unfinished).[49] The word *kexue* appears twenty-one times in this document (including in the title). Yan Fu pointed out that human thought can be divided into two types, rational thought and emotion, and that there is a difference between cerebral, rational thought and intuitive, emotional thought. He stated that moral education shapes the latter, and intellectual education the former, with science as its main instrument. Here Yan's "science" refers to the natural sciences, whose object is to discover the laws of nature. Thus, Yan agreed with Huxley that the purpose of education is to "clear the channels of the intellect, and broaden and deepen knowledge," and that the method of education should "broaden and deepen knowledge" through "clearing the channels of the intellect."[50]

---

49   The date must have been sometime before Yan Fu, "Jingshi Daxuetang yishuju zhangcheng" 京师大学堂译书局章程, *Ta Kung Pao* 大公报, August 29–31, 1903.

50   Yan 1986, vol. 2, 278–280.

Yan Fu pondered which of the sciences would be most effective to study to achieve this goal, given that time to study is usually limited. He asserted that the deductive sciences of mathematics and geometry and the inductive sciences of physics, chemistry, zoology, and botany would not only increase knowledge but even discipline the emotions and train the mind. He believed that the problem with Chinese education was that it "disproportionally stressed moral education at the expense of physical and intellectual education"—that is, that there was too much art and not enough physics, with an almost exclusively deductive approach that neglected inductive reasoning. In order to put forth arguments in the traditional manner, it was necessary only to think about an issue, not to gather facts. Thus, "mastering scholarship only leads to a slavishly dogmatic intellectualism." According to Yan, the antidote to this state of affairs was to increase the share of the physical sciences in the curriculum. He included physics, chemistry, zoology, botany, astronomy, geology, biology, and psychology in his "physical sciences." Most of these subjects are considered natural sciences today, but some of them are now considered human sciences. From the above analysis, it is clear that in 1903, Yan's understanding of science encompassed not only the concepts of dividing academic pursuit into separate fields, specialized inquiry, and academic fields but also those of the natural and human sciences, with a particular emphasis on induction.

Kexue *in "Lectures on Politics"*

The other major work by Yan Fu that contains the word *kexue* is "Lectures on Politics" (政治讲义).[51] This work was divided into eight lectures, and *kexue* appears thirty-seven times. The most significant instance occurs in the introduction of the first lecture, where he claimed that in the West, politics had already become a science. Politics that has "already become a science" must necessarily have adopted the fundamental principles underlying the practical physical sciences, in particular the division of subjects. After analyzing the contrast between observation (学) and experimentation (术), Yan remarked that when the scholars of old discussed politics, their arguments could not be considered "observation" when judged according to the standard of modern science. This was because "scholars, even if they seek the principles of physical things, examine only what is already known, while artisans know the path to morality when they have no work, but first ask what is appropriate." Yan further pointed out that "in order to learn science, it is necessary to start with correct names"; "what my generation calls politics is actually a science. If one

---

51   Ibid. (in Yan 1986, vol. 5, 1241–1316).

calls it a science, then whatever characters one uses must clearly distinguish fields, for if these standards are not properly observed, confusion will creep in." "In the vocabulary of science, whether a word has one meaning or two, it is necessary to ask whether the meanings are compatible," "because the vocabulary of science does not allow two separate meanings, and contradictions are even less acceptable." At the time, in 1906, specialized science terminology in Chinese was incomplete, so "in speaking of science, if my meaning and arguments differ from what is customary in our country, that creates difficulties; one problem is the seeking of clarity in names and meanings so as to avoid ambiguity, and another is the logical ordering of thought, since we are not used to it." Yan continued with great emotion: "Today we talk of science unabashedly with the nobility, but when we use our literary language to do so, it is just like a watchmaker using old-fashioned Chinese knives, saws, weights, and awls—I think such a watchmaker's difficulties are obvious to anyone. He can only make do with such tools: on the one hand, tinkering to make slight improvements and, on the other, shying away from using them at all, for he has no other art." Though Yan was very vocal in his belief that using Chinese to teach science was not a mistake,[52] at the same time he lamented that it would take twenty years before China reached that high level.[53]

V   Conclusion

To conclude, Yan Fu's view of SCIENCE can be summarized as follows: "science" (*xue*) and "art" (*shu*) are two opposing concepts, with the purpose of science being to pursue truth (in Yan's words, "the laws of nature"); "art" (*shu*) tends toward practicality, or what was described as "knowing the path to morality." "Art," however, can be elevated to science, the necessary requirement being to submit all observed phenomena to "systemization" (*tixihua* 体系化). As for science, in the past "academics" (*xue*) were classified as either "ethics of form or spirit" (形行气道德), which were metaphysics and "physics," respectively. Logic was one of the "physics," as a subcomponent of philosophy; since the modern era, however, the principles of "physics" (observation, generalization,

---

52   "Recently European theories have been flowing east, from statecraft to catching insects and fish, and they teach that it is not a mistake to conduct education in one's own national language" (方近欧说东渐，上自政法，下逮虫鱼，言教育者皆以必用国文为不刊之宗旨。). See Yan Fu, "Yan Fu zhi Wu Guangjian han" 严复致伍光建函 (Letter to Wu Guangjian), in Yan 1986, vol. 3, 586.

53   Yan 1986, vol. 2, 562.

experimentation) have been better appreciated, leading to the transformation of the old "ethics of form or spirit" into science (*kexue*), which was also a division of inquiry into separate fields. In particular, logic became a science par excellence, since it utilized both inductive and deductive reasoning. In Yan Fu's opinion, traditional Chinese scholarship "simply did not have the art of observation" "nor any appreciation for the necessity of proofs,"[54] so that "the knowledge of the people is underdeveloped, and the nation is poor and weak."[55] China needed to quickly adopt physics, chemistry, zoology, botany, astronomy, geology, biology, psychology, and other sciences. These "physical sciences," based on inductive reasoning, would raise the living standards and knowledge level of the people. The brand-new, highly systematized sciences would inevitably transform the old world, and they were also China's only route to salvation. This is the reason for Yan Fu's praise of "science," "logic," and the "physical sciences."

As for Chinese translations of scientific terminology, Yan Fu first used *kexue* in his translation of Smith's *An Inquiry into the Nature and Causes of the Wealth of Nations*. His decision to focus on the "study of a specialized field" portion of the meaning of "science" reflects the most common understanding in China at the time. What is original is that Yan infused the new Chinese word with a sense of the essence of what makes science scientific. An important caveat, however, is that he never abandoned the use of *xue* to mean the sum total of human knowledge and the systemization of scholarship. For example, starting in 1909 Yan Fu was chief editor at the Qing government's Bureau of Terminology (审定名词馆) in the new Ministry of Education (学部), where he was responsible for evaluating nearly thirty thousand new technical terms. The Chinese term for "science" adopted by this bureau as the national standard for use in education was *xue*. The entry for *kexue*, which had also been a strong contender, was simply defined as a neologism in wide use, showing that Yan Fu and his colleagues still had an ambivalent attitude toward it.[56]

From the above analysis, it is clear that Yan Fu himself thought deeply about the full meaning of SCIENCE, the necessity of adhering to the SCIENTIFIC method, the nature of SCIENTIFIC terminology, the fundamental chasm between Chinese and Western scholarship, and the attitudes of traditional society toward "science" and "art." How many people in Yan Fu's day truly, intimately understood science as presented in Yan's abstruse (by necessity)

---

54   Yan 1986, vol. 2, 281.
55   Yan 1986, vol. 2, 285.
56   Shen 2008.

writings?[57] Later, the leaders of the May Fourth Movement, who had included "Mr. Sai [Mr. Science]" (赛先生) among their slogans, spread scientific thought and even became intoxicated with the notion of the omnipotence of science. Thus, *kexue* developed along somewhat different lines than what Yan Fu had in mind when he wrote that science was "most effective in national enlightenment." As for what was gained and lost during the crucible years of modern China, that is not within the bounds of the study of the history of translation or of the history of individual words.

Bibliography

Dongyangshi yao 东洋史要 1899. Translated by Gezhi Xuetang 格致学堂. Shanghai: Dongwen Xueshe 东文学社.

Elman, Benjamin 2000. "Transition from the Pre-modern 'Chinese Sciences' to 'Modern Science' in China," *China Scholarship* 2: 1–43.

Fan Hongye 樊洪业 1988. "Cong 'gezhi' dao 'kexue'" 从 '格致' 到 '科学,' *Ziran bianzhengfa tongxun* 自然辩证法通讯 10, 3: 39–50.

Hida Yoshifumi 飛田良文 2002. *Meiji umare no Nihongo* 明治生まれの日本語. Kyoto: Tankōsha 談交社.

Thomas Henry Huxley 1902. *Evolution and Ethics and Other Essays*. New York: D. Appleton.

____ 1981. *Tianyanlun* 天演论 [Evolution and ethics], translated by Yan Fu 严复. Beijing: Shangwu yinshuguan, Xinhua shudian Beijing faxingsuo. [Originally published 1898.]

Jin Guantao 金观涛 and Liu Qingfeng 刘青峰, 2008. *Guannianshi yanjiu: Zhongguo xiandai zhongyao zhengzhi shuyu de xingcheng* 观念史研究：中国现代重要政治术语的形成. Hong Kong: Chinese University Press.

Jindai Zhongguo jiaoyu shiliao 近代中国教育史料 1928. Beijing: Zhonghua shuju.

Kato Shūichi et al., eds. 1991. *Nihon kindai shisō taikei 13 rekishi ninshiki* 日本近代思想大系13歴史認識. Tokyo: Iwanami shoten.

Liang Qichao 1899. "Lun Taipingyang zhi weilai yu Riben guoce" 论太平洋之未来与日本国策, *Qingyibao* 清议报 13 (March): 793.

Mill, John Stuart 1848. *A System of Logic*. New York: Harper and Brothers (http://www.archive.org).

____ 1949. *Ronrigaku taikei* 論理学大系, translated by Ōzeki Masakazu 大関将一. Tokyo: Shunjūsha.

_____

57   For the linguistic constraints Yan Fu faced, see Shen 1994, pt. 2, chap. 3.

___ 1981. *Mulu mingxue* 穆勒名学, translated by Yan Fu. Beijing: Shangwu yinshu-guan, Xinhua shudian Beijing faxingsuo.

Nakamura Masanao 中村正直, trans. 1874. "Seigaku ippan" 西学一斑, *Meiroku zasshi* 明六雑誌 10 (June 28): 341.

___, trans. 1879. *Shigaku* 史学1, pt. 1.

Nishi Amane 西周 1960. *Nishi Amane zenshū* 西周全集, edited by Ōkubo Toshiaki 大久保利謙, vol. 1. Tokyo: Nishi amane kinenkai.

Sanetō Keishū 実藤恵秀 1970. *Chūgokujin ryūgaku Nihonshi* 中国人留学日本史, translated by Tan Ruqian 谭汝谦 and Lin Qiyan 林启彦. Hong Kong: Sanlien shudian.

Shen Guowei 沈国威 1994. *Jindai RiZhong yuhui jiaoliushi* 近代日中语汇交流史. Tokyo: Kasama shoin.

___ 2003. "Kang Youwei yu Riben shumuzhi" 康有为与日本书目志, *Wakumon* 或问 5: 51–69.

___ 2008. "Guanhua (1916) ji qi yici—'xinci' 'budingci' wei zhongxin" 官话 (1916) 及其译词－'新词' '部定词' 为中心, *Ajia bunka kōryū kenkyū* アジア文化交流研究 3: 113–129.

Smith, Adam 1904. *An Inquiry into the Nature and Causes of the Wealth of Nations.* New York: G. P. Putnam's Sons.

___ 1981. *Yuanfu* 原富, translated by Yan Fu. Beijing: Commercial Press

___ 1995. *An Inquiry into the Nature and Causes of the Wealth of Nations.* London: William Pickering (http://www.archive.org).

___ 2008. *Guofulun* 国富论, translated by Xie Zujun 谢祖均. Beijing: Xinshijie chubanshe.

Sōgō Masaaki 惣郷正明, ed. 1986. *Meiji no kotoba jiten* 明治のことば辞典. Tokyo: Tokyodō Shuppan 東京堂出版.

Tsuji Tetsuo 辻哲夫 1973. *Nihon no kagaku shisō* 日本の科学思想. Tokyo: Chūkō shinsho.

Utagawa Yōan 宇田川榕庵 1980. *Shokugaku keigen/shokubutsugaku* 植学啓原／植物学. Tokyo: Kowa shuppan. [Originally published 1834.]

Xu Guangqi 徐光启 1965a. *Ke "Jihe yuanben" xu* 刻<幾何原本>序, *Jihe yuanben* 幾何原本 (1607), in *Tianxue chuhan* 天學初函, bk. 4. Taipei: Xuesheng shuju.

___ 1965b. *"Taixi shuifa" xu* <泰西水法>序, *Taixi shuifa* 泰西水法 (1612), in *Tianxue chuhan* 天學初函, bk. 3. Taipei: Xuesheng shuju.

Yamamuro Shinichi 山室信一and Nakanome Tōru 中野目徹, eds. 2009. *Meiroku zasshi* 明六雑誌. 3 vols. Tokyo: Iwanami shoten.

Yan Fu 严复1986. *The Collected Writings of Yan Fu* 严复集, edited by Wang Shi 王栻. 5 vols. Beijing: Zhonghua shuju.

Yuan Hanqing 袁翰青 1985. "Kexue, jishu liang ci suyuan" 科学，技术两词溯源, *Beijing Wanbao* 北京晚报, September 19.

Zhang Fan 张帆 2009. "Cong 'gezhi' dao 'kexue': Wan Qing xueshu tixi de duguo yu bieyi (1895–1905)" 从 '格致' 到 '科学' ：晚清学术体系的度过与别译(1895–1905), *Xueshu yanjiu* 学术研究 12: 102–114.

Zhou Cheng 周程 2009. "Jiujing shei shi zhongguo zuixian shiyongle 'kexue' yici?" 究竟谁是最先使用了 '科学' 一词?, *Ziran bianzhengfa tongxun* 自然辩证法通讯 4: 93–98.

Zhu Fajian 1899. *Dongyangshi yao* 东洋史要, translated by Gezhi Xuetang 格致学堂. Shanghai: Dongwen xueshe.

Zhu Jianfa 朱建发 2005. "Zao yinjin 'kexue' yici de Zhongguoren bianxi" 最早引进 '科学' 一词的中国人辨析, *Jishou Daxue xuebao (shehui kexue ban)* 吉首大学学报(社会科学版) 2: 59–61.

Chinese Scripts, Codes, and Typewriting Machines

Jing Tsu

Abstract

In the late nineteenth century, the Chinese writing system embarked on a path of unprecedented change. In the spirit of scientific thinking, language reformers, inventors, and pedagogues sought to abolish the character script and to replace it with other ideographic, numerical, and alphabetic systems. Swayed by the idea that a universal language in the modern world depended on quick access instead of cultural prestige, they aimed to forge a new script that would not only ease the process of translation but also match the alphabetic writing system in logic and efficiency. They expected the monumental feat to change the very terms on which China interacted with the world. From shorthand to Braille, universal alphabet to word/zi 字 segmentation, modernizing the Chinese language set the material and technological precondition for importing and transforming foreign knowledge. The result far exceeded its original conceit and entered the Chinese language into a race for global linguistic dominance and technology in the Cold War era. This essay discusses the various proposals that were put forth and their technological consequences, including a landmark invention of a Chinese-language typewriter in the 1940s.

Qian Xuantong, twentieth-century Chinese-language reformer and cultural critic, once recalled seeing an advertisement in 1920s Shanghai.[1] Amid the experimental cultural landscape of the Republican period, spotting an announcement for a public séance was nothing too extraordinary. But the growing credibility of such forums, Qian noted with disdain, was alarming. This particular one promised to deliver the spirits of past Confucian sages and literati. Instead of offering moral tales or glimpses into the future, however, the spirits were summoned from the netherworld to do one thing: expound on the virtues of classical phonology. Yet a return to traditional learning was not what the venerable apparitions urged. Nor did they denounce, in the expected tone of the day, the influx of Western knowledge and novelties that made China's own traditions look old, broken, and boring. The spirits rose above such quibbles by seizing on the changing signs of the times. No ghostly voice floated toward the audience from behind the curtains. Instead, they conveyed their messages tangibly in writing, using a mix of non-Chinese scripts, including the Latin alphabet and the Japanese kana.

Qian was not amused. The nationwide campaign to modernize the Chinese language had only just begun, he laments, and already people were ruining a

---

1   Qian 1999.

perfectly good science by turning it into a circus of gimmicks. Unlike what the charlatans tried to sell the masses, the national language project was based on empirical observation, phonological laws, and rational implementation. Although Qian was speaking more as a public ideologue than as a linguist, his observation highlights an important intersection between language, writing, science, and technology in modern China. The process was just about to gain nationwide momentum at the time of his writing.

The modern Chinese script reform movement was initiated by a few philologists and amateurs in the 1880s. With a mix of popular passion and ideological agenda propelling the movement, it continued under the state's auspices throughout the twentieth century. The net result was that the face of the script changed, as is well attested by the accomplishments of the simplification campaigns in mainland China in the 1950s and 1960s. A quieter revolution, however, was also afoot. The methods proposed for the script's technological delivery, in fact, have been undergoing continuous revision for more than 130 years. Garnering the most international attention, the simplification campaigns scored major political points at home, as they helped to reduce illiteracy and solidify national standardization. It was the longer scientific turn of the Chinese writing system since the late nineteenth century, however, that made all this possible. The script revolution, which was often pushed into the background while bigger political events seized the stage of twentieth-century Chinese history, turned out to be the lasting one. It irreversibly augmented the global influence and capacity of the Chinese language, thereby opening up a new space for competition and co-optation between the alphabetic and ideographic writing systems.[2] The rivalry reached a high point in the Cold War period, even though its impact is not widely known.[3] During the technological leap between the nineteenth century and the present, from telegraphy to automatic translation, the terms of the arms race between Chinese and English took shape between the 1880s and the 1950s.

In the spirit of the present volume, my analysis of the Chinese script as a transnational history of convertible technology builds on the recent compara-

---

2 I am well aware of the contended, if not obsolete, use of "ideograph" to designate the written Chinese script. The term has been unpleasantly associated with the philological bias of the Indo-European tradition, chinoiserie, missionary ethnocentrism, and European colonialism. The well-rehearsed critique, however, has been generated outside the context and materials I examine here. I therefore invoke "ideograph" in this essay as a historically laden project that was never stably fixed in itself but evolved with the different contexts of linguistic and technological standardization under examination. For a useful summary of the different ways of naming the Chinese script and their controversies, see DeFrancis 1984, 74–130.

3 Tsu 2010, 49–79.

tive studies of China's history of science, in relation to China's active response to Western science and technology, by taking the conversation in a new direction.[4] First, explicating Chinese science and technology on "Chinese" terms is often criticized for implicitly excluding the "Western," leaving the latter equally unqualified and misunderstood. Some view the polarizing tendency as a methodological problem intrinsic to comparison and suggest broadening the scope, as well as increasing the number of items to be compared. By breaking out of nation- and area-bound niches, it is hoped that no one measure of progress will be unilaterally imposed on different cultures and histories, regardless of their degrees of similarity or difference. On this view, attention to multiplicity and specific contexts helps to remove the Eurocentric lens that so often colors the comparative perspective.

What happens, however, when enforcing asymmetry—and imposing a standard interpretation of that comparison—is precisely the name of the language game? Between the alphabetic and ideographic systems, this question is reengaged at the center of a history of technological rivalry. The encounter between the two writing systems puts into play lasting dynamics of asymmetry and mutual dependence. Taking this as the point of departure, I identify linguistic mediality as a crucial material manifestation of how ways of writing—and the cultures to which they correspond—are made to be different, reciprocal, commensurate, and, finally, independent of these concerns. I take the script medium as an embodied indicator of a global struggle between Chinese and English to be the gold standard. In this case, standardization is more than just about enforcing a framework of normativity or top-down control. Rather, as we will see in the following, it opens the way to mutual accommodation and adaptability, giving currency to enhanced access rather than the exclusivity of accumulated power.

On this view, the terms of asymmetry by which we are accustomed to think about, and to interpret, China's modern history—belated, subjugated, catching up with a vengeance, and so on—are also out of step with how it has evolved on the material-technological front. Much of the existing interest in the nuts and bolts of the modern Chinese-language script has thrived primarily in the study of linguistics, state engineering, and planned education.[5] Seeking new connections that transcend this divided inquiry, I am interested in the popular innovations and unexpected subchannels that propelled the Chinese script into a dynamic role on the global stage. Without falling back on

---

4   See Elman 2005; Kurtz 2011; Sivin 1982; Lackner, Amelung, and Kurtz 2001; Pollard 1998; Xiong 2011; Guo 1998; Wang Hui 2004, vol. 4, 1107–1279.
5   DeFrancis 1989; Kaske 2008.

the rubrics of nationhood and state building as the predominant measures for the Republican era, I show how the technologization of the Chinese script simultaneously charted out a Chinese, as well as international, trajectory beginning in the nineteenth century. This process produced shared features that were later employed in the language campaigns during the Communist period. In response to the recent proposals to turn to nationalism as a new anchorage for the study of science in the modern period, then, this essay cautions restraint when it comes to the globalization of the Chinese script.[6] The question of modern national language standardization, as I have elaborated elsewhere in the context of Sinophone studies, has been held in constant tension with long-standing desires for transregional and global mediality.[7]

Second, my focus on script systems differs from an emphasis on translation in studies of European and non-European encounters.[8] While the study of how ideas flow and knowledge circulates helps move us past the impact-response model, here I am not interested in what happens to meanings when languages interact or what strategies empower or disempower acts of appropriation. The historical negotiations within the physical medium of script shift the emphasis from interpretive effects to technological materiality—that is, how the latter can structure the former in turn. The way in which writing systems, like the alphabetic and the ideographic, enter into polarity and disagreement raises questions about a priori differences or diverging civilizational mentalities, which are often implied as the deeper causes.[9] To keep the focal point accountable, I take legibility in a real and material sense, in order to gain traction on notions of civilizational difference that otherwise fall back on essentialist, culturalist positions or hide behind the familiar combat of China versus the West.

An inquiry into the Chinese script thus begins anew here. For most of the modern period, the Chinese logograph was singly pointed to as the writing system that was the least prepared for modern, scientific advancement due to its cumbersome physical shape. The suggestion has been provocative enough to incite heated and protracted debates. Without taking the same bait, however, one might ask how the fate of scientific thinking in China came to hang on a matter of a few less or a few more strokes. What kind of experiments were being carried out within the Chinese writing system that primed it for an alphabetic overhaul? What does it mean to refer to a "Chinese" or a "Western" writing system when such attributions are the effect, rather than the cause, of

6   Elman 2007; Fan 2007; Hu 2007; Schmalzer 2007; Shen 2007; Wang Zuoyue 2007.
7   Tsu 2010.
8   Huters 2005; L. Liu 2004.
9   Havelock 1987; Castells 1996; Lloyd and Sivin 2002.

a possible conversion between alphabetic and ideographic systems? Such a perspective articulates a view outside the silos of the "area studies" model that underlies parts of China studies, as well as a concern that bends the general disciplinary frame of the history of science, still thicker around the hard sciences and the European tradition than elsewhere, toward questions of the material, intercultural connections on the fringe. To build a possible bridge to a broader comparative conversation, then, I will examine three related contexts: (1) the debates on China's lack of scientific capacity due to its linguistic alterity; (2) the search for new Chinese writing systems that had little to do with Chinese; and (3) the attempted resolution to these historical debates by turning ideographicality into a virtual and global alphabetic medium.

I "Why Science Didn't Happen to Ideographic Writing—or Didn't It?"[10]

Few periods have witnessed more imaginative inferences about the Chinese script than the twentieth century. Gazing at it upside down—due to a typographical error or lack of linguistic knowledge—Marshall McLuhan saw "a vortex that responds to lines of force ... a mask of corporate energy."[11] Rivaling Ezra Pound, who once proposed that the Chinese script was "alive and plastic" and "not only the forms of sentences, but literally the parts of speech growing up, budding forth one from another," McLuhan belongs to a long line of illustrious commentators who treated the Chinese character as ideal alterity.[12] Going back to the seventeenth century, the Chinese ideograph, with designations ranging from "Real Character" to the "mother tongue" of God, was invested with the power of spiritual salvation and direct communion.[13] While it is easier to discredit, as many have, how McLuhan deployed the ideograph in his popular theory of the medium and the message, it is harder to dismiss the pleas voiced by the Chinese themselves. Chinese writing had long been consecrated with mystifying powers. From divination to recalling presence, the origin of writing partook in the formation of cosmic patterns in the universe.[14]

---

10 To a different end, I adapt this phrase from Nathan Sivin's provocative and seminal contribution to the pro and contra Needham paradigm debates. See Sivin 1982.

11 McLuhan and Parker 1968, 38.

12 Fenellosa 2008, 50.

13 Wilkins 1668; Webb 1669.

14 Lewis 1999. For a rich literary rendition of this myth against the backdrop of Chinese writing in diaspora in the early 2000s, see Malaysian Chinese writer Zhang Guixing's novel, *Qunxiang (Elephants)*

Given the long-standing significance and consecrated cultural status of the written Chinese script, any systematic change would be traumatic and was, for that reason, unthinkable for most of China's long history. Yet people started to contemplate just this possibility around the turn of the twentieth century. The survival of modern China, many urged, depended on what happens to its written script. Arguments of cultural heritage notwithstanding, the complexity of the character script, in contrast to the alphabet, was becoming a liability. While enforcing a unified script since the third century had served the purpose of centralization well, supporters and detractors alike noticed that the dogged adherence to character stroke orders, and the elitist cultural distinction reserved for the utmost mastery of its massive inventory, were getting diminishing returns. In the face of new international threats and old internal linguistic divides, China was plagued by the growing gaps in spoken topolects between the north and the south, on the one hand, and pressed by the outside world into international intercourse, on the other. To best interact with the world, one needed to, among other things, assimilate its modern forms of scientific knowledge. Developing a universalizable linguistic medium was vital. Influenced by missionaries' expressed woes of learning the difficult language, but adding their own sense of urgency, script inventors such as Lu Zhuangzhang, Shen Xue, and others treated the script question as a matter of life and death. Risking incarceration and sometimes even their lives, they set out to change the face of Chinese writing. Who had the time anymore, after all, to learn the right stroke orders when it was more pressing to make time for learning mathematics and physics? Common wisdom in the late nineteenth-century Chinese popular medical urbanscape had its own take on the subject too. Faulting the logograph for using up memory and clogging the brain, people welcomed and consumed brain tonics of different varieties in order to treat this very vulnerability that is particular to the modern age.[15]

If the logograph embodied for the Chinese a moment of widespread crisis at the close of a dynasty, it shouldered an even greater blame from the perspective of civilizational advancement. This argument resurfaced among Greek classicists in the mid-twentieth century, when the questioned relationship between orality and literacy invited speculations on whether a writing system like the alphabet was responsible for the advancement of philosophy and science in ancient Western civilization. Eric Havelock, once in the intellectual circle of McLuhan, and others argued that the advent of the Greek alphabet, superseding its Phoenician origins, was the first writing system to successfully

---

15   For a treatment of the growing phenomenon of neurasthenia in general, see Shapiro 2000.

reduce ambiguity between physically similar words by developing the capacity to represent any phoneme.[16] It was able to break down all semantic and phonetic units, then recombine them to represent any sound in spoken speech. This system of adaptation spurred the Greeks into developing higher and higher levels of abstractions that formed, in short, the prerequisite mental framework for science. While this view has been challenged by other classicists, it has nonetheless tapped into a long-standing popular prejudice.[17] That Chinese is not ideographic but, in fact, possesses both phonetic and pictorial components and is more binomial than monosyllabic remain nuances that are more important to the specialist than to the everyday reader.

One of the most concerted efforts to restore philosophical and scientific integrity to the Chinese language in its own right is presented in the two volumes of *Science and Civilisation in China* published in 1998 and 2004. Outlining the methodology, Christoph Harbsmeier gives an exclusive focus on language its due weight:

> The theory and practice of science and technology are inextricably bound up with language and logic. Scientific insights become transmittable cultural heritage to the extent that they are articulated in language. The insights add up to a scientific explanation to the extent that they are organized into a coherent argument. The explanations add up to a scientific system to the extent that they are organized into a general logical scheme.[18]

Picking up where Needham left off in volume 2 of *Science and Civilisation in China*, Harbsmeier points to language as the basic condition for articulating argumentation and explication.[19] This may at first appear as little more than an obvious fact. If the transmission of science from one cultural context to another means having to convey it in some written or verbal form, then any degree of language barrier could make the difference between having and not having this knowledge. Harbsmeier has something more specific in mind, however. Like pieces of a puzzle, language is further divisible into units of semantic conveyance, as in sentences or clauses, which can reflect the larger cognitive process in the Chinese language. Grammar, on this view, constitutes the diagrammatic logic of not only the structure of writing but also thinking

---

16   Havelock 1987.
17   Lloyd and Sivin 2002.
18   Needham and Harbsmeier 1998, i.
19   See Needham 1956, 199; Boltz 2000.

itself. Hence, for Harbsmeier, focusing on lexical changes in the act of translation would get at only the surface of the problem. He makes plain that the purpose is not to show how the Chinese have "logic" too, or that they have a theory of language that can be juxtaposed to European philology. Rather, he is careful to distinguish between complementary cultural differences and matching categories that do not exist. To this end, he uses a simple analogy for grasping the first misstep that can occur in approaching culturally distinct epistemic categories: though everyone can count, not everyone ends up developing a number theory. Just because logic exists, Joachim Kurtz's compelling study shows, it need not manifest along only one path of rational sense-making.[20]

While resonating with long-standing metaphysical questions about modes of perception and cognition, this important reminder goes beyond the assertion that culturally specific ways of thinking are "different" or on their own terms. It opens the way to the more critical path of asking how the idea of *different* writing systems, *different* thought processes, was emplaced as a cornerstone or standard in staging such evaluations. Any attempt at a comparative study of science in East and West inevitably comes up against the chosen terms themselves as a methodological constraint. This point was not lost on those who first labored over the question of the technology of Chinese writing on the ground. And they took a decidedly experimental approach to close the comparison gap.

II   How Chinese Almost Lost Its Script

Much of the scholarly debate on ideographic versus alphabetic writing systems, in fact, could have been preempted before the twentieth century got under way. In 1900, twelve years before Beijing Mandarin became the national language (*guoyu* 国语) of the new Republic of China, a wanted Chinese fugitive returned from Japan. Disguised as a Buddhist monk from Taiwan named Zhao Shiming, he stole across the border of the Qing Empire into Shandong Province, following a route south to Jiangsu before traveling back north to the city of Tianjin. All the while, he had with him a secret document: a draft proposal for a new phonetic writing system for the Mandarin dialect of Chinese, called the "Mandarin alphabet" (*guanhua zimu* 官话字母), which he had developed during his two years of exile in Tokyo.

While his story is more exciting than most, Wang Zhao—the fugitive's real name—is but one of more than a score of inventors and pedagogues who

---

20   Kurtz 2011.

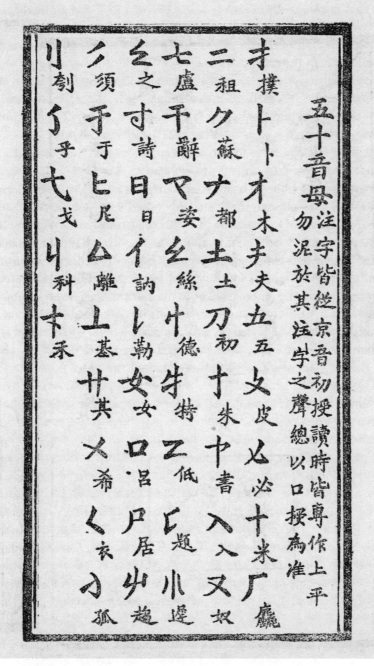

FIGURE 1   *The fifty vowels of the Mandarin alphabet.* (FROM WANG ZHAO 1903.)

attempted to change the Chinese writing system in the late nineteenth and early twentieth centuries. A largely overlooked group of impassioned pedagogues and practitioners, they held that the Chinese script was structurally inconsistent with the conditions of modernity. The amount of labor required to master the cumbersome writing system was to blame, according to one script inventor in 1908, for China's "evolutionary belatedness."[21] This complaint was, of course, not unprecedented. It followed a well-known trail of woes left by foreign missionaries since the time of Matteo Ricci and others.

Instead of relying on the Romanization schemes that the missionaries had developed for the purpose of proselytization, however, late Qing script reformers saw the Chinese script itself as having a decisive role to play. Left unimproved, they argued, the ideograph would stunt any form of modern learning—especially in the areas of technology, translation, commerce, and communication. If a key could be found, on the other hand, in the acoustic patterns beyond the written script to make it easier to learn, a whole new world would "open up" (*tong* 通) with it. Script inventors and language reformers, seeing an opportunity and infused with the spirit of science and empiricism, responded with an array of imaginative, at times esoteric, prescriptions. Some proposed replacing the logogram with alphabet letters, while others studied shorthand, notations for the deaf-mute, and numeral-based systems.

Ni Haishu estimates that, between 1892 and the 1910s, more than thirty script schemes were proposed. There were certainly more. Most of the ones Ni knew about were reprinted mainly in the late 1950s, in connection with the Chinese government's revived interest in simplified orthography. What has remained virtually unknown is that the proposals for a new orthography were being peddled abroad as well, bypassing the scrutiny at home altogether in attempts to reach a world audience.

One such example is a rare Cantonese phoneticization scheme in shorthand devised by a man from Hong Kong named Mok Lai Chi, who was a member of Pitman's Phonetic Society. Isaac Pitman—the inventor of modern phonography (shorthand)—published it in his *Phonetic Journal* in 1893. This was only one year after Lu Zhuangzhang's *A Primer at a Glance: Chinese New Phonetic Script in the Amoy Dialect* appeared.[22] Ni Haishu, and others citing his authority, believe that Lu was "the first person who had a concept of the pho-

---

21   Liu Mengyang 1957 [1908], 84. The acknowledgment of China's "evolutionary belatedness," however, was not always fatalistic, as it was often made in relation to the even less fortunate civilization that had fallen under the sway of Western imperialism. See Tsu 2005, 32–65.

22   Lu Zhuangzhang 1956 [1892].

netic script and devised a phonetic scheme in China."[23] While Mok did not use the Latin alphabet as Lu did, his phonography was a means to the same end. Mok was already working on his scheme prior to May 1892. This certainly disputes the commonly held belief that Cai Xiyong, who published his *Phonetic Quick Script* in 1896, was the first to develop a phonetic script for the Chinese language based on the shorthand system, and it puts into perspective other schemes that followed a different topolectal or tonal paradigm.[24]

The interest in phonography was widespread and surfaced beyond China's borders. In 1892, a Singaporean Chinese, Lim Koon Tye, printed a request for exchange and correspondence with an English-language phonographer in any part of the world in the *Phonetic Journal.* It is clear that change was in the air. Many contemplated the possibility of coordinating sound and script in the different Chinese dialects in a new way. Rushing toward this new frontier, they sought out different networks and resources to identify the appropriate audience.

Mok, for instance, had originally intended to limit his study to traditional rhyme dictionaries in order to figure out a similar system of phonetic classifi-

---

23   Ni 1948, 32.

24   Cai Xiyong 1956 [1896]. Cai was a translator who accompanied Emissary Chen Lanbing to the United States, Japan, and Peru in the 1870s. He spent more than a decade drawing up a tachygraphy-based Chinese shorthand system. Having witnessed the extraordinary efficiency (two hundred words per minute) of the use of Lindsley shorthand (*suohen* 索痕) in US congressional proceedings during his four years in Washington, D.C., Cai welcomed a similar prospect for the then roughly 40,000 Chinese characters. He encountered shorthand again in Japan as *shagenshu* and meticulously studied various manuals and handbooks related to the subject. Cai is often credited with the foresight of having developed a phonetic script using the Beijing-based Mandarin dialect, which was later chosen for the national language. Cai, however, had originally intended the quick script to serve as a tool for implementation after the national language had been chosen. The fact that the Latin alphabet can be used to spell and pronounce the different European national tongues inspired Cai to do the same for the various regional dialects in China by supplementing Sinograph recognition with an easy phonetic scheme that can notate several characters in one continuous stroke. Like many others, he first tested his ideas on his own family, who reputedly, within a month, mastered the scheme. His eldest son, Cai Zhang, carried on his research after his death. Improving upon it with Isaac Pitman's shorthand, and in collaboration with a Japanese stenographer, Cai Zhang published *Chinese Stenography* in 1934, which became the founding textbook for modern Chinese shorthand. Other similar shorthand schemes include Li Jiesan's *Min Dialect Quick Script* (1956 [1896]), which was a Min-dialect adaptation of Cai's northern-dialect formula, and Wang Bingyao's *Table for Phonetic Script* (1956 [1896]). Cf. Gitelman 1999; Downey 2008, 103–154; Kreilkamp 2005.

THE LORD'S PRAYER.

祈禱文

吾 父 在 天。 願 爾 名 聖。 爾 國 臨 格。 爾

旨 得 成。 在 地 苦 天。 所 需 之 糧。 今 日

錫 我。 我 免 人 負。 求 免 我 負。 俾 勿 我

試。 拯 我 出 惡。 爾 國 權 榮。 皆 爾 所 有。

受 及 世 世。 顧 所 願 也。

MOK LAI CHI.

Y.M.C.A., 31 Hollywood road,
Hong Kong.

FIGURE 2   *Example of "The Lord's Prayer."* (FROM MOK 1893, 470.)

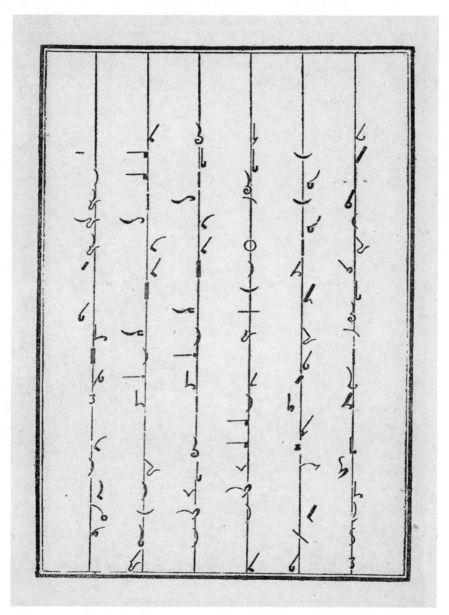

FIGURE 3   *"Imperial Edicts for General Instruction" in "quick script."* (FROM CAI XIYONG 1956 [1896], 36.)

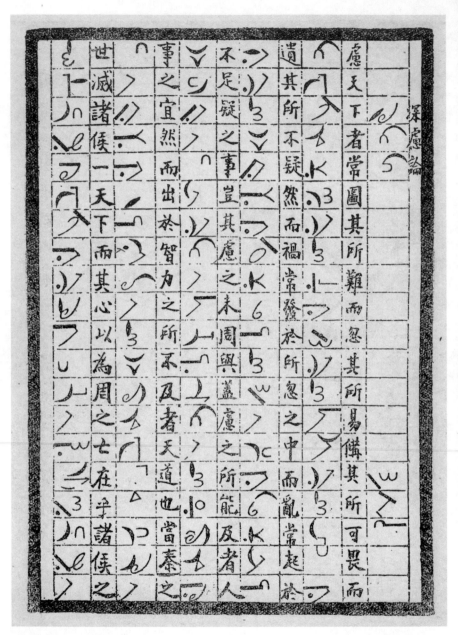

FIGURE 4   *Li Jiesan's "quick script in Min topolect," using as a teaching text Confucian scholar
Fang Xiaoru's (1357–1402) treatise "A Discussion of Profound Contemplation"* (SHENLU
LUN). (FROM LI 1956 [1896], 67.)

FIGURE 5   *Chen Qiu's "Seven-Tone New Script of Europe" was based on the five tones of the ancient pentatonic scale in Chinese musicology—gong* 宮, *shang* 商, *jue* 角, *zhi* 徵, *yu* 羽—*and two additional tones derived from gong and zhi. The five tones roughly correspond to the keys of C, D, E, G, and A in the Western diatonic scale. The phonetic scheme is called the new script of "Europe," because Chen viewed Asia and Europe as belonging to the same continent, with China at its eastern edge.* (FROM CHEN 1958 [1903], 64.)

cation. In May 1892, however, he read an article in the *Phonetic Journal* about a Chinese-language shorthand in the Pitman style created by the Reverend Alexander Gregory. Gregory was a missionary of the Presbyterian Church in Amoy who while learning Chinese conceived the idea that Pitman's shorthand "might be transferred [into Chinese] almost in its entirety, [character] strokes used for the consonants, and a somewhat increased number of vowel signs being put in round the outline thus obtained."[25] He further studied, as most missionaries did, local rhyme books in order to map the dialectal syllabic properties onto Pitman's shorthand system. With a few adjustments, he submitted his findings to the *Phonetic Journal*, hoping that it would "be useful as a starting point for others." Greatly inspired, in a letter to the *Phonetic Journal* in 1893 that was written in "beautiful phonography," Mok passionately voiced his own aspiration to follow suit and "to assist a phonographer to read, speak, and write Chinese by means of the simple phonographic signs." People found Mok's method so useful that, Mok describes, "schoolboys and clerks [were] asking me to open a shorthand class in the evening, which I intend to do."[26]

Four months later, Mok opened the Hong Kong School for Shorthand, attached to the Morrison English School, in which he taught, pro bono, a curriculum that included Pitman's shorthand, translation from English to Chinese, grammar, composition, and letter writing. It was designed for students interested in going into government service. His scheme, received as "a very ingenious adaptation of Pitman phonography to Chinese," comprised slightly fewer than fifty vowel signs and consonants, with provisions—light dots, lines, and word positions—for marking the nine tones in Cantonese.[27] At the time of his course offering in September 1893, his manuscript "Chinese Phonography, an Adaptation of Phonography to the Chinese Language in the Cantonese Dialect" was "in preparation."[28] In November of the same year, he posted a notice in the journal, seeking the help of potential collaborators and lithographers in producing the book.[29] It is unclear whether it was ever published.

From these examples, it is apparent that there were different developments in phoneticization going on at the same time. Some, like Mok, sought out support from Western phonographers and missionaries, while other Chinese script reformers saw their projects as distinctively Chinese and were implicitly disdainful of the missionaries' efforts. Regardless, the shorthand system

---

25   *Phonetic Journal* 51 (1892): 325–326.
26   *Phonetic Journal* 52 (1893): 290.
27   Ibid., 389, 470.
28   Ibid., 590.
29   Ibid., 722.

attracted a following and curiosity. In response to a reader's query in the February 4, 1893, issue of the *Phonetic Journal* regarding whether a phonographic system had been developed for the Cantonese dialect in China, a correspondent in Glasgow forwarded a letter from the Reverend W. H. Murray in which the latter described the use of an adapted version of the Pitman shorthand in Beijing for copying parts of the Bible for reading exercises.[30] Murray was a Protestant missionary who had been a resident of China since 1871 and had founded the School for the Blind in Beijing.[31] He spoke with authority on the subject of phonetic scripts, having invented in 1879 the Numeral-Type system for teaching literacy to the blind based on a classification of 408 distinct tones, or syllables, in Mandarin Chinese. He later modified the system for the purpose of general literacy, using black lines instead of the raised dots of the Braille system, to accommodate the needs of sighted but illiterate Chinese.[32] By 1895, it was reported that the use of Pitman's shorthand was evident in mercantile offices and schools in Shanghai.[33]

This broad view of the different motivations, the technical sources, and the international network for revamping the Chinese writing system pinpoints a new translocal and global locus for understanding the significance and ambition of the phoneticization movement of the late Qing dynasty. For one thing, it far exceeded the later scope of national standardization. Recent attention to this project has largely been restricted to the representative figures of the movement—such as Lu Zhuangzhang, Shen Xue, Cai Xiyong, and Wang Zhao—and has faithfully adhered to Ni's standard accounts.[34] Indeed, by the time Lu Zhuangzhang's *A Primer at a Glance* appeared in 1892, the question had already taken on a different color.[35] Unlike Mok's, Lu's scheme was designed to supersede missionary phoneticization in open rivalry.

Born in the first year of the First Opium War in 1840, Lu Zhuangzhang was raised in Xiamen, where missionary Romanized versions of the Bible were in circulation as early as 1852. He did not do very well under the traditional civil examinations system, which afforded him little prospect of official distinction. He converted to Christianity and sought out opportunities in the missionary community. While studying the Bible and learning about the Western sciences, his daughter later recounts, he became deeply involved in the question of

---

30 Ibid., 114.

31 Dennis 1906, vol. 3, 378.

32 Gordon-Cumming 1898, vii–x.

33 *Phonetic Journal* 54 (1895): 278.

34 Kaske 2008; Cheng 2001; Mair 2000.

35 Lu Zhuangzhang 1956 [1892].

alphabetic writing and its possibilities for reforming the Chinese ideograph.[36] At the age of twenty-one, he went to Singapore to study English for about four years. Afterward, he returned to Xiamen and worked as a language tutor for Chinese and foreigners before acquiring a post assisting John MacGowan of the London Missionary Society. Together they compiled the *English and Chinese Dictionary of the Amoy Dialect*, which appeared in 1883. While working under Macgowan, Lu had the chance to work extensively with the missionaries' system of "speech-sound script" (*huayin*话音), which used Latin letters to transcribe local dialects. The missionary Romanization schemes drew from local sources, in particular the fifteen tones already identified in the earliest extant rhyme book of the Zhangzhou dialect in southeastern Fujian Province.[37] Having perused the same sources and studied their transposition in the process, Lu came to believe that he could develop a better system.

Having gained more confidence from knowledge and exposure, Lu grew critical of the missionaries' endeavors. In the preface to *A Primer at a Glance*, he takes issue with the speech-sound script. Not only did he find the missionaries' reliance on the fifteen tones insufficient, but he also found their schemes structurally wanting. The speech-sound script required several letters to convey just one sound, leaving some words physically longer than others, uneven instead of aesthetically streamlined. To save space, Lu proposed a system of fifty-five *zimu* 字母 (alphabet letters), on the basis of which each character would be spelled out with exactly one letter for the rhyme vowel and another for the rhyme ending. Lu adapted the fifty-five *zimu* from the Roman alphabet, a method that a number of other reformers opted for as well. While some letters appeared to be Latin in origin, each had its own distinct pronunciation. Of the fifty-five letters, thirty-six were based on the Amoy (current-day Macau) pronunciation, nine were taken from the Zhangzhou and Quanzhou dialects, and the remaining represented composite tones from other regions. Lu used local rhymes and songs as practice lessons throughout the manual. Despite their exposure to foreign languages, grammar, and transcription systems, all script reformers cut their teeth on the well-established corpus of indigenous linguistic and phonological materials.

A major selling point of Lu's scheme was its purported ease of learning. It was designed to spare the brain unnecessary exertion, even dispensing with the presence of a teacher by allowing the student to recognize the intended Chinese character, now spelled out in Latin letters, on his or her own based on

36   Lu Tiande 2000, 77.

37   Van der Loon 1992, 15–57.

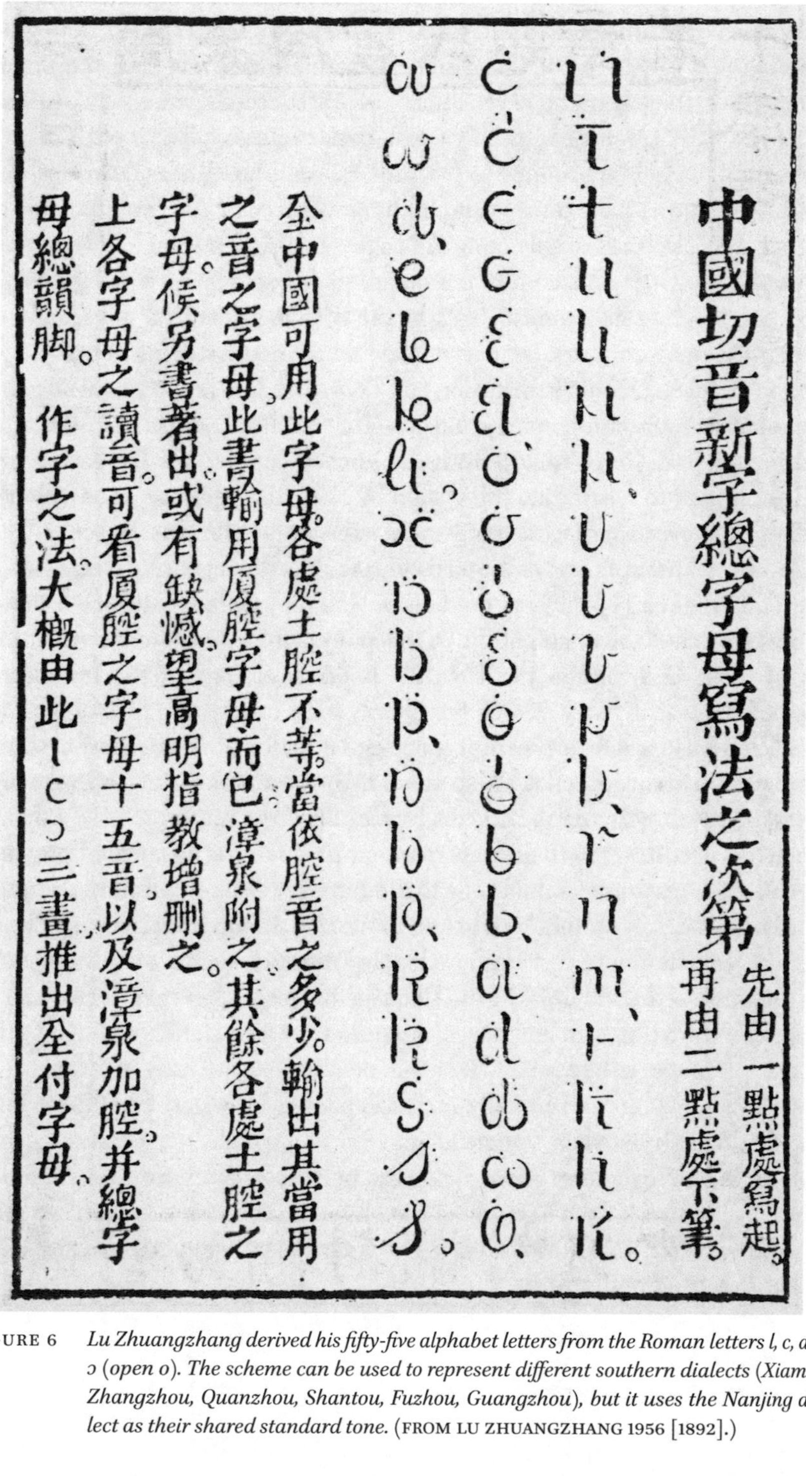

FIGURE 6   *Lu Zhuangzhang derived his fifty-five alphabet letters from the Roman letters l, c, and
ɔ (open o). The scheme can be used to represent different southern dialects (Xiamen,
Zhangzhou, Quanzhou, Shantou, Fuzhou, Guangzhou), but it uses the Nanjing dialect as their shared standard tone.* (FROM LU ZHUANGZHANG 1956 [1892].)

a few rules of thumb.[38] Lu anticipated the method's rapid spread throughout the nineteen provinces in China, its effect rippling through to the outside world. The orthography, however, did not at first succeed in being easy to learn. Other than the inventor himself (whose passionate familiarity with his own innovation made him immune to the growing complexity of added rules, principles, exceptions, and clauses) and his next of kin (who are generally the first test subjects and predictably compliant), few could pick it up in a few weeks, as advertised. This was a common problem that beset almost every new script proposal, sending its inventor back to the drawing board for a second, and often third and fourth, try. Even a trained linguist could still fail at grasping its basic principles. Linguistic historian Luo Changpei, in a not-atypical response, found Lu's scheme cumbersome and esoteric, "neither Chinese nor Western."[39] Later, while living in colonial Taiwan, Lu himself came to see the design flaws and attempted to recalibrate the system by using the Japanese kana syllabary. By then, however, there were many more new competitors on the scene.

Thwarted attempts and redoubled efforts aside, the ingenuity of the various script schemes can be gauged, not in how well they were received, but in how far they stretched the imagination. Of the many factors that could have doomed any of these innovations, however, the political climate of the tumultuous close of the last dynasty and of the bloody path toward the founding of the Republic was enough to preempt a definitive realization. The lasting significance of Lu's innovation lies not so much in the phonetic scripts as in the principles that went with them. Early on, he identified the importance of grouping characters according to their most frequent usage in the absence of punctuation marks, something comparable to modern-day segmentation in computational linguistics. A prerequisite to automatic translation from the early days of machine translation to current-day Google Translate, such a method of marking the basic semantic units of the Chinese language prepared the way for its conversion into different languages. Whereas the Latin alphabet allows for the separation of words by spacing, Lu explained, Chinese writing is composed of discrete characters, traditionally unaccompanied by visual cues that would help distinguish between semantic units (*ci* 词), which frequently consist of more than one logogram. Recognizing the need to account for the syntactical subunits, Lu used a dash to connect the phoneticized scripts within the same semantic cluster. Other script inventors followed suit by alternatively using parentheses and underlining.

---

38   Lu Zhuangzhang 1956 [1892], 3.

39   Luo 1934, 12.

Two lessons can be drawn from the late Qing script reform's ingenuity and failure. Despite their innovativeness and the significant advantage they offered in the long run, the various script schemes failed the practical test. The window for radical change—which some saw as the very abolition of the Chinese-character writing system itself—that opened with the cataclysmic fall of the dynasty was too small to allow gradual reform. The rise of nationalism essentially drew the movement to a close. It marked a conservative shift from fundamentally transforming the shape of the Chinese language to standardizing its geographically diverse pronunciations. Despite its general untimeliness, the late Qing script reform left unresolved certain issues that became new venues of pursuit. The recognition of a fundamental spatial disparity between segments of Chinese characters and segments of alphabetic words left behind a monumental challenge: the convertibility of the Chinese character script as a host medium for alphabetic languages, particularly English. This technical quandary was ingeniously confronted in the 1940s.

III   Alphabetizing Chinese

The concern during the late Qing period that the Chinese writing system was not conducive to modern thinking took a very different turn in the ensuing decades. On April 17, 1946, the Chinese writer Lin Yutang filed an application with the US Patent Office for a Chinese-language typewriter. The design, which took him fifty years to conceive and to build, realized the vision of the late Qing script reformers in typographic technology. His venture relied on an assemblage of different means of production from China to Europe and the United States. He put his idea to the test in 1927 by conducting an empirical study using an instruction manual on general mechanics and an English-language typewriter. In 1931, he spent time working with engineers in England and subsequently brought back an early template of his invention, custom-made in Xiamen. The casting mold for the type was customized in New York's Chinatown, and Lin found a small factory in the suburbs to make the special parts for his ideographic writing machine.

While Lin's literary success is well known, his technological experiments have won only passing mention. From 1930 onward, he authored numerous nonfiction and fiction works and introduced Chinese culture and civilization to the Anglophone audience. His commercial success in the United States further extended his reputation as one of modern China's best essayists. Nearly all his English-language works were published by the John Day Company with the help of Pearl S. Buck and her husband, Richard Walsh, both of whom were

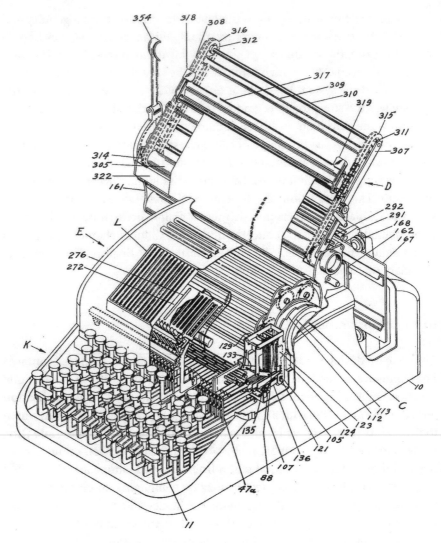

FIGURE 7 *"Chinese Typewriter" (US Patent 2613975, approved October 14, 1952).*

instrumental in persuading Lin to return to the United States in 1936. Their friendship did not survive the typewriter. To finance his typewriter, Lin exhausted nearly all the royalties from his English-language best sellers.[40] The 14- by 19-inch apparatus almost bankrupted Lin, and he tried to borrow money from Buck. She refused, and this reputedly precipitated their much publicized falling out. Against the odds, Lin's patent was finally approved in October 1952.

---

40   Lin 1994, 250–257.

By the mid-twentieth century, efforts to modernize the Chinese language were well under way. The new forms of its institutionalization, however, were far from uniform. Lin's typewriter played an important role in this process toward standardization and dissemination on an international scale. Though not the first Chinese-language typewriter, Lin's machine contributed greatly to the eventual digital globalization of the Chinese ideograph.[41] Details of his model served as a main reference in subsequent developments in electronic writing: multilingual electric typewriters, Chinese-language input in data processing, the encoding of Chinese characters into unique numerical codes for storage and transmission, and electronic software programs that use a phonetic version of Chinese.

All this was indebted to Lin's early interest in linguistics and phonology, which later took a back seat to his literary career.[42] Already twenty-three years before he filed the application at the US Patent Office in New York, Lin was developing important views on the history and taxonomy schemes of the Chinese language. After studying for a year at Harvard with Irving Babbitt and Bliss Perry and then earning a doctorate in historical phonology from Leipzig, he returned to Beijing in 1923 at the behest of Hu Shi, who offered him a professorship in linguistics and literature in the Department of English at Beijing University. Lin was a core member of the Committee for Research on the Romanized Spelling of the National Language that was appointed by the Ministry of Education in 1925. He strongly supported the use of the alphabet in Gwoyeu Romatzyh (National Romanization), a Mandarin Romanization system developed by the linguist Zhao Yuanren.[43] Debates about modern lan-

---

41　Lin made reference to three other Chinese-language typewriters in his patent application alone: "Type-writing Machine" (US Patent 1,247,585), by Pan Francis Shah of Tianjing, China, in 1916; "Apparatus for Writing Chinese" (US Patent 1,260,753), by Heuen Chi of New York in 1915; and "Chinese Language Typewriter and the Like" (US Patent 2,412,777), by Chung-Chin Kao of New York in 1943. Unlike other Chinese typewriters, which required typists to fill in the characters manually and memorization of characters, Lin's machine boasted a "self-evident" keyboard that required no training. See "New Typewriter Conquers Chinese Symbols," *Popular Science* 151 (November 1947): 137. One of the earliest Chinese-language typewriters was designed by the Protestant missionary Devello Z. Sheffield of the American Board mission. See Sheffield 1897.

42　Yet Lin acknowledged that his investment in the fate of the Chinese script remained a lifelong interest. His writings on the topic are collected in a volume separate from his other works. See Lin 1967.

43　DeFrancis 1950. For the history of and contemporary developments in Romanization in China, see the very useful website "Romanization Systems," www.pinyin.info/index.html (last accessed September 14, 2009).

guage reform reached new heights, stirring up controversies and oppositions that called for no less than a full-scale "Han script revolution" (*Hanzi gaige* 汉字改革).

Taking a more conciliatory approach, however, Lin advocated taking the best of both worlds.[44] Simplifying character strokes and developing a Romanization system for the Chinese language, Lin assured, were not mutually exclusive projects. It was necessary to pursue a parallel course. Lin thought it redundant to devise a new system of acoustic symbols when the alphabet had already proven its phonetic usefulness in the different Indo-European national languages. He reviewed other possibilities that were important in the discussions among European linguists and philologists on the correspondence between alphabetic notations and their symbolized sounds in the science of phonetics.

Otto Jespersen's Analphabetic System (later renamed antalphabetic) was one such "ultra-alphabetic" system. It used "half-mathematical" formulae to symbolize not sounds but elements of sounds and the positions of the various articulatory components of the speaking organ.[45] Lin thought the system, though devised with the precision of scientific transcription, bore no intuitive relation to everyday use. Alexander Melville Bell's Visible Speech, a second alternative, was similarly too intellectually detailed for the average language user. Bell wished to devise a system that would include all language sounds, from foreign to dialectal, as well as inarticulate sounds like sneezing and yawning, all by using iconic symbols that through their shape would indicate how the sounds were formed. Its classification of consonants and vowels was arbitrary and often disputed, undercutting its efficacy as a "Universal Alphabet." Neither did Lin find a simple shorthand system based on speed and accuracy—like Pittman's or Boyd's—entirely desirable. A common script, for Lin, needed to be not only clear and easy to use but also aesthetically pleasing. Only one scheme was agreeable to him. Henry Sweet's "organic alphabet" (derived from Bell's Visible Speech but replacing Bell's iconic symbols with Roman-

---

44   DeFrancis 1950. For the history of and contemporary developments in Romanization in China, see the very useful website "Romanization Systems," www.pinyin.info/index.html (last accessed September 14, 2009).

45   Jespersen 1889, 8–12. The technical precision with which Jespersen dissected the location and movement of sound made his system too abstruse even for the learned, and even more inappropriate for the audience Lin had in mind. Henry Sweet describes analphabetic type as "a group of symbols resembling a chemical formula, each symbol representing not a sound, but an element of a sound: the part of the palate, tongue, etc., where the sound is formed, the degree of separation (openness) of the organs of speech, and so on." As quoted in Henderson 1971, 255.

FIGURE 8   *The cover of a special issue of National Language Monthly (Guoyu yuekan 国语月刊) (August 1922) that features key essays and debates over the national language reform. Soldiers (center), wielding weapons of the phonetic alphabet, slaughter a horde of traditional Chinese characters in the ancient seal-script style (bottom right), while the masses coolly watch and stand united behind a row of Roman letters that spell out "Latin script" in Gwoyeu Romatzyh (middle left).*

alphabet-based notations) fitted his vision of a practical approach using the existing alphabetic system.

After careful study, Lin came up with his own solution in 1924. The breakthrough later became the cornerstone of the indexical system for his typewriter. Lin wanted to design a system that any user could "pick up without learning" (*bu xue er neng* 不学而能).[46] He proposed looking up any given character in a dictionary first by looking up the "top stroke" (*shoubi* 首笔) in the character's radical, or root, component.[47] The top stroke was further categorized into one of five stroke movements—straight across, straight down, down to the side, point, and hook—listed in the dictionary in that order. With the second stroke, the same order is repeated, thus narrowing the range of possible characters.

The idea was to classify the character according to its most identifiable component and then to index the character in a new order of progression. A complementary method was developed with reference to the "final stroke" (*mobi* 末笔). The exit stroke is generally the longest and thus most easily made out at the bottom portion of the character. The combined method, Lin boasted, was also greatly superior to those that came before, the majority of which depended on rhyme and vowels. "Reverse-cut" (*fanqie* 反切), for example, a method used in classical phonology since the late second century, indicates the pronunciation of one character by combining the opening consonant and closing vowel of two other characters. The cumbersome method, however, could not account for changes that took place in oral speech over the centuries. Even if one cuts correctly, the result may be far removed from its original pronunciation. As a lexicographical tool, the reverse-cut method ensures little inherent logic and systematization. In contrast, a system based on obvious top and bottom strokes, Lin noted, is "entirely based on shape and does not at all borrow from analytical methods, which are in any case not the strong suit of the Chinese."[48]

Lin's method, however, incorporated a more important mechanism. Although based on the graphic shape of the Chinese character, his system assimilated an alphabetic logic. The process of elimination by repetition of the five stroke types in fact had an augmenting effect. Lin likened it to the classifying order of aa, ab, ac, ad, and so on. Cai Yuanpei notes that Lin uses "the example of the alphabet and applies it to the strokes of the Chinese script," thereby

---

46   Lin 1967, 284.

47   Lin 1967, 273–274.

48   Lin 1967, 284.

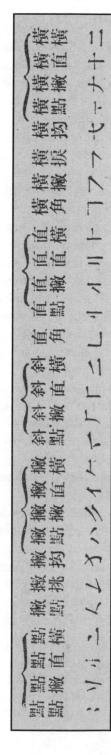

FIGURE 9   Lin Yutang's "top stroke," or first stroke, is categorized according to five stroke types: (1) straight across, or horizontal (heng 橫); (2) straight down, or vertical (zhi 直); (3) down to the side, or slanted (pie 撇); (4) point (dian 点); and (5) hook (gou 勾). After the first stroke is identified in this way, the process repeats with the second stroke, and so on. Lin likens it to the linear logic of alphabetism: aa, ab, ac, ad, etc. (LIN 1994, 273–74).

producing a veritable alphabetism of "aba, abb, abc, etc."[49] Taking on the alphabetic property of linear extension, the new method of assembling Chinese characters treated stroke order like the serial arrangement of the alphabet. Instead of a cluster of simple graphemic units, the ideograph was now conceptualized differently, linear rather than strictly combinatory. In short, Lin made it possible to "spell out" the Chinese character. Under this new configuration, the kind of cultural and philosophical difference that McLuhan and others saw in the ideograph would no longer be located on the level of visible arrangement. Instead, this cultural difference was converted into a new communicability between the ideograph and the alphabet. This transposition transformed the grammaticality—rather than the plain physical form—of the phonetic alphabet into the mechanization of the Chinese written language.

What had long distinguished the phonetic alphabet from the ideograph—combined syllabary, phonetic divisions, linearity—dissolved within a double frame of stroke and alphabetic index.[50] Lin's method shifted the frame of reference such that alphabetism could no longer be posed as the ideograph's lack. The idea that the ideograph is not phonetic or linear, nor the alphabetic pictorial and sensorial, one might recall, was never tenable or philologically sound. Yet Lin took the demystification a step further by reabsorbing that difference into the new classification system of the ideograph. By figuring out a new mode of accommodating and assimilating alphabetic languages, Lin fused what he thought were the best features of both languages. Behind the escalation of language wars between English and Chinese, a different kind of mutual governance came into play. Lin's pragmatic support of using the alphabet for Romanization, on the one hand, and innovative appropriation of its distinctive features to re-index the Chinese character, on the other, nullified the antagonism with strategic accommodation.

With this in mind, one can better appreciate Lin's design. The keyboard to Lin's typewriter displays not alphabetic letters but Chinese character radicals, separated and ordered in precisely the way he had outlined above. It has seventy-two keys, thirty-six of which represent the different top (upper left-hand) components, while the remaining twenty-eight represent the bottom (lower right-hand) components. When a top key and a bottom key are pressed simultaneously, the type roller matches the two together and prints a unit of eight

---

49   Cai Yuanpei wrote the preface to "Hanzi suoyinzi shuoming." See "Cai Jiemin xiansheng xu" (Preface by Mr. Cai Jiemin), in Lin 1967, 276–77.

50   Scholars now agree that the Chinese script is an imprecise syllabary that has both visual and semantic qualities. See Daniels and Bright 1996, 189–208. See also I. J. Gelb 1963, 85–88, 166–189.

possible combinations. An accompanying novel device is the method of displaying the qualifying characters. After the first round of selection, which is based on a match by radicals, a "magic eye," or projected window, appears above the keyboard. It allows the typist to see a maximum of eight characters displayed in a row. The typist then presses a key from another group of eight keys, each corresponding to a particular character in the viewer that is then finally printed on the paper.[51] With "reference to the shape or design of the strokes making up the character at the top and the bottom of the character," the machine can also be adjusted to transcribe other languages: "the same structure, but with modified key symbols and type arrangements, may be used to print other languages which are based upon the English alphabet and still other languages in which alphabets are not used."[52] Lin can truly be said to have developed an unprecedented Chinese writing machine that established a new logical parsing system of the ideograph, enabling its further use with other languages.

The convergence between mechanization and translation marked a new era for the ideograph and unexpectedly propelled the globalization of the Chinese script in a new direction. On May 18, 1948, Mergenthaler Linotype Company signed a contract to test Lin's prototype for a period of two years in order to evaluate its feasibility for mass distribution. The overhead cost, however, in manufacturing each typewriter and its customized parts was too high (about $1,000 each). In September 1951, Lin officially sold Mergenthaler the copyright for $25,000. At this point, the US Air Force embarked on a research project on "automatic translation," later known as machine translation. After multiple inquiries, the US Air Force concluded that they needed to use Lin's indexical keyboard as the prototype for their research on the Chinese language and gave it to the International Business Machines Corporation (IBM) for further development. Beginning in 1960, the IBM Research Center pursued the study, sponsoring projects conducted at various American universities. In summer 1963, IBM unveiled the "Sinowriter," which was jointly developed with the Mergenthaler Linotype Company. Gilbert W. King, the director of research at the center, led the project. With reference to a concurrent project on Russian-English

---

51   For a description of this process, see "New Typewriter Conquers Chinese Symbols," *Popular Science* 151 (November 1947): 137. Lin's magic window may be considered the prototype for the computer display of characters that share the same pinyin forms in contemporary Chinese-language software. As pinyin does not designate tones, all qualifying homophones are displayed in a rectangular window, which the user has to scroll through to identify the appropriate character.

52   Lin 1952a   3, 25.

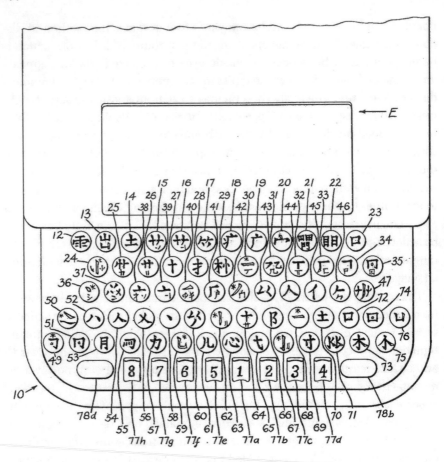

FIGURE 10   *The keyboard for Lin Yutang's "Minkuai" ("clear and quick") Chinese-Language Typewriter can generate up to ninety thousand characters from seventy-two keys (thirty-six top strokes, twenty-eight bottom strokes).*

machine translation, headed by Austrian Sinologist Erwin Reifler at the University of Washington in Seattle, King and his collaborators introduced the Sinowriter keyboard in a 1963 issue of *Scientific American.*[53]

Unlike its predecessor, Sinowriter was cost-efficient and put broad dissemination within reach. The news was picked up quickly by, among others, the *Armed Forces International Journal,* "almost an official organ of the armed

---

53   King and Chang 1963.

forces":[54] "The Sinowriter is an inexpensive machine which can be operated by typists who are not familiar with the Chinese language. After two weeks of training, a typist can obtain a speed of 40 characters per minute, or the equivalent of about 40 words a minute in English."[55] In designing the Sinowriter, King and his associates had developed a device for photographic storage and optical information retrieval that greatly improved upon the memory capacity for the number of characters in Lin's typewriter. The original search for "a keyboard that could be learned fairly quickly by people who are not necessarily able to read Chinese"[56] had led them to Lin's keyboard in the first place. Lin's "geometric-recognition scheme" provided a crucial missing piece.[57] Building on Lin's specific character index and display system, King's contribution, as he explained in a patent application in 1965, was to encode the Chinese input as punched holes on a Flexowriter tape, a conversion into binary codes that facilitates a faster storage and retrieval process.

Lin's upper and lower components, with the pressing of corresponding keys, could compose up to 90,000 characters based on a blueprint of 9,000 (he based the figure on the Chinese telegraph codebook). King's optical retrieval system, with the help of punched tape, called up and displayed the qualifying characters in "less than 100 milliseconds."[58] In terms of what a given operator had to do, the task was identical to Lin's prescribed steps: "the operator actuates a key having the desired upper segment configuration to thereby insert a binary X address code into an X address register and then actuates a key having the desired lower segment character configuration to thereby insert a binary Y address code into a Y address code register. These thereafter control the character plate having both the selected upper and lower character segments to be positioned within the retrieval area, which in turn enables the qualifying characters to be optically projected at a viewing area."[59] In effect, King explains, "the two keys activate a mechanism that projects onto a screen the whole family of characters sharing these particular configurations. The family may contain only one member or as many as sixteen. Each member of the family is numbered from one to sixteen, and the operator can easily identify the one that matches the desired character in the Chinese text."[60] This further paved

---

54   "Military Paper Willed to Club: Army, Navy Journal Worth Half Million," *Pittsburgh Press*, March 19, 1949, 17.
55   "X," *Armed Forces International Journal* 102 (1964): 15.
56   King and Chang 1963, 129.
57   King and Chang 1963, 130.
58   King et al. 1967, 3.
59   King et al. 1967, 2.
60   King et al. 1967, 2.

the way for Chinese-language machine translation, as the characters now "may be easily and quickly converted to a system for encoding and printing complex characters in a second language."[61]

King did not stop there and was about to take the project further. In 1962, he accepted a position at Itek Corporation, an important manufacturer of reconnaissance technology and a US defense contractor during the Cold War, and expanded research was undertaken both there and at IBM. In 1964, Itek came out with their own Modified Sinowriter, also known as Chicoder (Chinese Encoder). According to its public release statement in November 1966, it was capable of encoding 10,500 characters.[62] The method could be used for anything that required machine processing of large quantities of Chinese characters.

In retrospect, the impact of Lin's typewriter on the era of machine translation gave an unexpected twist to the original intent of machine translation. A historic memorandum by mathematician Warren Weaver on July 15, 1949 is credited with first launching machine translation as a scientific enterprise. Weaver, who was the director of the Natural Sciences Division of the Rockefeller Foundation and widely influential among major policy makers in US government agencies, was convinced that the success of cryptography during World War II had much more to say about the "frequencies of letters, letter combinations, intervals between letters and letter combinations, letter patterns, etc. *which are to some significant degree independent of the language used*" [emphasis in original].[63] His collaboration with Claude Shannon in pioneering the first introduction to information theory further convinced him to attempt a universal code for translating languages into one another.

Ironically, what helped to fulfill Weaver's vision was not the decipherment of a Chinese-coded English, or Basic English, but Lin's alphabetically coded Chinese. Insofar as machine translation involved the Chinese language as part of its universalizing project, the modern Chinese language reform met it halfway in its own quest for global mediality. Machine translation mechanized the cipher effect that was already in play when Lin embedded the alphabetic logic in the Chinese script. The process was well under way with China's late nineteenth-century script reform. Each system's race for its own distinction made possible, paradoxically, a new meeting ground. As the ideograph later joined the alphabet, the alphabet was also converted into one possible technological

---

61   King and Chang 1963, 130.
62   Eng 1966.
63   Weaver 1955, 16.

function of the ideograph.[64] As both orders of writing pursued their own visions of universal utility, their mutual material conversion formalized an implicit, shared desire to bring all languages under one roof—with one major difference. The alphabetic-ideographic conversion happened neither for the sake of common ancestry (a theory pursued by eighteenth- and nineteenth-century European linguistics under the sway of evolutionary theory) nor for the dream of a single lingua franca (in service of one nation's dominance over others during the era of Europe's self-ordained civilizing mission). What fundamentally changed in the twentieth century, if the Chinese script revolution has made its global mark, is how technological accommodation and practical hospitality came to define the true arena for the exercise, extension, and dissemination of world linguistic power.

Bibliography

Boltz, William 2000. "Logic, Language, and Grammar in Early China," *Journal of the American Oriental Society* 120, 2: 218–229.

Cai Xiyong 1956 [1896]. *Chuanyin kuaizi* [Phonetic quick script]. Beijing: Wenzi gaige chubanshe.

Cai Zhang 1934. *Zhongguo sujixue* [Chinese shorthand]. Shanghai: Zhonghua shuju.

Castells, Manual 1996. *The Rise of the Network Society: Information Age.* Cambridge, MA: Blackwell.

Chen Qiu 1958 [1903]. *Xinzi Ouwen qiyin duo* [Seven-tone new script of Europe]. Beijing: Wenzi gaige chubanshe.

Cheng, W. K. 2001. "Enlightenment and Unity: Language Reformism in Late Qing China," *Modern Asian Studies* 2: 469–493.

Daniels, Peter T., and William Bright, eds. 1996. *The World's Writing Systems.* Oxford: Oxford University Press.

DeFrancis, John 1950. *Nationalism and Language Reform in China.* Princeton, NJ: Princeton University Press.

_____ 1984. *The Chinese Language: Fact and Fantasy.* Honolulu: University of Hawai'i Press.

---

64 Interestingly, watching how his Chinese-index typewriter had transformed into the Modified Sinowriter, Lin did not seem to recognize the momentousness of his intervention. He was, in the end, awed by the photocomposition and retrieval system, like the one King designed based on his Mingkuai typewriter, which leaves the human operator jobless in the ever-refining technologization of mechanical printing.

____ 1989. *Visible Speech: The Diverse Oneness of Writing Systems.* Honolulu: University of Hawai'i Press.

Dennis, James Shepard 1906. *Christian Missions and Social Progress: A Sociological Study of Foreign Missions.* 3 vols. New York: Fleming H. Revell.

Downey, Gregory J. 2008. *Closed Captioning: Subtitling, Stenography, and the Digital Convergence of Text with Television.* Baltimore, MD: Johns Hopkins University Press.

Elman, Benjamin A. 2005. *On Their Own Terms: Science in China   1550–1900.* Cambridge, MA: Harvard University Press.

____ 2007. "New Directions in the History of Modern Science in China: Global Science and Comparative History," *Isis* 98: 517–523.

Eng, Albert 1966. "Experimental Modified Sinowriter (Chicoder)." Itek Corporation (November).

Fan, Fa-ti 2007. "Redrawing the Map: Science in Twentieth-Century China," *Isis* 98: 524–538.

Fenellosa, Ernest 2008. *The Chinese Written Character as a Medium for Poetry*, edited by Haun Saussy et al. New York: Fordham University Press.

Gelb, I. J. 1963. *A Study of Writing.* Rev. ed. Chicago: University of Chicago Press.

Gitelman, Lisa 1999. *Scripts, Grooves, and Writing Machines: Representing Technology in the Edison Era.* Stanford, CA: Stanford University Press.

Gordon-Cumming, Constance F. 1898. *The Inventor of the Numeral-Type for China, by the Use of Which Illiterate Chinese Both Blind and Sighted Can Very Quickly Be Taught to Read and Write Fluently.* London: Downing.

Guo Yanli 1998. *Zhongguo jindai fanyi wenxue gailun* [Introduction to translated literature in modern China]. Hankou: Hubei jiaoyu chubanshe.

Havelock, Eric 1987. "Chinese Characters and the Greek Alphabet," *Sino-Platonic Papers* 5: 1–4.

Henderson, Eugénie J. A., ed. 1971. *The Indispensable Foundation: A Selection of the Writings of Henry Sweet.* London: Oxford University Press.

Hu, Danian 2007. "The Reception of Relativity in China," *Isis* 98: 539–557.

Huters, Theodore 2005. *Bringing the World Home: Appropriating the West in Late Qing and Early Republican China.* Honolulu: University of Hawai'i Press.

Jespersen, Otto 1889. *The Articulation of Speech Sounds Represented by Means of Analphabetic Symbols.* Marburg: N. L. Elwert.

Kaske, Elizabeth 2008. *The Politics of Language in Chinese Education   1895–1919.* Leiden: E. J. Brill.

King, Gilbert W., and Hsien-Wu Chang 1963. "Machine Translation of Chinese," *Scientific American* 208: 124–135.

King, Gilbert W., et al. 1967. "Character Printer." US Patent 3,330,191 (July 11).

Kreilkamp, Ivan 2005. "Speech on Paper: Charles Dickens, Victorian Phonography, and the Reform of Writing," in Leah Price and Pamela Thurschwell, eds., *Literary Secretaries / Secretarial Culture*. Aldershot, Hants, UK: Ashgate   13–31.

Kurtz, Joachim 2011. *The Discovery of Chinese Logic*. Leiden: E. J. Brill.

Lackner, Michael, Iwo Amelung, and Joachim Kurtz 2001. *New Terms for New Ideas: Western Knowledge and Lexical Change in Late Imperial China*. Leiden: E. J. Brill.

Lewis, Mark E. 1999. *Writing and Authority in Early China*. Albany: State University of New York Press.

Li, Jiesan 1956 [1896]. *Minqiang kuaizi* [Script for the Min Topolect]. Beijing: Wenzi gaige chubanshe.

Lin Yutang 1946. "Invention of a Chinese-Typewriter," *Asia and the Americas* 46: 58–61.

——— 1952a. "Chinese Typewriter." US Patent 2,613,795 (October 14).

——— 1952b. "Visual Selecting Device for Chinese Typewriters and the Like." US Patent 2,613,794 (October 14).

——— 1967. *Yuyan xue luncong* [Collection of philological essays]. Taibei: Wenxing shudian.

——— 1994. *Lin Yutang zizhuan* [Autobiography of Lin Yutang]. Beijing: Zhongguo huaqiao chubanshe.

Liu, Lydia H. 2004. *The Clash of Empires: The Invention of China in Modern World Making*. Cambridge, MA: Harvard University Press.

Liu Mengyang 1957 [1908]. *Zhongguo yinbiaozi shu* [Chinese phonetically indexed script]. Beijing: Wenzi gaige chubanshe.

Lloyd, Geoffrey, and Nathan Sivin 2002. *The Way and the Word: Science and Medicine in Early China and Greece*. New Haven, CT: Yale University Press.

Lu Zhuangzhang 1956 [1892]. *Yimu liaoran chujie: Zhongguo qieyin xinzi Xia qiang* [A primer at a glance: Chinese new phonetic script in the Amoy dialect]. Beijing: Wenzi gaige chubanshe.

Lu Tiande 2000. "Zhonghua shouchuang yinzi zhi yuanzu Lu Zhuangzhang xiansheng" [Mr. Lu Zhuangzhang: The original pioneer in creating a phonetic script in China], reprinted in Xu Chang'an, ed., *Yuwen xiandaihua xianqu* [Pioneers in language modernization]. Xiamen: Xiamen shi yuyan wenzi gongzuo weiyuanhui bangongshi   76–80.

Luo Changpei 1934. *Guoyin zimu yanjinshi* [A history of the development of the national tonal alphabet]. Shanghai: Shangwu tushuguan.

Mair, Victor H. 2000. "Advocates of Script Reform," in William Theodore de Bary and Richard Lufrano, eds., *Sources of Chinese Tradition*. 2nd ed. 2 vols. New York: Columbia University Press, vol. 2, 302–307.

McLuhan, Marshall, and Harley Parker 1968. *Through the Vanishing Point: Space in Poetry and Painting*. New York: Harper & Row.

Mok Lai Chi 1893. "An Adaptation of Phonography to the Chinese Language in the Cantonese Dialect," *Phonetic Journal*, July 29.

Needham, Joseph 1956. *Science and Civilisation in China: History of Scientific Thought.* Cambridge: Cambridge University Press.

Needham, Joseph, and Christoph Harbsmeier 1998. *Science and Civilisation in China: Language and Logic.* Vol. 1. Cambridge: Cambridge University Press.

Ni Haishu 1948. *Zhongguo pinyin wenzi yundong shi jianbian* [A short edition of the history of China's phonetic script movement]. Shanghai: Shidai shubao.

Pollard, David 1998. *Translation and Creation: Readings of Western Literature in Early Modern China 1840–1918.* Philadelphia: J. Benjamins.

Qian Xuantong 1999. "Suigan lu" [Random notes], in *Qian Xuantong quanji.* 6 vols. Beijing: Zhongguo renmin daxue chubanshe, vol. 2, 1.

Ramsey, S. Robert 1987. *The Languages of China.* Princeton, NJ: Princeton University Press.

Schmalzer, Sigrid 2007. "On the Appropriate Use of Rose-Colored Glasses: Reflections on Science in Socialist China," *Isis* 98: 571–583.

Shapiro, Hugh 2000. "Neurasthenia and the Assimilation of Nerves into China," paper presented at "Symposium on the History of Disease," Academia Sinica, Institute of History and Philology, Nankang, Taiwan, June 16–18.

Sheffield, Devello Zelotos 1897. *The Chinese Type-Writer, Its Practicality and Value.* Paris: Imprimerie nationale.

Shen, Grace 2007. "Murky Waters: Thoughts on Desire, Utility, and the 'Sea of Modern Science,'" *Isis* 98: 584–596.

Sivin, Nathan 1982. "Why the Scientific Revolution Did Not Take Place in China—or Didn't It?" *Chinese Science* 5: 45–66.

Tsu, Jing 2005. *Failure, Nationalism, and Literature: The Making of Modern Chinese Identity 1895–1937.* Stanford, CA: Stanford University Press.

_____ 2010. *Sound and Script in Chinese Diaspora.* Cambridge, MA: Harvard University Press.

Van der Loon, Piet 1992. *The Classical Theater and Art Song of South Fukien: A Study of Three Ming Anthologies.* Taibei: SMC Publishing.

Wang Bingyao 1956 [1896]. *Pinyin zipu* [Table for phonetic script]. Beijing: Wenzi gaige chubanshe.

Wang Hui 2004. *Xiandai Zhongguo sixiang de xingqi* [The rise of modern Chinese thought]. 4 vols. Beijing: Sanlian.

Wang Zhao 1903. *Chongkan guanhua hesheng zimu xuli ji guanxi lunshuo* [Reedition of the Mandarin combined tonal alphabet and an exposition of its principles and relations]. Beijing: Biaobei hutong yishu cangban.

Wang, Zuoyue 2007. "Science and the State in Modern China," *Isis* 98: 558–570.

Weaver, Warren 1955. "Translation," in William N. Locke and A. Donald Booth, eds., *Machine Translation of Languages: Fourteen Essays.* Cambridge, MA: Technology Press of MIT; New York: Wiley 15–23.

Webb, John 1669. "An Historical Essay Endeavoring a Probability that the Language of the Empire of china is the Primitive Language." London: Printed for Nath. Brook.

Wilkins, John 1668. *An Essay Towards a Real Character, and a Philosophical Language.* London: Printed for S. Gellibrand [etc.].

Xiong Yuezhi 2011. *Xixue dongjian yu wan Qing shehui* [The dissemination of Western learning and late Qing society]. Beijing: Zhongguo renmin daxue chubanshe.

Semiotic Sovereignty

The 1871 Chinese Telegraph Code in Historical Perspective

Thomas S. Mullaney[1]

Abstract

In 1871, a cable between Shanghai and Hong Kong connected the Qing Empire to the rapidly expanding international telegraphic network. Morse code—the underlying semiotic architecture of telegraphy—was thereby brought into contact with a writing system it was ill-equipped to handle. The product of this convergence was the Chinese telegraph code of 1871, in which approximately seven thousand common-usage Chinese characters were assigned a series of nonrepeating, four-digit numerical codes. To send a Chinese telegraph, one first had to translate Chinese characters into numerical sequences and then transmit these numbers using standard Morse pulse patterns. Because of its reliance on numbers, the code of 1871 set in place a condition of ambiguity that haunted Chinese telegraphy for over a century. Often conflated with ciphers used to send secret or reduced-cost transmissions, even regular Chinese transmissions could be subject to the steeper tariffs and restrictions established to regulate clandestine communication. The particular form of mediation at play in Chinese telegraphy rendered it the one telegraphic language for which no clear boundary existed between "plaintext" and "secret" transmission. Within this context, a loose network of political authorities, entrepreneurs, codebook publishers, and everyday users dove headlong into the realm of mediation and began to play with it. They did not change the four-digit telegraph code itself, but rather they engaged in complex remediation through which they reimagined and configured anew their relationship to the emerging global information infrastructure.

In 1871, the growing network of telegraphic communication reached the shores of the Qing Empire, with a single line opened between Shanghai and Hong

---

1   The research in this paper was supported by the Hellman Faculty Scholar Award and the Stanford University Freeman Spogli Institute China Fund and was made possible by the assistance and generosity of archivists, librarians, and private collectors in China, Denmark, the United Kingdom, and France. In particular, special thanks goes to Alan Renton at the Cable and Wireless Archives in Porthcurno, to Taoshelene Hani at the Bibliothèque Nationale de France, and to both Henning Høeg Hansen and Daniel D. Beattie in Copenhagen for letting me view their private Chinese telegraph book collection (as well as to Kurt Jacobsen and Erik Baark for their invaluable assistance in tracking down codebooks in the Danish National Archives). Earlier drafts of this chapter received essential critical feedback from the volume editors Jing Tsu and Benjamin Elman, as well as Matthew Sommer, Leon Rocha, Kristen Haring, Wolfgang Behr, Yakup Bektas, Roger Thompson, Endymion Wilkinson, Steve Bellovin, Scott Klemmer, Roy Chan, and Mathias Crawford. All errors and omissions remain the sole responsibility of the author.

Kong in April of that year. Carried out by two foreign companies—the Great Northern Telegraph Company of Denmark and the Eastern Extension A&C Telegraph Company of the United Kingdom—the installation of this line marked an early step in the construction of an empire- and subsequently nationwide communications web, woven one filament at a time. A line was installed between Saigon and Hong Kong in June 1871, another between Shanghai and Nagasaki in August, and a third between Nagasaki and Vladivostok in November. In the ensuing years, this network expanded to encompass Amoy, Tianjin, Fuzhou, and multiple cities throughout the empire.[2] Chinese authorities and companies would steadily gain ownership of this web and expand it to a total length of approximately sixty-two thousand miles by the middle of the Republican period (1911–1949).[3]

The history of telegraphy in China has attracted a modicum of attention, primarily because of the significance of this particular form of information technology within wider histories of imperialism, the nation-state, and modernity more broadly. Historians have long referenced the telegraph as a time-eradicating and space-warping mode of communication that was central to the formation of the imagined community of the modern nation-state and to the projection of modern colonial power across spectacular distances. Ironically, however, the history of this Chinese information technology has thus far been narrated with practically no attention to the basic fact that the "information" itself was encoded in Chinese. Prioritizing questions of treaty relations, the negotiation of international landing rights, the resistance or co-optation of regional and local authorities in the construction of telegraph lines, and the role of telegraphy within key diplomatic and military episodes, the history of the Chinese telegraph has been reconstructed as if it were a technology scarcely different from those of railways, arsenals, mining equipment, and so forth. When "information" per se enters the frame, it has done so as a kind of language-invariant, knowledge-bearing fluid that flowed (or perhaps failed to flow) through the branching conduits of the network.[4] Readers in the history of Chinese telegraphy have rarely if ever been invited to reflect on the rudimentary fact that Chinese telegraphy was a *Chinese-language information technology*.[5]

To the extent that we consider telegraphy through the lens of political economy, its history can be narrated and debated according to any one of a number

---

2   Bull 1893, 7–10.
3   Zhu 1937, 149.
4   Baark 1997; Yang 2011.
5   One critical exception is the recent work by Jing Tsu (2010).

of conventional rubrics. As an assemblage of cables and poles, and perhaps an abstract "method," telegraphy lends itself to discussions of diffusion and technology transfer, wherein we chart the dissemination of materials and methods from loci of invention to loci of adoption and adaptation, and often from the West to the non-West.[6] As one of the critical infrastructures of the nation-state—alongside railways, mines, deep-water ports, and arsenals—Chinese telegraphy and its history would seem to fall squarely within the broader discussion of sovereignty and "rights recovery."[7] The primary concerns here pertain to the financial and legal control of this critical infrastructure: to what extent was ownership in Chinese hands, to what extent was ownership held by foreign parties, who were these foreign parties, what impact or influence did their ownership afford them in geopolitical terms, and what efforts were mounted by Chinese political authorities toward the redemption of sovereignty?

Such approaches to Chinese telegraphy are vitally important to our understanding of the nineteenth- and twentieth-century political and economic situation, and yet they disregard at their own peril the unavoidable place of language within this history. In this chapter the baseline objective is to demonstrate how and why the linguistic dimension of Chinese telegraphy cannot be bracketed or jettisoned from our discussion but must be kept in play—and even central—as we reflect upon questions of materiality, finance, and legal frameworks. When we begin to chart out intersections of telegraphic technology and Chinese language and culture, we rapidly discover that the Chinese telegraph was unlike other infrastructural technologies with which it is so often paired. It was neither a new type of mining drill, a new type of warship, nor indeed any type of machine that, at some rudimentary level, could be "switched on" upon reaching Chinese soil. The most basic functionality of Chinese telegraphy depended upon the creation of a novel information-semiotic system—that is, a Chinese telegraph code. Whereas other parts of the world faced the same prerequisite, of course, the challenge was significantly steeper in the Chinese case. In Japan, for example, the entire *kana* syllabary could be outfitted with Morse code–like pulse sequences, as could Russian Cyrillic, Arabic, Hebrew, and so on. There were, that is to say, *prêt-à-porter* methods by which to map the specific modularity of Morse code upon the modularity of these different scripts, be they alphabets or syllabaries. But how was one to build a telegraph code for Chinese, a script neither alphabetic nor syllabaristic? How would Chinese as a language be absorbed into this rapidly

---

6   Rogers 2003 [1962]; Headrick 1988.
7   Lee 1968; Esherick 1976; Hedtke 1977; Rankin 2002; Zheng 2009.

globalizing technolinguistic form whose own creation was foundationally connected to alphabets? From its inception, Chinese telegraphy demanded critical semiotic reimaginations in ways not true of other languages, a historical problem that has received scant scholarly attention.

To consider the "language problem" in the history of Chinese telegraphy is not a question of "more history"—it demands *different* history. This chapter will complicate the question of "telegraph sovereignty" by dwelling upon a rather straightforward fact: namely, that even after the material, financial, and legal infrastructures of telegraphy were brought under Chinese ownership, the problem of the Chinese telegraph code remained unresolved. That is, an enduring parallel problem of sovereignty over the symbolic or semiotic remained whose solution proved elusive. I offer the term "semiotic sovereignty" as shorthand, by which I mean the prerogative to craft a Chinese telegraph code with an attention to the particularities of the Chinese language itself, akin to the ways in which Morse went about developing a code carefully attuned to the affordances, limitations, possibilities, and problems that he and his collaborators encountered within the English language and the Roman alphabet. Semiotic sovereignty is thus an admittedly ambiguous concept through which we imagine sovereignty, not merely over the poles and wires of information networks but also over the notational, representational, and mediating logics by which one designs symbolic systems and paralanguages that make it possible to transmit language at great speed over great distances. Morse code, as we will consider shortly, was fashioned out of a detailed, sensitized conversation with the English language and the way in which it employed the Roman alphabet. What, then, might a telegraph code look like that was similarly attuned to the affordances and limitations of Chinese script? In the contemporary period, we might ask parallel questions of computational character-encoding schemes such as ASCII, reflecting on what such a scheme might look like had it been developed, not outward from an English-language or alphabetic starting point, but in direct conversation with the Chinese language from the outset. One might go so far as to ask: what might a computer-programming language look like that employed Chinese, rather than English, as its terminological and syntactical foundation? To reflect on questions such as these is to come into conversation with the concept of semiotic sovereignty as I will attempt to develop it here.[8]

---

8   As evidenced by the foregoing discussion, and further developed in the sections that follow, my concern with sovereignty departs from that of Lydia Liu in her work *Clash of Empires* (2004). Rather than centering on particular moments of semantic hegemonization, as in her much-discussed case of *yi/barbarian*, I am more concerned with the hegemonization of an

At first blush, the very notion of semiotic sovereignty might seem to place an impossibly steep demand on the history in question and invites—even demands—reasonable objections of many sorts. Control over the semiotic is enjoyed nowhere and by no one, humans everywhere being born into preexisting webs of signification. If we restrict our focus to telegraphic communication, any duly trained ditty bopper does not enjoy a prerogative over the codes in which he or she traffics but instead must internalize established conventions—these very conventions being the sine qua non of communication. What is more, even if we did wish to reflect upon the hegemony of Morse, one could just as easily focus on France and Russia as on China. French telegraphers blistered against the onetime exclusion of accented letters from the internationally recognized set of permitted codes. Russian telegraphers likewise had to push for the expansion of the internationally accepted conventions of telegraphy to include Cyrillic. The internationalization of Morse was a long, conflict-ridden process. Even when we account for important qualifications such as these, the history nevertheless shifts in unmistakable ways when we move into the domain of Chinese telegraphy. Simply put, we realize that, if semiotic sovereignty is by its very definition impossible, "impossibility" is itself a wide terrain and that not all impossibilities are created alike. In the domain of alphabets and syllabaries, the inclusion of the French *é* into the repertoire of signifiers that would be considered permissible for transmission by the international body governing telegraphic communication—to cite one example that we will revisit later—was not a technological problem but a political one. This symbol could be included, and quite easily—the question was *whether* or not it *should*. By contrast, upon venturing into character-based Chinese script and its relationship with Morse code, we begin to encounter questions that encompassed but were not limited to politics—a historical episode that involved, not the question of "whether or not" to include Chinese characters within the repertoire of acceptable and transmittable telegraphic signifiers, but rather an overloading encounter between a pseudo-universal nineteenth- and twentieth-century information infrastructure and the very script that, through its seeming incommensurability with this system's underlying semiotic architecture, stood frustratingly beyond that system's reach, revealing its pretense to universalism as imperfect, bogus, hubristic. Chinese simply could

entire field of signification wherein the expansion of colonial power at once advanced and was advanced by the globalization and normalization of an information infrastructure that structurally disadvantaged Chinese writing and further denormalized it, not only within and among the colonial powers but within China as well. I am concerned with a crisis of semiotic sovereignty in which the very semiotic capacity of Chinese was called into question.

not be "added" in the same way as French, Russian, Arabic, and the list goes on—its full inclusion would seem to require a fundamental reimagination and reorganization of the system itself.

As we will see, no such world-changing moment materialized. The inclusion of Chinese script was undertaken in a way that left the international system largely unaltered and placed the Chinese language—and, with it, China—in a position of structurally embedded inequality in the emerging global information infrastructure.[9] In particular, the first commercially used Chinese telegraph code—the code of 1871, which provides the focus of this chapter—bore the signature of the deeply unequal conditions within which China first entered the global telegraphic community. The Chinese telegraph code of 1871 continued to embed China and the Chinese script within a place of comparative disadvantage long after "telegraph sovereignty" (*dianxin zhuquan*) had been redeemed in a political economic sense. Long after Chinese authorities took possession of the materiality of cables and towers, as well as financial and legal authority, the question of the Chinese telegraph code persisted.

The narrative of the period in question is not one of an acquiescent or supine China, however. While Chinese telegraphers were not in a position to effect a transformation of the global information infrastructure itself, they did undertake a radical transformation of their *relationship* with this infrastructure and in the process set out on what will be described here as the redemption of semiotic sovereignty by other means. As will be charted out in the closing sections of this chapter, a loose network of political authorities, entrepreneurs, codebook publishers, and everyday users engaged in innovative and thoroughgoing experimentations with telegraphic transmission—not to *change* the system and the code but rather to establish novel relations of agency vis-à-vis both. Through their incremental efforts, these individuals began to take symbolic possession of their own inequality and alienation, engaging in a complex play of mediation through which they reimagined and reconfigured their relationship to the global information infrastructure.

---

9   In the longer book-length study to which this abridged chapter belongs, I consider in great depth a small coterie of nineteenth-century Chinese, French, and American thinkers who were, in fact, inspired by the Chinese language to reimagine telegraph codes at a root level. For reasons of space limitations, however, and because the writings of these avant-garde thinkers did not alter the historical trajectory of the problem of "semiotic sovereignty" considered herein, I have chosen to focus on the 1871 code and its long line of descendants.

The Chinese Telegraph Code of 1871

Samuel Morse was a decidedly cosmopolitan and international figure, dedicating considerable energy to the promotion of his invention in Russia, western and southern Europe, the Ottoman Empire, Egypt, Japan, and parts of the African continent.[10] At the same time, Morse conceived of his invention in deeply specific and particular terms, referring to it both as "the American telegraph" and, even more intimately, "my telegraph."[11] Affective relationships are no doubt common between inventors and the products of their labor, and yet Morse's choice of words here were more than mere sentimentality. Morse's telegraph code was deeply—we might say foundationally—connected to the Roman alphabet and the English language, which wove the fabric of Morse's linguistic world, an inextricable relationship that persisted even as Morse and his machine traveled the globe. This connection was not only an incidental one, moreover—not limited, that is, to the mere fact that the preponderance of messages that first passed through the nascent metallic web were sent and received by English speakers—but also a deeply entrenched linguistic one. With the short "dot," the long "dash," and code sequences ranging primarily from one to four units in length (with an optional but less economical fifth), the entire code could accommodate around thirty discrete units: just sufficient to encompass the twenty-six English letters, with four code spaces remaining. Essential symbols—such as Arabic numerals and a select few punctuation marks—could then be relegated to the less efficient domain of five-unit code sequences (later expanded to even less efficient six-unit sequences in "Continental Morse").[12] While ideally suited to handle English, the same could

---

10   Bektas 2000.

11   Bektas 2001, 206.

12   Between a four- and five-unit pulse sequence, one faced an increase of anywhere between 25 to 75 percent in the time required for transmission. Because of its inefficiency, this five-unit code area—which contained 2^5, or 32, additional spaces—was originally limited to numerals and special symbols (including punctuation). Connections to the English language went further than this, however, and included subtle considerations of the particular types of ambiguities that might surface during the transmission of letters. Within Morse code, the allocation of pulse patterns also had to take into account the co-occurrence of letters and the ambiguities that could arise if a sequence of two originally distinct patterns was partitioned incorrectly, thereby leading to erroneous transcription. Confronted with the most common two-letter sequences ("digrams"), then, the pulse patterns of such letters had to be sufficiently distinct so as to prevent misreading, even though this meant allocating pulse patterns whose energetic efficiency fell short of what a given letter's frequency might otherwise demand.

not be said for other languages—even alphabetic ones. With its thirty letters, German chaffed against the limits of the code's capacity, while French and its variety of accented letters spilled out beyond it. Such English-language centrism was further embedded with the original repertoire of acceptable signifiers. At the time of the International Telegraphic Union conference in Vienna in 1868, the collection of acceptable symbols was confined to the twenty-six unaccented letters of the English language, the ten Arabic numerals, and a small group of sixteen analphabetic symbols (the period, comma, semicolon, colon, question mark, exclamation mark, apostrophe, cross, hyphen, accented é, fraction bar, equal sign, left parenthesis, right parenthesis, ampersand, and guillemet).[13] The expansion of this authorized repertoire was an extremely conservative and slow affair, moreover. Seven years later, at the Saint Petersburg conference of 1875, this list of twenty-six letters was finally expanded to include a twenty-seventh: the accented e (é), now no longer sequestered to the specialized list of "signes de ponctuation et autres."[14] The 1875 conference further stipulated that, for those using Morse code, it would now be possible to transmit six other special, accented symbols: Ä, Á, Å, Ñ, Ö, and Ü. These would not be available, however, to those using "Hughes Signals," an alternative system to Morse. It was not until the London conference of 1903, almost two decades later, that this supplemental list of accented letters was granted admission into the "standard" semiotic repertoire.[15]

As it ventured into the second half of the nineteenth century, the international telegraphic community approached a crossroads. The 1860s witnessed the accelerated expansion of modern colonialism and with it the beginning of a rapid expansion of the telegraphic network far beyond the Roman alphabetic world. In 1864, cables were laid in the Persian Gulf, which, when connected to the existing landline system, put India into direct telegraphic connection with Europe.[16] In 1870, a rapid expansion saw cables laid from the Suez to Aden and Bombay, and from Madras to Penang, Singapore, and Batavia.[17] Such proliferation brought telegraphy into contact with scripts it had not been originally designed to handle, thereby raising the question: would the inclusion of new languages, scripts, alphabets, and syllabaries prompt a radical reconsideration of telegraphy itself, or would their absorption be performed via a subordina-

---

13   *Convention télégraphique internationale* 1868, 58.
14   *Convention télégraphique internationale* 1876, 22; *Convention télégraphique internationale* 1885, 15.
15   *Convention télégraphique internationale* 1903, 16.
16   Headrick 1988, chap. 11.
17   Bull 1893, 4.

tion of these languages to the logic and syntax of existing approaches (some more comfortably than others)? When considering the deeply conservative nature of the semiotic infrastructure of telegraphy—conservative even when viewed from the perspective of the very European signatories to these conventions—it is perhaps unsurprising that the latter path was ultimately followed.

When China and the Chinese language entered the domain of international telegraphy starting in 1871, what ensued was not a reimagination of the modes or syntax of telegraphic encryption and transmission but the subordination of Chinese within a system uniquely ill-equipped to handle it. The product of this convergence was the first commercial Chinese telegraph code, developed by a Danish professor of astronomy named H. C. F. C. Schjellerup and formalized in 1871 by the French harbormaster in Shanghai, Septime Auguste Viguier.[18] In this system, approximately 6,800 common-usage Chinese characters were organized according to the Kangxi radical-stroke system and then assigned a series of distinct, four-digit numerical codes running from 0001 to 9999. Approximately 3,000 blank spaces were left at the end of the codebook, and a few blank spaces were left within each radical class, so that individual operators could include otherwise infrequently used characters essential for their work.[19] To transmit a Chinese telegram using this system, one would begin by enciphering the message into a sequence of four-digit codes, which would then be transmitted using standard Morse signals (figs. 1–2). As Viguier originally imagined it, Chinese telegrams would be "taxed at the same price as those written in any European language, that is, in proportion to the number of characters contained therein."[20]

Given the encipherment process fundamental to the 1871 code, it is tempting to assume that it might have been strikingly less efficient than those employed elsewhere in the world. By this point in history, however, the use of encoded and enciphered transmission was practically ubiquitous, being the rule rather than the exception in global telegraphy.[21] Beginning decades earlier, telegraphers who were concerned with secrecy and, much more so, with saving money began to develop an immense variety of telegraph codes that could be used in coordination with Morse.[22] As the creator of one particularly popular codebook from 1870 phrased it evocatively, his system was designed with the goal of "almost entirely preventing the possibility of a message being

---

18   Jacobsen 2001.
19   Viguier 1871a.
20   Viguier 1871b, preface.
21   Standage 1999.
22   Bellovin 2009.

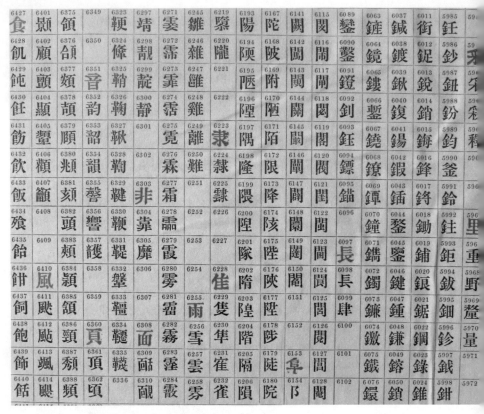

FIGURE 1   *1871 Chinese telegraph code (sample).*

deciphered by those skillful investigators who by long practice have acquired wondrous facility in reading off messages which the senders had hoped were concealed from all the world but those they were intended for."[23] Authored by Robert Slater, the secretary of the Société du Cable Transatlantique Français Ltd., the code was targeted against the threat of government interference. Starting on February 1, 1870, Slater explained in his opening message to the reader, telegraph systems throughout the United Kingdom were to be turned over to the government and, with that, to government clerks. Slater distrusted these clerks, describing them as "prying spirits, curious as to the affairs of their neighbours."[24]

The dominant objective of codes, however, quickly became that of cutting costs: saving money on transmission by creating code words that represented longer sequences, including entire sentences. In a codebook used as early as

---

23   Slater 1870, vi.
24   Slater 1870, iii.

FIGURE 2   *Sample of Chinese telegram and encryption process. To transmit a message in Chinese (left), the telegrapher would first translate the message into a series of four-digit codes (right).*

1888, and well into the 1920s, single words could specify, not only particular commodities, but also specific types and quantities of commodities. While the code word "battled" referred to "beer," for example, "battlement" indicated "beer in bottles," and "batty" indicated "beer in kegs."[25] A single word could often be used to transmit orders or requests for particular commodities, in certain quantities, shipped at specified times, and to exact destinations. An even more vivid example comes from *The Theatrical Cipher Code* of 1905, in which the code word "filibuster" was used to represent the message "Chorus girls who are shapely, good looking and can sing."[26] When compared to telegraph trans-

25   Shay 1922, 22.
26   Bellovin 2009.

International Morse Code

1. The length of a dot is one unit.
2. A dash is three units.
3. The space between parts of the same letter is one unit.
4. The space between letters is three units.
5. The space between words is seven units.

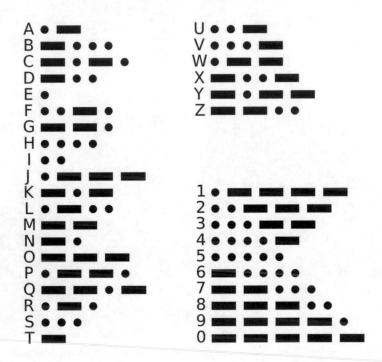

FIGURE 3   *Morse code. By Rhey T. Snodgrass & Victor F. Camp   1922 [Public domain], via
Wikimedia Commons.*

missions in which words like "filibuster," "battled," and many thousands of others lived such strangely double lives, then, there was nothing especially peculiar, or indeed inherently inefficient, about the idea of a Chinese telegraph code in which the four-digit sequence 6255 represented "rain" (*yu*) and 6351 represented "sound" (*yin*).

Whereas the process of encipherment itself did not distinguish Chinese markedly from other telegraphic environments, other properties of the Chinese code had immense and negative implications for China's starting position within the international community.[27] First, the basic (and only) units of the

---

27   Viguier's original list of characters was later slightly adjusted by De Mingzai 德明在, atta-
      ché to the Qing mission to France in 1871, who detected flaws, he explained, in the

Chinese telegraph code were Arabic numerals and thus were among the longest code sequences within Morse. As shown in figure 3, the shortest numerical code (the number 5, with its five short pulses) was already five times longer than the shortest letter (*e*, with its single short pulse). The numeral 0 was the longest transmission sequence in the entire code, requiring five long pulses. Overall, then, the Chinese telegraph code's status as a purely numerical system handicapped it from the outset.[28]

More significant than the inefficiency of transmitting numerals rather than letters, however, was the unforeseen and disadvantaged position that the Chinese telegraph code came to occupy within the web of existing international telegraph laws. A maze of regulations designed to police and limit the use of encrypted transmissions was in place circa 1870. Confronted with the midcentury spread of coded languages, governments and cable companies had pushed for amendments to the international telegraphic system, in terms of both permission and pricing. Codes and "numbered languages" (*langue chiffrée*) were repeatedly barred from use in the context of telegraph addresses and in many other types of communications. In terms of pricing, "secret" transmissions were almost universally assessed at significantly higher per-word costs, so that cable companies could make up for the money they were losing thanks to code words such as "battlement" and "filibuster." As a practical consequence of such regulations, telegraphers operating in English, French, German, or any of the other telegraphic languages were thus presented with a constant choice: namely, whether to transmit a message in code or "in the clear."

When China entered the international telegraphic community under the aegis of the 1871 code, it instantly became entangled within this web of rules and regulations. Unlike telegraphy elsewhere, however, the international community did not recognize any distinction between Chinese "code" and Chinese "plaintext." By virtue of being fundamentally a *langue chiffrée*, the Chinese telegraph code—and with it the Chinese language—was understood as being an *inherently secret language* in which transmissions "in the clear" were consid-

---

Frenchman's stroke-count sequence (Baark 1997, 85). Following the 1949 revolution, we witness the further creation of two versions of the code, one in the mainland and one in Taiwan, which both made use of four-digit codes but with different code assignments. Even when accounting for these changes, we see that the basic model of Viguier's system became the industry standard in China for over a century.

28   No mention is made in Chinese, Danish, or British sources as to whether Chinese telegraphers made use of the "abbreviated numerals" in Continental Morse (also known as International Morse Code). If so, it is possible that telegraphers were able to bypass the inefficiency of standard Morse codes for numerals, while still facing the limitation inherent to using only numerals and never letters.

ered impossible by definition. Rather than deciding between coded and plaintext transmissions like their counterparts elsewhere, Chinese telegraphers circa the 1870s operated within a legal framework in which their language in its entirety was considered "clandestine" by its very nature and thus subject in its entirety to the same punitive tariffs and restrictions of usage that, in other contexts, were meant to counteract solely cost-saving and covert communications.

The Impossibility of Chinese Plaintext

It was the international cable companies themselves that first confronted the challenges associated with the Chinese telegraph code whose creation they sponsored, and which they helped establish as a global convention—a central irony of our story. The Great Northern and the Eastern Extension A&C were acutely aware that this disadvantageous pricing regime and coding system threatened to make the transmission of telegrams too expensive for average Chinese users, and thereby obstruct cable companies' goals of penetrating the interior market. To accommodate this situation, these companies focused their energies on compensating for the disadvantaged position of Chinese script by establishing preferential transmission rates. In particular, they attempted to exploit the inherently ambiguous relationship between "characters" and "words"—a fluctuating relationship that, as Lydia Liu has argued, was the subject of deeply politicized efforts at commensuration as part of the "clash" between the Qing and British Empires (and, later, American and western European powers).[29]

The precise relationship between characters and words was never fixed but instead fluctuated, not unlike like the exchange rate between two currencies. This exchange rate could be manipulated using both legal and financial levers. Legally, regulations could be adopted that established a definitional relationship or ratio between "characters" and "words," such as defining a "word" as any combination of two characters, or the like. Insofar as international regulations assessed telegraph tariffs on per-word bases, such legal classes of equivalency between characters and words could be compensatory vis-à-vis Chinese, all without resolving the underlying system inequality that made such compensations important. The second lever was financial. Significant reductions of per-word tariffs for Chinese transmissions created de facto revaluations of these same classes of equivalency that could attenuate or even entirely neutralize or

---

29   Liu 2004.

invert de jure ratios. If a law set every two characters equal to a word, while at the same time tariff regulations assessed Chinese "words" at half the price of their English counterparts, the effective equivalency was 4:1. From the establishment of the first telegraph line in China through the twentieth century, a complex assemblage of corporations and political authorities attempted to manipulate each of these levers, all while leaving the foundational inequality untouched.[30]

Legal attempts to improve the disadvantaged position of Chinese became more sweeping and complex over the course of the late nineteenth and early twentieth centuries, particularly as representatives of the Qing and the Chinese Republic began to participate directly in the proceedings of the international telegraphic community. Invited to join the conference in 1883, the Qing eventually dispatched a representative to the Lisbon conference of 1909.[31] From that moment on, Chinese representatives were an active part of the international organization, doggedly attempting to renovate the international system in incremental ways that would remove the disadvantages to their language.[32] Moving beyond questions of differential pricing, Chinese representatives—

---

30 Perhaps the earliest advocates for change were the monopolists themselves—Great Northern Telegraph Company of Denmark and Eastern Extension A&C Telegraph Company of the United Kingdom. In an 1887 agreement between the Imperial Chinese Telegraph Company, the Great Northern Telegraph Company, and the Eastern Extension, Australasia and China Telegraph Company, the cost of Chinese telegrams was valued at less than half of foreign transmissions between key nodes in the network. Between Shanghai and Fuzhou, foreign telegrams were priced at 0.33MD (Mexican dollar) per word, while Chinese telegrams were priced at 0.15MD per word. Between Shanghai and Hong Kong, foreign telegrams were assessed at 0.44MD, with Chinese telegrams at 0.19MD. All the other transmission lines listed in the agreement featured the same price differential. None of this was undertaken out of altruism, it bears emphasizing, but as part of the effort to make telegraphy more affordable for subjects of the Chinese empire. "An Agreement" 1887, 7.

31 *Documents de la Conférence télégraphique internationale de Lisbonne* 1909, 482.

32 In the wake of the 1911 revolution, and the collapse of the Qing, the newly established Republican regime continued to push for preferential tariffs in the international community and to establish more attractive tariff policies at home. In 1933, for example, the Ministry of Communications promulgated new regulations that transferred the financial burden of encipherment and decipherment—that is, of translating Chinese characters into four-digit codes and vice versa—away from clients and to the telegraph offices themselves. Thereupon, messages were to be "translated free of charge by the office of destination before delivery." In the same year, special rates were assigned to telegraphic "greetings" and "social letters," such as New Year's messages, birthday cards, and condolences. Zhu 1937, 166–168.

trained telegraphy engineers with extensive experience in the industry—
slowly began to carve out a space of legal exception for the four-digit Chinese
telegraph code itself. In a moment both sublime and brilliant, Chinese representatives pushed for recognition of the four-digit code as the *artificial equivalent* of Chinese plaintext—to exempt this particular *langue chiffré* from the
status of "secret" transmission and treat it as the effective equivalent of "in-the-
clear" Chinese transmission.

Chinese and other representatives were largely successful on this front, as
they fashioned a kind of legal biosphere in which Chinese could survive rather
than suffocate in a hostile and insupportable environment. Under new international guidelines appearing as early as 1893, any numbered transmission recognized as a "bona fide" Chinese message was to be counted as a single word
when transmitted between stations in China, and at a slightly higher rate when
transmitted to Japan or further abroad.[33]

Owing to the complex, variegated structure of the conventions and amendments that governed telegraphic tariffs, however, the goal of ameliorating the
position of Chinese could not be achieved through any single victory. Each
gain was incremental and easily rolled back. With the addition of each new
subtechnology or practice to the expanding telegraphic repertoire—whether
it be preferential pricing for holiday telegrams, novel methods of addressing
telegrams, or the emerging coordination of telegraphy and surface post in the
form of "letter-telegrams"—Chinese representatives could be certain that any
such amendment would undoubtedly include legal language pertaining to
restrictions on encoded and numbered language. Christmas telegrams were to
be assessed at a lower rate than regular telegrams, and yet such telegrams had
to be written in plaintext. Likewise, it was stipulated in multiple agreements
that telegram addresses had to be written in plaintext as well and could not be
written in either codes or ciphers. In the 1932 conference in Madrid, the con-

---

33   Bull 1893, 115. Even victories such as these carried unforeseen and sometimes negative
      repercussions, however. In this case, the general lack of familiarity with the Chinese lan-
      guage, and thus inability to authenticate a given transmission as "bona fide," meant that
      China would be required to "deposit" official copies of the codebook in multiple foreign
      countries, to serve as the gold standard. In an act not unlike the ritual interment of the
      "standard meter" and the "standard kilogram" in the French town of Breteuil in 1889—an
      act designed to preserve and render sacred a system of weights and measures for eter-
      nity—the standard Chinese telegraph code would have to be "buried" in sites across the
      globe to ensure uniformity and temporal persistence. Such perpetuity, while it was to be
      desired in the case of weights and measures, could in the case of Chinese telegraphy only
      restrict future efforts to redesign the code. *Fawen yi Huayu dianma zihui*, n.d.; Galison
      2003, 91.

gress outlined more detailed rules governing the practice of both "télégrammes de félicitations" and "letter-telegrams," stipulating that such transmissions had to be written entirely in "clear language."[34]

If regulations pertaining to discounted Christmas telegrams might appear utterly unrelated to the concerns of China, they were—and it is precisely this irrelevance that lies at the center of our ongoing consideration of semiotic sovereignty. China's peculiar disadvantage within the international telegraphic community manifested itself, not by means of calculated, premeditated infringements upon China's interests by European and American powers, but through processes decidedly more aloof and unaware. In far-off convention halls in Lisbon, London, and the like, it was rather a largely indifferent Euro-American community that passed one regulation after the next that, despite having nothing to do with Chinese per se, in fact exerted measurable influence upon China's interests, pricking, bumping, and bruising China practically each time. (One's mind turns to thoughts of communities residing along the flight paths of commercial jets or the routes of commuter trains, having to rearrange and tidy up their homes each time elites rumbled by, distracted and oblivious.) Each time that "clear language," "plain language," or analogous terms were invoked, representatives of China once again found themselves having to push for adjustments, lest China find itself once again excluded from one or another part of the ever-expanding and diversifying complex of global telegraphy.

In 1925, during the meeting of the International Telegraphy Conference in Paris, Chinese representative Wang Jingchun and his associates successfully advocated an amendment to the accepted methods of writing addresses, which had long prohibited the use of coded or numbered language. As the Chinese delegation argued at the conference, an estimated three thousand telegrams had of late failed to be delivered because of ambiguities and errors in the Romanized addresses. Once again because of the exceptional, disadvantaged position of China within this seemingly innocuous formulation—Chinese being an inherently "encoded" language within the telegraphic infrastructure—Wang and his colleagues pushed through to adoption a powerful, single-sentence adjustment of this long-standing article: "Pour les télégrammes à destination de la Chine, l'emploi de groupes de quatre chiffres est admis pour désigner le nom du destinaire. ADOPTÉ."[35] Similar amendments were successfully adopted with regard to "télégrammes de félicitations" as well. Perhaps emboldened by these not insignificant victories, the Chinese delegation also filed a statement of broader reach. Beginning in the 1920s, representa-

---

34   *Documents de la Conférence télégraphique internationale de Madrid* 1933, 429.

35   *Documents de la Conférence télégraphique internationale de Paris* 1925b   422.

tives of China began to comment upon the entire semiotic system of the international telegraphic infrastructure, seeking to win, not merely these minor battles, but the wider war as well. The delegation made the following statement:

> Without a doubt, the "differentiated international" service has been introduced and maintained by the different countries within the International Telegraphic Union in the interests of all the people of European and non-European regimes. *At the same time, this service has not afforded any benefit whatsoever to the Chinese within these regimes. ... For this reason, we request that the prohibition against the use of numbers be reduced so as to place the Chinese public on the same footing as the nationals of other countries* [emphasis added].[36]

By way of this complex manipulation of the legal definition of characters, the rates at which Chinese characters were assessed, and persistent and incremental interventions at international telegraphic conventions, the status of the Chinese telegraph code was ameliorated but never equalized over the course of its first five decades. Long after the material infrastructure of telegraphy was brought under Chinese control, then, the problem of the semiotic remained unresolved.

The Mediation of Mediation

In 1926, the renowned Chinese linguist and language reformer Qian Xuantong railed against the inefficiency of the Chinese telegraph code, as well as the broader Chinese information infrastructure. "[I]f we use characters," he pondered "what other choice do we have than that painfully ridiculous fantasy of '0001, 0002 ...'?" "Using phonetic characters, sending one letter at a time," he continued, "as soon as a sentence has been uttered, it has already been transmitted. This is the only proper definition of a real telegraph."[37] Notwithstanding Qian's omission or perhaps unawareness of the secret commercial codes that were ubiquitous by the 1910s—a ubiquity that disrupts the myth of instantaneity and transparency that he considered the sine qua non of "real" telegra-

---

36   *Documents de la Conférence télégraphique internationale de Paris* 1925a   427. Wang
        Jingchun would become one of the most influential figures in pinyin telegraph transmission, a subject I address at great length in my larger project.
37   Qian 1918.

phy—Qian's damning critique captures what would seem to be the central theme of the foregoing discussion. Short of a radical and revolutionary overhaul of the global information infrastructure itself—a utopian future wherein the rules of the game were reorganized so as to place every language and every people on "the same footing"—it would seem that the very idea of semiotic sovereignty was quite simply beyond China's reach. To partake in and benefit from modern systems of communication, China circa the early twentieth century was presented with no other option than to relinquish its pursuit of equal linguistic footing and abandon the Chinese script altogether.

Over the course of the late Qing and Republican periods, however, there emerged a vibrant, local-level sphere of innovation within Chinese telegraphy that was distinct yet inseparable from the vicissitudes of the international telegraphic community. In this parallel domain, a loose network of telegraphers, codebook publishers, and entrepreneurs dedicated themselves to the development and refinement of a wide array of experimental methods designed to make the Chinese telegraph code faster and more efficient. To an even greater extent than their elite and metropolitan counterparts, such individuals lacked the political power to transform the thick and alienating medium that was built into the technological and legal framework of telegraphy, and so instead they focused their energies on developing and refining a wide array of experimental methods designed to make this alienating medium less of a liability. Much like the community of late Qing and early Republican language experimenters examined by Jing Tsu, this local level network was the site of a "largely overlooked group of pioneering pedagogues and avid practitioners," ones "undeterred by thwarted experiments" (Tsu 2010, 18). The shared objective of their otherwise diverse experiments was that of *mediating mediation*—that is, to insert novel practices and devices of their own design between themselves and the Chinese code so as to render their relationship with the code (which was already a form of mediation, as we have seen) more favorable, workable, memorable, suitable, or otherwise advantageous. As we will see, these mediations of mediations—or *hypermediations* for short—were decidedly personal, corporeal, local, and immediate, encompassing a variety of novel aides-mémoires, forms of training and bodily practice, and reorganizations and repaginations of the Chinese telegraph codebook, among other strategies. Each in their own way, then, these individuals inserted *additional layers of mediation* between themselves and the four-digit telegraph code—additions that, while they would seem prima facie to increase the time and effort needed to use the code, often rendered it faster, less expensive, and yet fundamentally unchanged in terms of its root-level semiotic architecture.

As we examine some of the most important examples of such experimental forms of mediation, our primary concern is not with purely technical questions of speed or cost, however. We are more concerned with the broader symbolic effects of these experiments, and the way in which they aggregated to form the *redemption of semiotic sovereignty by alternative means*. Unable to develop a Chinese telegraph code from scratch that would be deeply attuned to the affordances and limitations of Chinese script—that is, unable to reperform the experimental process by which Morse created his code—these individuals approached a similar destination but via the long way around. They left the underlying semiotic architecture of the Chinese telegraph code untouched, while at the same time engaging in radical reimaginations of their *relationship* with, and pathways through, that architecture. Through the gradual establishment of a novel relationality with and agency toward the code—by playing with it, mediating it, suiting it to their own linguistic and physical preferences, capacities, and limitations—these individuals gradually took symbolic possession of a system and a condition that were, at their inception, alienating and unequal. The balance of this chapter will explore this argument and close with an even more far-reaching one: namely, I will argue that the uniquely *hypermediated* practices one encounters within modern Chinese information technologies (including but not limited to telegraphy) were not merely means by which Chinese and other inventors went about "reconciling" the incongruities that, as seen above, placed Chinese at a distance from the alphabet-centered information technologies of the nineteenth and twentieth centuries. Far beyond this immediate objective, hypermediation became a zone of experimentation unto itself, a fundamental strategy and property of modern Chinese information technology from the late nineteenth century onward. Indeed, it could be argued that everyday users of Chinese information technologies in the late nineteenth and early twentieth centuries were developing and refining hypermediated semiotic strategies that, elsewhere in the world, would not take hold to any critical degree until the turn of the twenty-first century.

Even when limiting our focus to telegraphy, there are too many examples of experimental mediation to address within the confines of the present chapter. Among the wide variety of systems, this section will consider four (table 1). Each involved the development of a different practice and/or device that operators could use to mediate their relationship with the four-figure telegraph code in a favorable or advantageous way. These experiments, as we will see, focused on both the physicality and materiality of telegraphic operation, both with regard to the bodies of telegraph operators and to the artifacts used by operators (most notably the codebook).

TABLE 1   Systems of Telegraphic Mediation Developed in China   1871–1940s

Year	System of mediation
1871	Telegraph stamps (S. A. Viguier)
1872	Chinese vs. Arabic numerals
1881	Chinese character mediation of Roman alphabet (used to exploit three-letter code)
1937	Repagination of telegraph books

The earliest example of experimental mediation of the Chinese telegraph code was a set of stamps invented by Viguier himself to expedite the processes of encipherment and decipherment.[38] Referred to as "tables d'expédition et de réception," these were two-sided stamps, with Chinese characters on one face and their corresponding four-digit codes on the other. For those enciphering and transmitting messages, the stamps were organized with the characters face up and arranged according to the Kangxi radical-stroke system. For reception and decipherment, the numerical code was oriented face up, running from 0001 to 9999. The process of encipherment and decipherment was thus transformed into a process of locating, grasping, inking, stamping, and replacing each stamp, with the end result being a fully translated message. As Viguier imagined it, this parasystem would release telegraph operators from practically all need to use his codebook, save for occasional reference, and any need to memorize the codes for characters. To further facilitate usage, moreover, Viguier later outfitted his codebook and stamps with Chinese instead of Arabic numerals.[39] Presumably, the operators could memorize the electrical-pulse patterns for *yi* (one), *er* (two), and *san* (three) rather than 1, 2, and 3 and thereafter transmit these codes to a receiving agent who would then decipher the transmission back into Chinese numerals. As Viguier explained, "In order to facilitate the introduction of telegraphy to the interior of the empire, I Chinese-ized my system by replacing Arabic numerals used to represent the character with Chinese numerals."[40] At its very inception, then, what began seemingly as a system of direct, one-to-one mediation between Chinese characters and four-digit Arabic numeral codes was premised from an early stage upon a diverse array of metalevel mediations that could either cooperate with, sup-

---

38   Viguier 1875, 6.

39   "T'een-piao-hsin-shu" 1872.

40   Viguier 1875, 7.

plement, or bypass one another and certain dimensions of the underlying code.

Chinese telegraph operators maintained Viguier's Chinese numeral equivalents and yet ultimately made little, if any, use of the stamp system. In the decades following, Chinese operators instead developed their own techniques of mediating the codebook. One of the earliest involved retrofitting a relatively new technique of three-letter, or trigraph, coded transmission. Developed in the United Kingdom and elsewhere, this code sought to exploit the Roman alphabet as a base-26 system, wherein a sequence of three letters could be used to transmit a total of 26^3, or 17,575, units. Such a system in theory proved far more efficient than the four-digit system, insofar as the trigraph system required one fewer code unit (constituting a 25 percent reduction at the outset), and because it made use of letters rather than numerals (with letters being faster to transmit than numerals within Morse code, as explained above). In a newly published codebook from 1881, steps were taken to employ this alternative encoding system by pairing each of the conventional four-digit sequences with a unique three-letter code. The code 0001 was paired with AAA; 0002 with AAB; and so forth. A second layer of mediation acted in tandem to this one, being a set of twenty-six Chinese characters used by the editors to represent the Roman alphabet and thereby simplify and render more approachable graphemes that were still foreign to most circa the 1880s (figs. 4–5).[41] Using this second mode of mediation in concert with the first, a Chinese telegrapher could in theory carry out his work *entirely in Chinese* even as he operated with and through the Roman alphabet. Phrased differently, if we imagine being able to wiretap the electrical pulses that coursed through telegraph cables in China at this time, what might have appeared to us as the three-letter code sequence D-G-A would, for both the sending and receiving agents, have been understood in terms of the Chinese-character sequence 諦基愛 (*di-ji-ai*). Even "alphabetic order" itself could be emulated using such a mediation technique, with telegraph operators being able to memorize the

---

41   The system was further mediated, moreover, through what appears to have been one or another Chinese dialect—perhaps Cantonese. That is, while the table in figure 5 lists only the pinyin pronunciation of each character in *putonghua*, a close consideration of particular letter-character combinations indicates that the pairings were quite likely made with a Chinese dialect in mind. The pairing of *zai* and *j* only makes sense, for example, when we consider its Cantonese pronunciation *joi*. The same is true for the *w*, whose accompanying character is pronounced *wu* in Cantonese; and the letter *y*, whose character is pronounced *wai*. Other elements within the list suggest a dialect other than Cantonese, however, such as the character used to represent *k* (凱, which is pronounced *hoi* in Cantonese). My thanks to Roy Chan for bringing this wrinkle to my attention.

sequence 愛、比、西、諦、依 ... (*ai, bi, xi, ti, yi,* ...). Within this complex and multilayered interplay of mediations, the very dynamics and valences of signification could thus be reimagined, all while leaving the underlying architecture of the telegraphic code unchanged.[42]

As familiarity with Arabic numerals and the Roman alphabet increased in China, Chinese-character mediation schemes such as these began to disappear from codebooks. Being more comfortable with the likes of 1, 2, 3, ... and a, b, c, ... the average telegrapher presumably no longer needed or found useful the mediations of 一二三 and 愛、比、西. Mediations of other sorts soon replaced them, however, one of the most subtle and successful centering on a clever repagination of the Chinese telegraph codebook itself. In contrast to early codebooks, in which each page started with the code sequence ---1 and concluded with ---0 (e.g., 0101 to 0200 with codebooks that featured one hundred characters per page, or 0101 to 0300 in those that featured two hundred characters per page), publishers in the Republican period began to reorganize the codebook such that each page ran from code sequence ---0 to --99 (e.g., from 0100 to 0199, 1200 to 1299, etc.) (fig. 6).[43] Although seemingly trivial, this alteration produced a secondary effect eminently useful within the everyday work of the Chinese telegrapher, as the page numbers of the codebook itself were hereby transformed into aides-mémoires that corresponded to the first two digits of the codes featured on any given page. Repagination of the codebook made possible a new "memory practice," one in which a telegrapher using a codebook circa 1946 knew, for example, that the code 1289 could be found on page 12, code 3928 on page 39, and code 9172 on page 91—with the corresponding page numbers being featured in bold red letters at the top and bottom of each page.[44] To understand how telegraphers engaged with the Chinese code, mediating this relationship in novel and experimental ways, we therefore must also pay close attention to the social history of the codebooks themselves.[45]

---

42   Changing regulations governing international telegraphic communication often closed pathways of mediation. Trigraph encryption was eventually limited by the very same corporate and statist watchdogs that policed money-saving and/or clandestine communication. By the early nineteenth century, all telegraphic transmissions had to be "pronounceable"—a somewhat vague designation that mandated the inclusion of a minimum number of vowels within any coded transmission. Long sequences of consonants would no longer be permitted, thereby greatly reducing the number of code spaces in the original trigraph system.

43   Jiaotong 1946.

44   Jiaotong 1916; Bowker 2005.

45   In my analysis of the materiality, structure, and organization of Chinese codebooks, as well as the somatic interplay between the telegraphers and the codebooks, I am inspired

FIGURE 4   *Chinese-character-based mediation of Roman alphabet (letter/character/pinyin).*

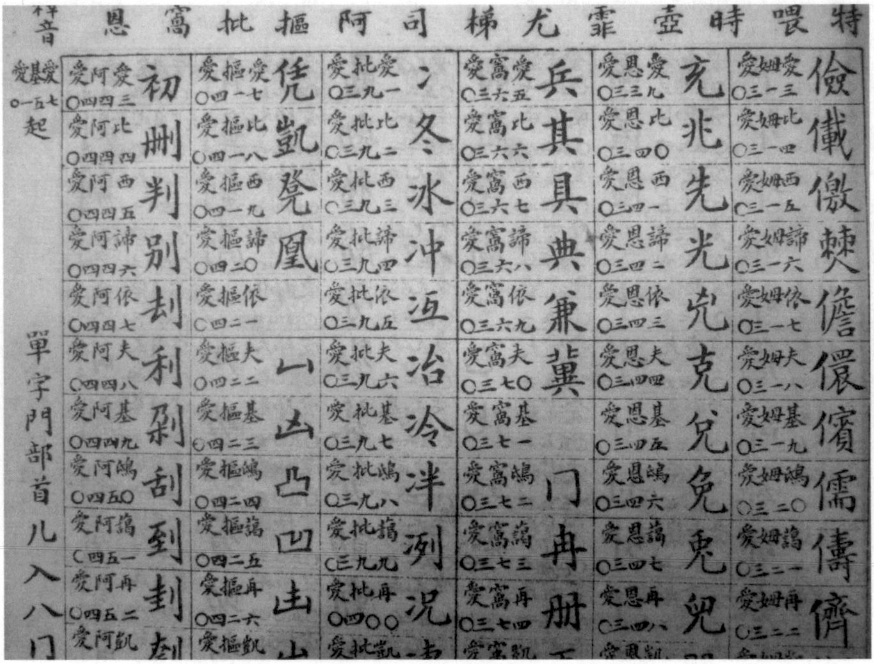

FIGURE 5   *Sample page from character-mediated version of codebook.*

and guided by the work of scholars such as Roger Chartier, Robert Darnton, and Adrian Johns. In particular, changes within the telegraph codebook bring to mind the work of Chartier and the ways in which the emergence of new linguistic technologies and material forms in Europe enabled and constrained the body in unprecedented ways. The codex, as a form, permitted modes of pagination impossible with the roll, which in turn rendered possible the creation of indexes, concordances, and other referential paratechnologies. Within this new technosomatic ensemble, the reader was now able "to traverse an entire book by paging through" (Chartier 1995, 19). Moreover, unlike the roll and its accompanying requirement of two-handed reading, the codex "no longer required participation of so much of the body" (19), liberating the reader's hand to, among other

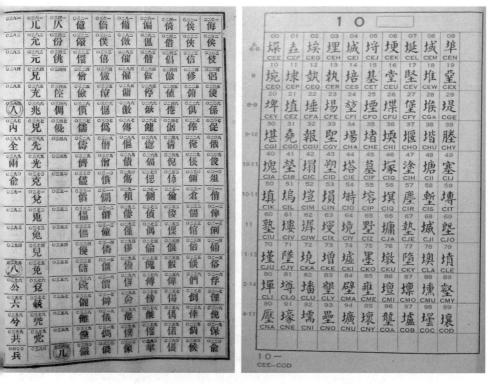

FIGURE 6   *Repaginated codebook in which page numbers serve as aides-mémoires.*

At the microhistorical level, each of these efforts was oriented toward a variety of facilitations: enabling Chinese telegraphers to work within a foreign and unfamiliar alphanumeric context and accelerating the process by which telegraphers could track down a given character or code, among other objectives not considered here. On the macrohistorical level, these local-level efforts added up to something much broader: a historical process connected to what has been theorized in this chapter in terms of semiotic sovereignty. Through the creation of these many different mediations of mediations, telegraphers were not only undertaking pragmatic processes pertaining to time and money but also establishing a novel relationship with an information infrastructure that, as examined in the preceding section, placed Chinese within a position of structurally enforced inequality. Telegraphers were taking "symbolic possession" of the four-figure code—again, by playing with it, mediating it, suiting it

things, take notes. Dynamics not unlike this were central to the types of experimentation and innovation found within Chinese telegraphy as well.

to their own linguistic and physical preferences, capacities, and limitations. With these incremental, highly localized acts, telegraphers began the process of surrounding a system that had once surrounded them.

Reflecting on Linguistic Hypermediation in Modern China

The histories outlined here do not resolve in the Republican period. Reverberations of these processes continue to be felt in the historical development of pinyin and, perhaps more clearly, in more recent histories of code schemes such as ASCII and Unicode. In these and many other cases, many of the same features of Chinese technolinguistic modernity considered here are discernible: first, the persistent and elusive problem of semiotic sovereignty; and second, the way in which complex, interoperating, multilayered practices of mediation have become arguably *the* defining strategy within modern Chinese information technology. If the Chinese telegraph code was the first example of an information system in which Chinese script was relegated to a condition of structurally embedded inequality, a remarkably similar story of disadvantage and alphanumeric-centrism is discernible in the history of computing as well. ASCII and other codes have consistently been designed with the alphanumeric environment in mind—with "just enough" spaces for letters, numerals, and a handful of symbols. Through encounters with Chinese script, such systems have not been fundamentally rethought so much as stretched and "extended" (with Chinese being relegated, yet again, to the less efficient "extensions" that result). The early history of dot-matrix printing and monitors witnessed a strikingly similar phenomenon as well, with the earliest x-y matrices encompassing an array of nodes that, while sufficient for the expression of graphemes such as A, 1, and ?, lacked the density of plot points necessary to express Chinese characters such as 電 and 信. The persistence of examples such as these bespeaks the persistence of a historical trajectory that originated in the closing decades of the nineteenth century.

With regard to the second feature—what we might call the markedly hypermediated quality of modern Chinese-language information technology—this too finds its origins in the ongoing efforts of Chinese telegraphers, inventors, linguists, publishers, dictionary compilers, and others to deal with, and relate to, the deep structural inequalities that the Chinese language faced within the nineteenth- and twentieth-century "clash" of information systems. Hypermediation became an avenue, as it were, to eke out more empowered relationships with—and even semiotic sovereignty over—systems that otherwise alienated the Chinese script. Operating with limited power to change the underlying infrastructures, an impressive spectrum of creative energy was focused instead on reimagining and redefining the user's *relationship* with this otherwise-alienating infrastructure. These experiments involved drawing upon the resources of Chinese characters, foreign scripts, and invented symbolic languages to create complex webs of mediation in which, depending upon the situation, a Chinese character might be used to represent a letter, a wooden stamp to represent an entire code sequence, a Roman letter to represent a particular tone, and so forth, in increasingly rich and complex chains of signification. In but one of the examples outlined above, we find telegraphers using Chinese characters to represent letters, which in turn were used to represent numerals, which in turn were encoded into a sequence of electrical pulses, which then underwent a process of decoding in the opposite, stepwise, hyper-mediated direction. Taken together, this long cascade of mediations was itself, collectively, a kind of aggregate signifier ultimately used to transmit one specific Chinese character, which in turn was itself also a signifier in the more conventional sense of the term.

For anyone who has used a computer in China, moreover, it becomes clear that this reliance upon hypermediation has become increasingly sophisticated, rich, and fast. Sitting down in front of a QWERTY keyboard, an average Chinese operator is comfortable with a semiotic interface in which the "law of identity" simply does not apply. The letter Q need not signify the letter Q, for example, but, more often than not, is used in accordance with one or another "input system" (*shurufa*) in order to signify perhaps the pronunciation of the operator's desired character, the tone of the desired character, the radical of the desired character, or some other graphemic or phonetic feature thereof. In recent years, this basic condition of flexible mediation has become richer, owing to the near ubiquity of "predictive text" applications that employ algorithms to offer probabilistic guesses as to what the user is typing as he or she types. Whereas predictive algorithms are in large part limited to text messaging in much of the world, in China they have become a fundamental and basic part of word processing, web applications, search, and so forth, thereby adding even further possibilities of mediation. With the declining cost of memory and the meteoric rise of processing speeds, moreover, such strategies of hypermediation have been arguably the most profitable space of potential when it comes to the acceleration of information technology. The first step toward understanding this complex condition—one both historical and contemporary—is to sensitize ourselves to the linguistic dimensions of Chinese information technology, a dimension that has so often been bracketed or jettisoned in our considerations thereof.

Bibliography

"An Agreement entered into the 10th day of August   1887 between the Imperial Chinese Telegraph Company and Great Northern Telegraph Company of Copenhagen and the Eastern Extension, Australasia and China Telegraph Company, Limited" 1887. Cable and Wireless Archive DOC/EEACTC/1/304 E.Ex.A&C.T. Co. Ltd Agreements with China and Great Northern Telegraph Co. etc., pp. 185–195.

Ahvenainen, Jorma 2004. *The European Cable Companies in South America before the First World War*. Helsinki: Finnish Academy of Sciences and Letters.

Baark, Erik 1997. *Lightning Wires: The Telegraph and China's Technological Modernization   1860–1890*. Westport, CT: Greenwood Press.

Bays, Daniel H. 1978. *China Enters the Twentieth Century: Chang Chih-tung and the Issues of a New Age   1895–1909*. Ann Arbor: University of Michigan Press.

Bektas, Yakup 2000. "The Sultan's Messenger: Cultural Constructions of Ottoman Telegraphy   1847–1880," *Technology and Culture* 41: 669–696.

_____ 2001. "Displaying the American Genius: The Electromagnetic Telegraph in the Wider World," *British Journal for the History of Science* 34, 2 (June): 199–232.

Bellovin, Steve 2009. "Compression, Correction, Confidentiality, and Comprehension: A Modern Look at Commercial Telegraph Codes," paper presented at the Cryptologic History Symposium, Laurel, MD.

Biaozhun zhengkai mingmi dianma xinbian 標準正楷明密電碼新編 [Standard regular script plaintext and secret telegraph codes—new edition]. 1937. Shanghai: Da bei dianbao gongsi.

Bowker, Geoffrey C. 2005. *Memory Practices in the Sciences*. Cambridge, MA: MIT Press.

Brokaw, Cynthia, and Christopher Reed, eds. 2010. *From Woodblocks to the Internet: Chinese Publishing and Print Culture in Transition, circa 1800 to 2008*. Leiden: E. J. Brill.

Brokaw, Cynthia, and Kai-wing Chow, eds. 2005. *Printing and Book Culture in Late Imperial China*. Berkeley: University of California Press.

Bull, W. 1893. "A Short History of the Shanghai Station," Shanghai. Handwritten manuscript. Cable and Wireless Archive DOC/EEACTC/12/10.

Chartier, Roger 1995. *Forms and Meanings: Texts, Performances, and Audiences from Codex to Computer*. Philadelphia: University of Pennsylvania Press.

Chia, Lucille 2003. *Printing for Profit: The Commercial Publishers of Jianyang, Fujian*. Cambridge, MA: Harvard University Asia Center.

Chow, Kai-wing 2004. *Publishing, Culture, and Power in Early Modern China*. Stanford, CA: Stanford University Press.

Convention télégraphique internationale (1885): Documents de la Conférence télégraphique internationale de Berlin 1885.

Convention télégraphique internationale de Paris, révisée à Vienne (1868) et Règlement de service international (1868)—Extraits de la publication: Documents de la Conférence télégraphique internationale de Vienne 1868. Vienna: Imprimerie impériale et royale de la Cour et de l'État.

Convention télégraphique internationale de Saint-Pétersbourg et Règlement et tarifs y annexés (1875) Extraits de la publication—Documents de la Conférence télégraphique internationale de St-Pétersbourg. Publiés par le Bureau International des Administrations Télégraphiques 1876. Berne: Imprimerie Rieder & Simmen.

Convention télégraphique internationale et Règlement et tarifs y annexés révision de Londres (1903) 1903. London: Electrician Printing and Publishing Co.

"Denshin jigô" 電信字号 [Telegraph code]. "Extension Selskabet—Japansk Telegrafnøgle" 1871. Arkiv nr. 10.619. In "Love og vedt{ae}gter med anordninger." GN Store Nord A/S SN China and Japan Extension Telegraf. Rigsarkivet [Danish National Archives], Copenhagen, Denmark.

Documents de la Conférence télégraphique internationale de Berlin; Bureau International des Administrations Télégraphiques 1866. Berne: Imprimerie Rieder & Simmen.

Documents de la Conférence télégraphique internationale de Lisbonne; Bureau International de l'Union Télégraphique (1909) 1909. Berne: Bureau International de l'Union Télégraphique.

Documents de la Conférence télégraphique internationale de Madrid (1932)—Tome I 1933. Berne: Bureau International de l'Union Télégraphique.

Documents de la Conférence télégraphique internationale de Paris (1925)—Tome I 1925a. Berne: Bureau International de l'Union Télégraphique.

Documents de la Conférence télégraphique internationale de Paris (1925)—Tome II 1925b. Berne: Bureau International de l'Union Télégraphique.

Esherick, Joseph 1976. *Reform and Revolution in China: The 1911 Revolution in Hunan and Hubei.* Berkeley: University of California Press.

Fawen yi Huayu dianma zihui 法文譯華語電碼字彙; *Dictionaire télégraphique officiel, Chinois en Français* n.d. Shanghai: Dianhouzhai. 10619 GN Store Nord A/S. 1870–1969 Kode- og telegrafbøger. Kodebøger 1924–1969. Rigsarkivet [Danish National Archives], Copenhagen, Denmark.

Galison, Peter 2003. *Einstein's Clocks, Poincare's Maps: Empires of Time.* New York: W. W. Norton.

Headrick, Daniel 1981. *The Tools of Empire: Technology and European Imperialism in the Nineteenth Century.* New York: Oxford University Press.

_____ 1998. *The Tentacles of Progress: Technology Transfer in the Age of Imperialism   1850–1940.* Oxford: Oxford University Press.

_____ 2002. *When Information Came of Age: Technologies of Knowledge in the Age of Reason and Revolution   1700–1850.* Oxford: Oxford University Press.

Headrick, Daniel, and Pascal Griset 2001. "Submarine Telegraph Cables: Business and Politics 1838–1939," *Business History Review* 75, 3: 543–578.

Hedtke, Charles H. 1977. "The Sichuanese Railway Protection Movement: Themes of Change and Conflict," *Zhongyang yanjiuyuan jindaishi yanjiusuo jikan* [Bulletin of the Institute of Modern History, Academia Sinica] 6: 353–407.

International Telegraph Convention with Berlin Revision of Service Regulations and Tariffs 1885. London: Blackfriars.

Ismail, Ibrahim bin 1984. "Samuel Dyer and His Contributions to Chinese Typography," *Library Quarterly* 54, 2 (April): 157–169.

Jacobsen, Kurt 2001. "A Danish Watchmaker Created the Chinese Morse System," *NIASnytt (Nordic Institute of Asian Studies) Nordic Newsletter* 2 (July): 17–21.

Jiaotong bu 1916. *Ming mi ma dianbao xinbian—jingxiao xiushen ben* 明密電報新編 [Plaintext and secret telegraph codes—new edition]. Shanghai: n.p.

_____ 1946. *Ming mi dianma xinbian* 明密電碼新編. N.p.: Jiaotongbù.

Lauture, comte D'Escayrac de 1862. *On the Telegraphic Transmission of the Chinese Characters*. Paris: E. Brière.

Lee, En-han 1968. "China's Response to Foreign Investment in Her Mining Industry," *Journal of Asian Studies* 28, 1 (November): 55–76.

Legrand, Marcellin 1845. *Tableau des 214 clefs et de leurs variantes*. Paris: Plon frères.

_____ 1859. *Spécimen de caractères chinois gravés sur acier et fondus en types mobiles par Marcellin Legrand*. Paris: n.p.

Liu, Lydia 2004. *The Clash of Empires: The Invention of China in Modern World Making*. Cambridge, MA: Harvard University Press.

Macgowen, D. J. 1851. *Bowu tongshu* 博物通書 [Philosophical almanac]. Ningbo: Zhen shen tang.

McDermott, Joseph P. 2006. *A Social History of the Chinese Book: Books and Literati Culture in Late Imperial China*. Hong Kong: Hong Kong University Press.

Pauthier, Guillaume 1838. *Le Tào-te-Kîng, ou le Livre de la raison suprême et de la vertu, par Lao-Tseu, en chinois, en latin et en français, avec le commentaire de Sie-Hoèi, etc.* Paris: F. Didot Frères, Libraires.

Qian Xuantong 1918. "Zhongguo jinhou de wenzi wenti" 中國今後的文字問題 [China's script problem from now on]. *Xin qingnian* 4, 4.

Rankin, Mary Backus 2002. "Nationalistic Contestation and Mobilization Politics: Practice and Rhetoric of Railway-Rights Recovery at the End of the Qing," *Modern China* 28, 3 (July): 315–361.

Review of *Dianbao xinshu* 1874. *Chinese Recorder and Missionary Journal* 5 (February): 53–55.

Rogers, Everett M. 2003 [1962]. *Diffusion of Innovations*. New York: Free Press.

Shay, Frank 1922. *Cipher Book for the Use of Merchants, Stock Operators, Stock Brokers, Miners, Mining Men, Railroad Men, Real Estate Dealers, and Business Men Generally.* Chicago: Rand McNally.

Slater, Robert 1870. *Telegraphic Code to Ensure Secresy* [sic] *in the Transmission of Telegrams.* London: W. R. Gray.

Standage, Tom 1999. *The Victorian Internet: The Remarkable Story of the Telegraph and the Nineteenth Century's On-line Pioneers.* New York: Berkeley Books.

"T'een-piao-hsin-shu, Nouveau code du télégraphie chinoise" 電報新書. April 1872.

Tsu, Jing 2010. *Sound and Script in Chinese Diaspora.* Cambridge, MA: Harvard University Press.

Viguier, Septime Auguste. 1871a. *Dianbao xinshū* 電報新書. In "Extension Selskabet— Kinesisk Telegrafordbog." Arkiv nr. 10.619. In "Love og vedt{ae}gter med anordninger." GN Store Nord A/S SN China and Japan Extension Telegraf. Rigsarkivet [Danish National Archives], Copenhagen, Denmark.

———— 1871b. *T'een-piao-shu-tsieh* (*Code de télégraphie chinoise*). Shanghai: n.p. In private collection of Henning Høeg Hansen, Copenhagen, Denmark.

———— 1875. *Mémoire sur L'établissement de lignes télégraphiques en Chine.* Shanghai: Imprimerie Carvalho.

Wang, Chin-chun 1929. "The New Phonetic System of Writing Chinese Characters," *Chinese Social and Political Science Review* 13.

Yang, Daqing 2011. *Technology of Empire: Telecommunications and Japanese Expansion in Asia   1883–1945.* Cambridge, MA: Harvard University Asia Center.

Zheng, Xiaowei 2009. "The Making of Modern Chinese Politics: Political Culture, Protest Repertoires, and Nationalism in the Sichuan Railway Protection Movement in China," Ph.D. dissertation, University of California–San Diego.

Zhongguo dianbao xinbian 中國電報新編 [Chinese telegraph code—new edition]. 1881. 10619 GN Store Nord A/S. 1870–1969 Kode- og telegrafbøger. Kodebøger 1924–1969. Rigsarkivet [Danish National Archives], Copenhagen, Denmark.

Zhu Jiahua [Chu Chia-hua] 1937. *China's Postal and Other Communications Services.* Shanghai: China United Press.

Proofreading Science

Editing and Experimentation in Manuals by a 1930s Industrialist

Eugenia Lean

Abstract

In the 1930s, leading industrialist and famous romance novel writer Chen Diexian (1879–1940) compiled and published technological treatises and collectanea on household science, medicine, and industrial technology, ostensibly for the general reader. An examination of the texts reveals how Chen mixed long-standing practices of collecting and compiling knowledge with new forms of industrial and commercial pursuits to popularize and legitimate the incorporation of modern science into daily life and industrial endeavors. In an era when words, texts, and things were easily mass-produced and often copied and pirated, Chen Diexian's practices of editing sought to bring order to a world of material and textual abundance by serving to authenticate certain regimes of knowledge and practices, as well as objects. His texts promoted the virtue of shiyan, or vetting through practice and experimentation, to arm readers with strategies by which to identify false goods, nonnative commodities, and the inauthentic, even while seeking to identify certain forms of commerce as patriotic, technology and industry as honorable, and native production and products as virtuous. By focusing on these manuals, this article illuminates the role of commerce, maverick entrepreneurs, and popular print culture in endowing science with cultural authority and value in the Republican period.

The preface to the 1935–1941 domestic-technologies collectanea *A Collection of Household Knowledge* (*Jiating changshi huibian* 家庭常識彙編) listed several principles of compilation, one of which included the following excerpt:

> If one can use boric acid (*pengsuan* 硼酸) to gargle but commits a simple mistake in miswriting a single character, you could write "sulfuric acid" (*liusuan* 硫酸) or "sodium nitrate" (*xiaosuan* 硝酸), which can cause one to burn one's tongue and lips, and even death. You can see how dangerous it is. Once there was a funny contributor who mixed in a satirical piece with the other articles that he contributed. In it, he wrote, "In order to stop a severe stomachache, you can use 礜石 (*yushi*)." In fact, *yushi* is [similar] to the word for "arsenic" (*pishi* 砒石), which certainly treats human life as a joke! Also, the character 礜 (*yu*) is similar to the character 礬 (*fan*). Among our (country's) old methods, we once used *fanshi* 礬石 [alunite] (and even *mingfan* 明礬 [alum]) to cure stomachaches. If there was one way to cure gas pain (*shaqi futong* 痧氣腹痛), it was grinding *fanshi* and ingesting it with water. But if I wrote by mistake the character

礜, and the proofreader missed this, it would be disastrous. Thus, we must prohibit the pirating (*fanyin* 翻印) of this book in order to avoid misleading people. If you want a reprint, you can always order by mail from me.

In this passage, the editor of the collectanea, Chen Diexian 陳蝶仙 (1879–1940), warned against the dangers of false, misprinted words, substandard medicine, faulty chemical knowledge, as well as the pirating of texts. All such falsities—false knowledge, bad editing, and questionable epistemological regimes—had the potential to result, Chen asserted, in the loss of human life. His point here was not to discredit modern science or industrialism per se. As a modern entrepreneur and inventor, Chen was highly invested, both literally and figuratively, in promoting industrial modernity and its attendant objects and knowledge regimes (such as modern science). Yet, in this text, he nonetheless acknowledged, if not actively generated, unease about industrialism and modern biomedicine and chemistry, if only to legitimate the reason behind the publication of his text. One slip of the brush could transform the compound *fanshi* to *yushi* and, thereby, turn a beneficial medicine into an agent of death. Or, as Chen put it, "one word difference [could] reverse benefit to harm." The difference between life and death then was not simply dependent on the appropriate use of modern chemistry or the right application of new forms of medicine; it also turned on authoritative, reliable editing of scientific and medicinal knowledge. Dependable methods of compilation would guarantee trustworthy information on industrial arts and modern science and enable authentic production. Chen's personal care in vetting and editing such information was thus crucial.

By examining two collectanea compiled by Chen Diexian, this chapter inquires into anxiety about inauthentic knowledge and false things in the 1930s, exploring in particular what the exact relationship was among sincere and trustworthy editing, authoritative scientific and technical knowledge, and authentic production in a modern age. To be sure, a primary motivation behind the publication of such collectanea was fairly straightforward: namely, to promote Chen's own industrial products and confer legitimacy on himself. Yet, in an era when words, texts, and things were easily mass-produced, often copied, and inevitably pirated, Chen's practices of compilation and editing were also offered as means to authenticate the regimes of knowledge and practices that constituted industrial modernity. In these texts, Chen mixed long-standing textual strategies of compiling information and knowledge with new goals of industrial and commercial pursuits. His collectanea promoted certain forms of commerce as patriotic, proper technology and industry as honorable, and native Chinese production and products as ideal. In an era when markets were perilous, when knowledge could be false and science questionable, Chen's series emphasized empirical and textual verification to offer a sense of security and strategy for survival.

Using textual strategies to bring order to a world of material and textual abundance was not new to the twentieth century. Since at least the seventeenth century, Chinese literate elites compiled lists and edited catalogs to bring order to an era of overwhelming commercialism and materialism, to restore the moral fabric of society, as well as to reaffirm their literati identity vis-à-vis newly powerful merchants.[1] The mere ability to acquire luxury objects did not define status. Rather, the ability to investigate things and, by extension, the natural world through textual knowledge was crucial in establishing taste.[2] The practice of investigating things (*gewu zhizhi* 格物致知) generated an abiding interest in ordering everyday life through texts such as daily encyclopedias and the pharmaceutical world through comprehensive texts such as the *Compendium of Materia Medica* (*Bencao gangmu* 本草綱目).[3] Such textual strategies were to persist into the late Qing, when collectanea (*congshu* 叢書) on modern science and technology were published as part of the Foreign Affairs Movement (Yangwu yundong 洋務運動) to bring order to the sudden profusion of new forms of knowledge (as well as new and foreign things).[4] Literally referred to as *xinxue* 新學, or "new knowledge," this knowledge included modern science and Western technology, as well as Western philosophies of law and governance.

By the third decade of the twentieth century, Chen Diexian found similar textual strategies useful in bringing order to an era when anxiety about newfound commercial materialism and modern science was deepening. Like his late imperial predecessors, Chen turned to strategies of compilation. What was unprecedented was the context in which the need to identify authentic things and words emerged. Mass production, the unsettling integration of Chinese

---

1   See, e.g., Clunas 2004, 2007.

2   Clunas 2007, 112–113, has nicely linked this encyclopedic and classificatory interest in collecting information on the pharmacopoeia of the Ming to the popularity of the medicine cabinet and its multiple drawers that physically separated elements of the world of pharmacology into sections. Such a desire to order the material and natural world stemmed from the intellectual trend of *gezhi* 格致, or "investigation of things," which was itself an abbreviation of *gewu zhizhi* 格物致知, an "investigation of things and an extension of knowledge."

3   For more on the vernacular interest in knowing the natural world in the late imperial period, see Elman 2004.

4   For more on the late Qing textual classification of new knowledge, including science and technology, see Elman 2004.

markets into global capitalism, a highly politicized "buy native" movement, growing ambivalence toward the regime of Western science, along with the vexed question of the modern relevance of native Chinese and local knowledge systems, were unprecedented factors shaping anxiety about what constituted modern technology. An unruly abundance of slick words and misleading signs in a new era of mass media was moreover cause for concern. It was in this context that Chen sought to guarantee in these publications both trustworthy textual strategies and authoritative technological know-how. It was precise and capable editorial work that could provide the reader with the practical and trustworthy knowledge needed to navigate a world of inauthentic goods—from counterfeits to nonnative products—and, by extension, maneuver both the promise and peril of China's industrial modernity.

By examining how industrial manuals sought to authenticate know-how as "scientific" and promote the spirit of *shiyan* 實驗 (investigating through practice), this chapter moves beyond the Needham paradigm that seeks to recover an authentic "Chinese science" in contradistinction to "Western" modern science. It eschews the question of why modern science appeared in Europe first and examines, instead, the conditions under which Chinese actors strategically decided to deem certain ways of knowing and engaging with the material world (as well as authenticate certain things and objects) *as* scientific and modern. By extension, the chapter does not assume a fixed notion of "science" but treats "science" as something whose definition and articulation are inextricably linked with and, indeed, mutually constitutive of the making of social identities (e.g., the industrialist, the modern connoisseur of scientific knowledge, the authoritative compiler of industrial know-how), the establishment of new epistemological regimes and practices (e.g., empiricism, experimentalism), and the formation of a new society and nation that is patriotic and productive. Efforts at authenticating know-how and practice as scientific and modern were hardly unique to China. They were part of a larger global phenomenon in which complex cultural and social strategies were deployed for and by ordinary citizens to grapple with a new material world shaped by science and industry in their everyday lives. Yet, what our local case study demonstrates is that although Chinese actors like Chen Diexian were engaging in practices that characterized societies worldwide, the particular strategies deployed and the historical conditions shaping that deployment were strikingly local and highly specific.

Chen Diexian: Editor, Science Advocate, and Industrial Entrepreneur

Who was Chen Diexian, and why was he invested in using long-standing compilation techniques to authenticate scientific and technological knowledge in the modern period? Chen Diexian has been best known as either a writer of popular romance novels or a patriotic industrial captain who founded a regional (China and Southeast Asia) pharmaceutical empire, Household Industries (Jiating gongye she 家庭工業社). Less well known is how Chen had a lifelong interest in the promotion of modern science and industrial technology. He was part of a new generation that sought to promote science not through officialdom or state-sponsored institutions but in and through the worlds of publishing and commerce. Chen had started to merge his scientific interests with commerce early on. In the first decade of the twentieth century, he founded the Gather Profit (Cui Li 萃利) Company, a Hangzhou shop that sold books and imported scientific instruments and appliances. While serving in an official staff position early in his career (1909–1913), he not only wrote one of his most famous serial romance novels, *Money Demon* (*Huang jin sui* 黃金祟), but also began to experiment with an inexpensive ingredient of tooth powder. Chen's later entrepreneurial success stemmed directly from this early "amateur" scientific experimentation.

In the 1910s, Chen spent a considerable amount of time disseminating scientific knowledge and industrial know-how through editorial activities. With his move to Shanghai after 1911, Chen became involved with editing and writing for some of the most powerful journals and newspapers of the day, and it was in those various capacities that he promoted science and industry. By December 1914, he was chief editor of *Women's World* (*Nüzi shijie* 女子世界), a pioneering journal that was devoted to poetry and fiction but also included a variety of practical household information and technological pieces. This included the column "Warehouse for Manufacturing Cosmetics" (Huazhuangpin zhizao ku 化妝品製造庫), which was published in the Industrial Arts (Gongyi 工藝) section of the journal and did not merely provide household tips but also promoted industrial knowledge to would-be industrialists and connoisseurs of technology. From 1913 to autumn 1918, Chen was actively involved in the production of the *Shenbao* 申報 literary supplement "Free Talk" (Ziyoutan 自由談) and served as its editor from late 1916 to September 1918. He not only published his novels in serial installments in *Shenbao* but also contributed regular entries on medicine and industry to the "Household Knowledge" (Jiating changshi 家庭常識) column.[5]

By the end of the 1910s, Chen turned to actual industrial pursuits. By the middle of 1917, Chen had perfected his tooth-powder formula and begun its production. In May 1918, he listed his Household Industries as a joint-stock company, whose most notable product was Wudipai 無敵牌 (Peerless) tooth powder, a product uniquely characterized by its ability to double as face cream. By the 1930s, Household Industries had grown into a regional pharmaceutical empire, and its tooth powder was one of the most popular toiletry items in China and Southeast Asia.[6] At this point, Chen turned his attention away from literary pursuits to focus on building his pharmaceutical empire. Nevertheless, he was not entirely willing to forgo the bully pulpit of print media or the editorial techniques of ordering and authenticating knowledge. As a leading "industrialist" (shiye jia 實業家), he continued to compile and publish (often through his company's in-house press) a range of technological treatises and collectanea on household science, medicine, and industrial technology.[7]

Two of his collectanea will be examined in more detail below. First, the collectanea *A Collection of Household Knowledge*, mentioned above, consisted of a series of booklets on domestic matters published from 1935 to 1941. This was actually an eight-volume compilation of the column of the same name published in *Shenbao*'s "Free Talk" literary supplement when Chen was its editor during the late 1910s. Second is Chen's 1933 *Collectanea on How to Acquire Wealth in Industry* (*Shiye zhi fu congshu* 實業致富叢書). Targeting industrialists and manufacturers, the *Collectanea on How to Acquire Wealth in Industry* was a multivolume compilation of the 1916–1917 Agriculture and Commerce Ministry serial publication *Industry in Laymen's Terms* (Shiye qianshuo 實業淺說). It was printed by Shanghai's Xinhua Press, under Chen's pen name Tianxuwosheng 天虛我生.

Beyond the stated desire to make available "articles on household practice in order to contribute to society," these collectanea sought to reassure readers in a period of growing apprehension regarding inauthentic science, false

5   The specific column "Household Knowledge" was published only until May 10, 1917, even though "Free Talk" was printed until 1918. For this information, see Chen 1935–1941, vol. 1, 1.

6   For biographical information on Chen Diexian in English, see, e.g., the introduction to Chen 1999.

7   Chen continued, albeit on a smaller scale, to contribute to newspapers and journals, including household science columns, into the wartime period. As noted by Cochran (2006, 112), Chen contributed the column "Common Sense about Family Experiments" (Shiyan jiating changshi 實驗家庭常識) in the wartime journal *Healthy Home* (*Jiankang jiating* 健康家庭).

goods, and misleading words and information.[8] The 1930s was a period when mass production in China had developed at an unprecedented scale and when industrial science and the practices of modern capitalism had substantially matured.[9] Yet these developments hardly signaled straightforward progress. The specter of faulty industrial goods loomed large, and doubts about modern science and industry were growing. In the aftermath of World War I and with the start of yet another world war in Europe, intellectual ambivalence about Western science and technology in China had deepened considerably.[10] As science spread more widely, the general population also found the newness and esoteric nature of science to be a source of considerable unease. The ability to distinguish between good and bad science and to identify false promises, quackery, and misinformation became increasingly important. As was the case in other countries where modern advertising resulted in the sudden rise and tremendous popularity of patent medicines, the explosion of patent medicines in China's urban markets engendered apprehension about quackery and the subpar quality of medicinal items.[11] These new domains and products of science and industry coexisted, often confusingly, with indigenous technologies and knowledge regimes, including traditional Chinese medicine, local technologies, and other "folkloric knowledge."

Mass production and modern capitalism similarly generated unease. A shadow economy of broadly circulated untrustworthy items and pirated and foreign goods accompanied the flow of "legitimate" commodities and objects. Both domestic brands and international products were widely copied, and in certain circles, such copying was increasingly deemed a problem.[12] As twenti-

---

8   Chen 1935–1941, vol. 1, 1.

9   For example, by the 1930s, China's light industry had become relatively robust and competitive. By 1934, there were fifty-eight soap factories in the province, with a majority of them concentrated in Shanghai. By 1931, there were about fifty cosmetic manufacturers (*China Industrial Handbooks Kiangsu* 1933, 499–501, 509–511). China Chemical Industries Co. was the most successful, especially its "Three Star" brand products, and drew most of its customers from Sichuan and the Yangzi delta (ibid., 514). Chen Diexian's company, Household Industries, was next in importance. Its products, especially its "Butterfly" brand tooth powder, were distributed throughout the country and even penetrated the Southeast Asian market.

10   An example of this intellectual ambivalence can be found in the famous 1924 debates on science and metaphysics that involved May Fourth thinkers such as Ding Wenjiang, Liang Qichao, Hu Shi, Zhang Junli, and others.

11   For more on patent medicines in China's urban print media in the early twentieth century, see Lean 1995.

12   See Dikotter 2006 on counterfeits in China's modern markets. He makes an interesting argument that counterfeited goods piggybacking on brand names often resulted in the

eth-century China saw its markets being integrated into a global system of capitalism characterized by a developing legal and commercial regime of intellectual property rights, a copy culture that had long thrived in Chinese commercial and cultural markets was becoming increasingly unacceptable.[13] Concerns about trademark and pirating developed, and trademark infringement cases engendered legal and even diplomatic entanglements.[14]

Finally, the 1930s were characterized not merely by a profusion of mass-produced things but also a profusion of mechanically reproduced texts and mass media. New print technologies and mechanized means of reproducing texts meant the proliferation of print material and knowledge. As a result, more and more people had access to the production of knowledge, which rendered many intellectuals, once the primary arbiters of culture, highly anxious. Just as the sudden proliferation of things generated anxiety and concern about fake goods, the spread of new print technologies engendered the circulation of illicit texts and a growing concern about this circulation. To add to these problems, with the influence of international ideas of copyright and intellectual property rights, the notion of owning ideas and words and the rise of the notion of the modern "author" meant that copying texts without attributing one's source, once a relatively unproblematic and common practice in China, had become problematic.[15] Pirated texts and the spread of false information became social and cultural concerns.

As an editor *and* entrepreneur, Chen Diexian had a particularly good vantage point from which to grasp the "problem" of copying and uncertain goods in both the world of culture and the world of commerce. In his capacity as a captain of industry, Chen saw his own brand-marked commodities stolen and

---

popularization of the brand-name goods.

13   For more on the history of copy cultures in the imperial period and the rise of a modern global regime of copyright and intellectual property in China, see Alford 1995.

14   See, e.g., Lean 2013, which explores trademark infringement cases involving copycat Chinese companies counterfeiting Lux Soap and Hazeline Snow Face Cream, globally popular products of two British pharmaceutical empires, Unilever and Burroughs, Wellcome, and Co.

15   New ideas about intellectual property rights started to circulate in China in the late nineteenth and early twentieth centuries. By the Nanjing Decade, the Nationalists (GMD) had promulgated the first part of a new legal regime on intellectual property, its Copyright Law, which was modeled after the German example as filtered through Japan (e.g., Alford 1995, 50). Authors could register with the Ministry of Internal Affairs to protect their literary and artistic works, and if their rights were infringed upon, they could bring civil action to seek damages. The question remains as to how effective the enforcement of this law was on the ground.

counterfeited by smaller-scale, crafty Chinese merchants. As a result, he became invested in eradicating piracy and counterfeiting and in rendering China's commercial and legal practices "modern." He tried to achieve these goals by aggressively pursuing counterfeiters of his Wudipai tooth powder in the market and the courtroom.[16] In his role as editor and commercial writer, he was also no doubt fully cognizant of the widespread plagiarization of literary products and dissemination of false knowledge. It was in this 1930s context, when the mass production of things and the mechanized reproduction of texts had started to generate concern about the counterfeiting and piracy of both objects and words, that Chen displayed a commitment to fight such textual inauthenticity. He did so not just through legal means but also with strategies of compilation and editing.

Household Knowledge: Compiled Authenticity in an Age of Industrial Science

Chen Diexian's faith in using editorial strategies to give order to industrial knowledge is evident in his *Collection of Household Knowledge*. While a compilation, the 1930s publication nonetheless sought to assuage newfound anxiety regarding inauthentic things and words at a heightened stage of industrialization in China. The publication was, on the one hand, invested in promoting modern industrial goods and scientific know-how and yet, on the other hand, exhibited a degree of apprehension about industrial modernity. It was only through the authority of Chen's compilation practices that industrial goods and scientific know-how could be disciplined and authenticated. To accomplish this, the serial publication presented a variety of technologies, many of which were informed by modern science and industry. It guaranteed the authenticity of the information by assuring readers that the material had been thoroughly "vetted and tested empirically in practice" (*shiyan*). Once armed with such texts, readers would, the collectanea seemed to promise, be better able to navigate a dubious world of misinformation, counterfeits, fakes, and copied commodities.

A *Collection of Household Knowledge* presented information in an organized fashion and aimed to disseminate such information for the sake of strengthening the nation and making China's households and industries more competitive. Content-wise, the collectanea provided what might be characterized as

---

16   As early as 1921, Chen Diexian's Wudipai tooth powder generated copycat products. For
      evidence of this, see Zhang 1921.

accessible technical know-how, even practical "tips." Sections retained from the original *Shenbao* column included those on clothes and tools, food and drink, the body, animals, and plants. The 1930s reprint involved some recataloging and the addition of new bulletins. These included a bulletin on new recipes for Chinese and Western foods; a bulletin on "things one must know for industry" that included various manufacturing methods that should be used in industry; a section on collected bits of beneficial wisdom, which specially posted scholarly research and included a Q&A forum; and finally, a section called "miscellaneous jottings" on medical prescriptions.

Who might have read this series? Actual readers of the earlier newspaper column and the 1930s collectanea most likely shared a similar background, though the imagined reader of the 1930s collectanea might have been new. One of China's major commercial papers, the *Shenbao* had a circulation of thirty thousand by the early 1920s, which suggests a considerable circulation for the paper by the late 1910s.[17] The paper's audience had grown somewhat beyond its late Qing clientele, which had generally been restricted to literati-officials, merchants, and industrialists. By the turn of the century and into the first decades of the twentieth century, readership had expanded beyond the local and regional to encompass the nation, including the rural arena and women.[18] However, those reading "Household Knowledge" specifically most probably remained a fairly exclusive group of lettered elites and men of commerce. Despite the fact that its name might appear to imply that the general reader was the target audience, "Household Knowledge" was part of the "Free Talk" literary supplement that Chen edited from 1916 to 1920, which was characterized by a highly literary sensibility of playfulness and connoisseurship. Such a literary style appealed to China's professionals and merchants, as well as its men-of-letters.[19] These same readers would have read the column "Household Knowledge."

As collectanea, the 1930s compilation had a far smaller circulation than the earlier column in the *Shenbao*. Actual readership is difficult to determine, though much of the textual evidence suggests that the readership was not a general one but a fairly tight-knit group of like-minded connoisseurs of tech-

---

17   For circulation figures, see Narramore 1989, 373. Keep in mind that these numbers did not accurately reflect actual readers, who were much greater in number, as it was common practice for the newspaper to be posted on public reading boards for readers to read communally.

18   For a discussion of late imperial and early Republican rural elites having access to modern newspapers like *Shenbao*, see Harrison 2005, 105. For women readers as ideal and in reality, see Mittler 2004, 245–311.

19   For more on readers of "Free Talk," see Meng 1994.

nology or industrialists. The target audience likely had followed Chen's writings on technology in newspapers over the years and had written letters to the editor with personal observations and feedback. It was this feedback, we will see, that became an integral and guiding part of the 1930s publication.

From the content of the series, one further gets the sense that the reader would be someone who could appreciate modern science and industrial technology. To be sure, not all entries concerned modern science, and some read very much like entries from late imperial "everyday encyclopedias" (*riyong leishu* 日用類書). The section on skin that was part of the larger section on the "human body" dedicated a full two pages to a variety of cures for the affliction of *luoli* 瘰癧, a skin disease that causes the upper torso and neck to be covered in rotting sores and pustules. Cures ranged from Chinese herbal-medicine concoctions to local remedies. Remedies included a soup made of cat meat and a paste to rub on sores that required small pellets of white-mice feces.[20] At the same time, other entries were clearly informed by modern science and included a considerable amount of chemical information.[21] An entry in the subsection on teeth and the mouth explained how to treat tooth rot with chemical products.[22]

Yet, if parts of the compendium were eclectic in their range of knowledge, industrial knowledge and attendant goods nonetheless loomed large, and the prominence of industrial knowledge in the publication suggests that a large portion of the audience consisted of industrialists.[23] The original newspaper column had already identified *gongyi*, or industrial arts, as one of the funda-

---

20   Chen 1935–1941, vol. 2, 74.

21   The excerpt from *A Collection of Household Knowledge* on using *fanshimo* (alunite powder) to cure stomach pain, discussed at the start of this essay, is one such example.

22   See, e.g., Chen 1935–1941, vol. 7, 55 or 5.

23   On a related note, contrary to what we might expect, industrial know-how sat comfortably with domestic technologies and was presented as meaningful and useful for the domestic domain. Chen similarly links the domestic realm with industry in his *Collectanea on How to Acquire Wealth in Industry*, the collectanea that I discuss next. Chen comments in the preface that "we must allow every family and household to know of this [knowledge]" in order for the gospel to spread to industrialists. It is an interesting choice of words, and hardly arbitrary (Chen 1933, 1). It spoke to an early twentieth-century reality in which, although industrialization had indeed started to develop in cities, much of it remained family and household based rather than corporate or factory based. It also spoke to a larger cosmological correspondence between the microcosmic realm of the household and the macrocosm of the newly industrializing nation and the belief that bettering society, industry, and, ultimately, the Chinese nation not only started with the household but was indeed rendered virtuous by being associated with domestic production, a strategy that had a long history in China.

mental organizing categories of the series, and the category included entries on the industrial uses of water and how to make one's own soap, artificial ivory tusks, grass paper, different kinds of perfumed oils, and potassium nitrate. The 1930s compilation included a new bulletin, "things one must know for industry," suggesting the continued importance of industrial know-how. The preface to the compilation points out that one of the reasons for the reprinting was the need to verify the information in the section on industrial arts in particular.

This emphasis on industrial knowledge was in part indicative of the 1930s, a period when industrial development had advanced considerably, and the concept of *gongyi* had gained currency. In her research on Republican era art magazines, Carrie Waara has found that the term *gongyi* had multiple connotations in the High Republic, referring to technical arts like printing or textile manufacture, decorative art and design (as in *gongyi meishu* 工藝美術), craftsmanship and technique, as well as industrial design and tool and machine use in the production of consumer goods.[24] In Chen's *A Collection of Household Knowledge,* the meaning of *gongyi* referred specifically to industrial arts, which connoted knowledge and practices involving the use of original ingredients to make something new, often through chemical processes. It might even be argued that *A Collection of Household Knowledge* rendered *gongyi* into something close to "industrial folkloric craft" with its emphasis on methods of self-manufacturing industrial items. Some of these *gongyi* entries appear to promote goods that Chen's company, Household Industries, sold as premade commodities, and overall, the publication sought to promote these sorts of industrial products as daily necessities, which certainly was in Chen's commercial best interest. Other entries, including how to self-manufacture objects like tooth powder, coincided more with Chen's long-term commitment to spreading science and modern production know-how to his compatriots. The knowledge gained from self-production, it was implied, would enable the reader to discern false from true, good from bad, and copy from original, a necessary skill in this era of instability and overwhelming materialism.

The introductory sections to the segments on industrial arts (*gongyi bu* 工藝部) state explicitly that, for these segments at least, industrialists were the intended audience. For example, in the introduction to the industrial arts segment of volume 4, Chen Diexian notes that this bulletin was dedicated to his "good friends in the world of industry, who help each other."[25] Chen would also often dispense advice directly to the world of industry. At one point, after he had provided entries on how to manufacture soap, Chen urged China's

24   Waara 1999.
25   Chen 1935–1941, vol. 4, 121.

industrial world not to spread itself too thin but to specialize in the production of one kind of product and excel in it. He writes that "a common problem with the industrial sector is that it wants quantity but it doesn't specialize. ... If you make one thing and can sell it far and wide, then you can enjoy endless benefits. If you look at one thing and, then, try another thing, then it is 'in the morning it is three and in the evening it is four,' and I can predict you will not succeed."[26]

In addressing fellow industrialists, Chen expended energy in the preface reviewing at great length the publication's compilation principles to demonstrate that the information was properly cataloged and, thus, that the publication overall was transparent, authoritative, and authentic.[27] In these principles, Chen featured the notion of *shiyan* as a guarantor of the authenticity of the material presented in the collectanea. In modern Chinese, *shiyan* means scientific experimentation, but in this preface the term assumes more the meaning of being based in practice or having been tested against or confirmed by reality.[28] The knowledge in the serial, Chen specifically contended, had been vetted along the lines of whether it qualified as *shiyan*, and, as such, worthy of compilation.

The preface suggested that the editorial compilation process was itself a means to vet and verify and, in effect, qualify as a textual form of *shiyan*. In the

---

26   Ibid.

27   It might be apt to describe prefaces in Chinese literary history as being similar to what Gérard Genette (1997) has described as "paratexts." For Genette, paratexts, or elements in a published work that accompany the main body of the text, including prefatory material, provide a space or zone where editors or authors can assume a strong authorial voice with the intent to intervene in and shape the reading process. Chen Diexian treats his prefaces in these collectanea much in the way that Genette describes paratexts. However, while much of his treatment conforms to Genette's definition, certain aspects of his preface seem unique to the Chinese literary tradition, including the use of principles of compilation to authenticate a publication's content. Another characteristic of the Chinese preface that might be more specific to the Chinese tradition is its inclusion of material not only by the author or editors but also by the author's friends or network of famous associates, to authenticate a text.

28   At the time, the concept of *shiyan* had multiple connotations. According to the *Comprehensive Chinese Word Dictionary* (*Hanyu dacidian* 漢語大詞典), Lu Xun 魯迅, for example, used the term to connote something based in reality or practice that is able to disprove previous assumptions. Liang Qichao 梁啓超, Hu Shi 胡適, and Guo Moruo 郭沫若 all used the term to refer to the scientific process of forming a hypothesis and carrying out an activity to interrogate or examine that hypothesis. The entry lists examples of how this connotation is used in particular phrases, including "carrying out experiments" and "chemical experimentation."

second principle of compilation, Chen articulated the ideal of *shiyan* as practice or reality-based know-how rather than erudite book knowledge. But, then, he proceeded to point out the importance of the compilation process in identifying such knowledge and in rendering this publication unique in the print market. He wrote, "All entries in this edition are based in practice and proven to be efficacious (*shiyan youxiao* 實驗有效) ... these [entries] were collected from earlier publications, and I, the editor, relied upon my knowledge to discriminate and select accordingly. I believe it is impossible to speak of similar books on the market in the same breath."[29]

Chen spent a considerable amount of time detailing exactly what editorial practices were undertaken. In the first principle, he wrote, "Originally there were several articles, all including names of those who contributed the articles. ... Those with mistakes have been corrected. ... There were some cases where I discovered repetitions when I was proofreading. Also, I may have supplemented certain articles according to my own knowledge [of the topic], ... which I indicate with the word 'supplemented' (*bubai* 補白), but I have not added my name."[30] By keeping the names or pen names of the original contributors of entries (the names or pen names are given at the ends of the entries), Chen acknowledged the collaborative effort of the compilation, credited individual contributors, yet upheld his own authority as the ultimate editor and compiler of information. Despite the possibility of mistakes on the part of individual contributors, it was his meticulous review of each entry that guaranteed the trustworthiness of the information presented.

It is in this section, too, that an ideal reader emerged in the text: one who engaged in *shiyan*. As I mentioned earlier, the assumption that editing information could serve to verify and give order to the knowledge and, by extension, the material world had a long history in imperial China, and Chen Diexian continued this tradition to order and discipline industrial science and related technological know-how textually in the modern moment. Chen's practice also converged with earlier literati traditions with respect to the faith he placed in the authority of his readers. This is evident in Chen's prefatory statements, where he starts by making clear that public demand had warranted the reediting. He notes that he had received letters complaining about the difficulty of reading the original *Shenbao* column and encouraging him to reedit the information to fit into one single booklet. "Each day many made the request (for reediting); thus, ... we have taken the above-mentioned bulletins and the already-published material and have compiled and collated the entirety to

---

29   Chen 1935–1941, vol. 1, 1.

30   Ibid.

print a single book in order to meet the needs of daily household use."[31] As compilers in the past had done, such a mention of readers helped establish a sense of a network of like-minded readers who had faith in the compilation.

What was relatively unprecedented, however, was the faith placed in the ability of readers to vet information through practice. In this preface, it is claimed that reader participation did not stop with this initial demand but also played a crucial role in verifying the knowledge being presented. Chen made a point of noting that in addition to his own personal verification, the content was scrutinized by a number of readers who had put such knowledge into practice, and the entries were confirmed through their positive results and reports of satisfaction. He wrote that "the part on the industrial arts (*gongyi*), in particular, has undergone the test of application in practice (*shiyan*) by multiple readers and each entry has been confirmed by reports of success and accomplishment."[32] Thus, whereas in the past, the role of demanding publication of a particular text would have been ascribed to a friend or a select group of like-minded people, here the reader-participant—who was somewhat more anonymous—did not simply demand republication but also participated in the vetting of the know-how.

In the third principle of compilation, Chen provided a particularly illuminating explication of the ideal of *shiyan*. In what is a remarkable passage, Chen validated certain skills of practice once associated with a different class from himself and his readers, including the ability to improvise and adapt:

in these entries, the manufacturing procedures and the medicinal portions need to be adapted to circumstance. Thus, at different times, depending upon the particular conditions, decide what is appropriate ... such an approach is similar to using a recipe in cooking. For example, if you have to show how to fry fish, sauté fish, and determine how much oil, salt, sauce, and vinegar you need to use, there are not, I feel, very precise portions [that you can identify]. In fact, if you follow a recipe mechanically, this does not live up to what your kitchen maidservant does. That is because the heat of the stove always needs to be adjusted accordingly, and one cannot be a stickler for rules. If, in music, you can only record its scoring and identify the beat in a scholarly manner, that is not enough to play music. You need to have technique in your fingers and the beat needs to be alive and nimble, and that is where good musical technique is sepa-

31   Ibid.
32   Chen 1935–1941, vol. 1, 2.

rated from the bad. Thus, it is not simply a matter of memorizing the
score but depends on one's ability to feel the music.[33]

In this excerpt, Chen promoted willingness to use the compilation in a daring,
improvisational manner, flexibly adapting the information to new environments. In the face of an ongoing series of political and national failures, a critical discourse of effete, ineffectual, and even corrupt literati holding back
China's progress had gained remarkable currency in the early twentieth century. One of the most trenchant elements of the critique was how classical
literati culture had focused myopically on textual moral knowledge and how,
in their reverence for the Classics, traditional literati had passively received
knowledge and failed to look forward and engage in new, practical, and progressive knowledge *and* practice. In this context, Chen's statement can be
understood as a challenge to elite readers to appropriate strategies and skill
sets from an entirely different class of people. He drew an analogy between
how one should approach the collectanea's knowledge with adaptability and
the willingness to be flexible, with the skill used in cooking, a skill long associated with the lower classes, as the mention of the maidservant makes clear.
The following reference to the ability to *feel* the music, rather than merely
mechanically execute music, only further reinforced the idea that improvisa-
tion and adaptability in application and practice were virtues.

    A Collection of Household Knowledge provided Chen Diexian with the oppor-
tunity to bring a sense of order to a new stage of industrial modernity, authenticate industrial goods and scientific know-how, promote virtues (including
that of *shiyan*), and provide trustworthy know-how in an era characterized by
fakes, copies, and epistemological uncertainty. The publication exhibited a
degree of apprehension about industrial science while at the same time promoting it. The use of this tension was quite clever. By generating unease in
readers, the series became a necessary source of information and comfort with
its promise to assuage any ambivalence through its editorial feat of compilation. Such authentication, preached the text, rested in hands-on practice and
the ability to manufacture things oneself. Finally, Chen's promotion of an ethos
of self-sufficiency was in line with the promotion of autarkic production at a
national scale.[34] As a leading merchant in the National Products Movement, a

---

33   Ibid.

34   According to Zanasi 2006, industrialists were working in a fraught alliance with the GMD
      state at the time. The Wang Jingwei 汪精微 faction's push for autarkic productivist eco-
      nomic policies sought to create an alliance between the state and Jiangnan industrialists,
      moving resources away from the rural landed elite. Specifically, the Wang group organized

movement that had begun in the early twentieth-century boycotts of foreign goods but reached its apogee in the 1930s with the support of the Nationalists, Chen's advocacy of self-manufacturing for the sake of nationalism was hardly surprising and, as we will see next, was inextricably linked to his agenda to increase native manufacturing at the expense of foreign goods.[35]

Wealth in Industry: Native Production and National Authenticity

If *A Collection of Household Knowledge* points to the unsettling influence of insincere knowledge, *Collectanea on How to Acquire Wealth in Industry* reveals a deep concern with false, nonnative industrial goods and issues a call to arms for industrialists to engage in a form of virtuous and masterful copying in order to strengthen national industry. Part of the imperative to authenticate "practical knowledge" through strategies of compilation stemmed from a concern with the influx of foreign goods into China's markets. Over the course of the early twentieth century, treaty port industry and commerce became fraught with political meaning, and commodities and commercial practices grew increasingly inflected with nationalistic and anti-imperialist connotations. Chen Diexian's own patriotic consciousness grew gradually over the course of his personal career, paralleling this general growth of Chinese anti-imperialism. By the 1930s, Chen came to understand his various enterprises, commercial and otherwise, in highly patriotic terms. He had become a leading native industrialist and architect of the "Buy Chinese" campaign, the National Products Movement (Guohuo yundong 國貨運動).

In such a capacity, Chen sought to mobilize through his publications widespread patriotic consumption and to promote domestic production against fraudulent enemy commodities and imperialist merchants. In *Collectanea on How to Acquire Wealth in Industry*, he explicitly targeted industrialists and drew sharp lines between authentic and virtuous native goods versus politically fraught and dubious nonnative "enemy commodities" (*dihuo* 敵貨). Yet even as he demarcated enemy goods from authentic, native products, Chen promi-

---

the economic administration of the state along corporativist lines, centered on a controlled economy. This structure institutionalized at the national level the informal networks of cooperation between reform-minded industrialists and local authorities that had developed since the Self-Strengthening period. Tension, however, existed between a wish to support development of a private industrial economy and the idea of a strong interventionist state.

35   For more on the National Products Movement, see Gerth 2003.

nently included foreign technologies without hesitation. In fact, a rich discourse on "copying" (*fangzao* 仿造) from foreigner sources emerged in this text that was indicative of its implicit agenda of import substitution for the purposes of building national industry.

The content of the publication primarily consists of concrete advice and tips for China's industrial development. It was divided into several parts, with the first being "How to Acquire Wealth in Agriculture." With their titles following the format of "How to Acquire Wealth in *X*," the remaining parts were on forestry, industry, commerce, mining, husbandry, fishing, and the silkworm industry. The preface of the publication explicitly linked virtuous commerce and industrial development to the betterment of society and strengthening the nation, making explicit that the information in the text was compiled for patriotic purposes. Chen noted that the appeal to have industry save the nation was hardly unique to this text, citing the 1916–1917 Agriculture and Commerce Ministry textbooks upon which the current compendium was based.[36] He then detailed how he improved the original textbooks to generate the current compendium. The relevance of the compiled information was made evident in the following statement: "in order to have the whole country be rich and strong, we can use industry to save the nation. To ensure that this does not become hot air, this text provided the techniques for us to be rich."

The following section of the preface is particularly illuminating in how it deems the material presented to be "public knowledge" meant to improve China's national fate:

> The original book was published in 118 volumes. ... I compiled and categorized each volume. ... At first, it was simply for my own purposes. ... But because the original book is not easy to obtain, the full picture remains incomplete. ... We should treat this [republication] as public gospel. We need to let every family and household know of this in order to promote general interest [in industry]. If industrialists of all ranks know what [knowledge] to select, ... people will have jobs and we can progress [as a country]. Our masses will work hard and resist imports. If we had had the support of the masses, would we have been attacked by outside goods

---

36   Again, while I note above that Chen Diexian and others worked outside the official arenas of commerce and publishing to promote science and industry, I do not mean to say that the central government or civic governments did not play a role in developing commerce (see n. 34). Here, with these textbooks, we see that the central government, even during a period of weakness and disarray, was still committed to promoting commerce and industry.

and, now, be heading each day toward extinction?! ... In the past two years, our economy has been spiraling down; the unemployment rate is at a frightening level, impacting the entire nation. ... There are many roads to prosperity. ... It is up to those who have determination.

This rather lengthy excerpt asserted the importance of compilation in making the crucial information "public gospel" for all industrialists. Chen noted how he had first cataloged the information for his own purposes but then realized its value. Compilation became the means for him to verify the information and make it accessible for Chinese manufacturers to study and "select." He then called upon industrialists to use this information to provide the masses with jobs, something necessary, he reminded readers, given China's worsening national and economic situation. To underscore the urgency of the matter, he used the metaphor of war to describe economic competition and the threat of foreign commodities and, thereby, effectively raised the specter of national and economic extermination in the charged atmosphere of the National Products Movement.

A fundamental impulse behind the compilation was the desire to present a narrative of national authenticity. Notably, authentic national production did not necessarily hinge upon the use of indigenous Chinese production methods but, instead, could rely on the capacity to imitate cutting-edge production skills from abroad. An understanding of authentic production resting on emulation rather than invention can be found, for example, in entries such as how "to manufacture an imitation (*fangzao*) of the small foreign knife." The contributor of this particular entry was Zhang Yingxu, who noted that locally made knives were vulgar in their shape and had thus failed to attract buyers. As a result, consumers were flocking to buy Western-produced knives. To turn the flow of profit back to his fellow countrymen, Zhang promised to present a method "that does not require an engineer, nor the building of a factory." It was "effortless and simple to accomplish" and, "in less than ten days," would result in "locally made, new-style 'small foreign knives' sold on the street."[37] What that method entailed was building on preexisting industry, if only to turn it into something new. He used a neighborhood in Beijing where knives and scissors shops were plentiful to make his point. He reasoned that existing blade stores could copy the Western-style blades, shops that make traditional handles could produce the handles for the new knife, copper artisans could make the copper nails and springs, and electroplating factories could assume the responsibility of electroplating the nickel. He urged patience, warning that at

---

37   Chen 1933, vol. 1, 8.

first such an approach might require an investment. He appealed to his readers' sense of nationalism: "this [method] will count on patriotic gentlemen ... For those who are manufacturing this kind of thing you must not become discouraged halfway because there is no immediate profit. Those who buy, you cannot refuse its purchase simply because the price is a bit more expensive. You must be willing to improve industry along the way and have the will to promote national products (*guohuo* 國貨)."[38] Patience, determination, and skillful copying were of the essence.

Precedents for Chen's promotion of innovative nativist imitation of foreign technologies can be found in modern Chinese history. In an article on the late nineteenth-century Jiangnan Arsenal, Meng Yue notes the importance of the concept of "adapting" or "duplicating" (*fangzhi* 仿制) in the process of weapons production, which was focused on technological learning and training with the ultimate goal of cultivating technological creativity.[39] To be sure, Meng's approach is revisionist, and as many earlier scholars writing on the arsenal have noted, there was considerable failure at the arsenals, with bombs notoriously failing to explode.[40] Yet, where Meng's contribution lies is precisely in her eschewing the issue of success versus failure in our assessment of the arsenal. In doing so, she is able to investigate what did occur, including the rise of a discourse and practice of *fangzhi*. She shows that creativity in technological production at the arsenal was heavily dependent upon duplication of Western knowledge and technology. The process of *fangzhi* was complex and included research, sample making (of parts of a weapon), and testing for efficacy (*yanshi* 演試); a crucial practice that created the space for creativity in the process of *fangzhi* was *gaizao* 改造, or "remaking." As Meng notes, it was in the adaptation of Western technology that products and models were remade.

Chen Diexian's collectanea demonstrate that, several decades later, the notion of adapting Western technological knowledge for creative and nationalistic industrial endeavors remained relevant, even urgent. Indeed, Frank Dikotter has discussed how China's copy culture in the Republican era was characterized by a great deal of imitation of Western products to serve the large swath of the population that would never have been able to purchase

---

38   Ibid., 9.

39   Meng 1999, 20–23.

40   In response to these earlier scholars, see Elman 2004, 355–395, for a reassessment of their narratives of "failure." He argues that while the Jiangnan Arsenal was considered rather innovative in its day, it was the Sino-Japanese War in 1894–1895 in particular that "refracted" the story of the arsenal, and the Self-Strengthening Movement more generally, into narratives of failure.

expensive imports.[41] Such copying of commodities, he notes, was not seen as problematic. Nor was it identified as a legal infringement. In fact, it served to bolster the National Products Movement. As one of the leading participants of the movement, Chen not only had little problem with masterful copying but also actively encouraged it in his technological treatises. Chen's endeavors as a pharmaceutical giant owed their success to a form of undeclared import substitution by his company—the substituting of "Western goods" (*yanghuo* 洋貨) such as soap and tooth powder with native versions. Just as his industrial endeavors helped recategorize *yanghuo* into quintessentially native products, his editorial activities in these domestic manuals and industrial treatises presented masterful copying of foreign know-how as the foundation of authentic nativist production.

Conclusion

By compiling technical and industrial information into a single series, or *congshu*, Chen engaged in an interesting maneuver, promoting the long-standing collectanea genre, which collects knowledge and preserves it in a series format, over the newer format of the newspaper column, which prioritizes the constant production of new knowledge to be consumed (and presumably jettisoned) on a daily or weekly basis. His intention was to offer a more systematic treatment and ordering of production and technological knowledge to the industrially oriented reader-participant. Such collectenea on science and technology were not published in a vacuum. Large-scale print projects to give coherence to new terms and new ideas were under way in the 1930s. Encyclopedia projects and the publication of *congshu* were in vogue. As Robert Culp has shown, Wang Yunwu 王雲五, the editor of the Commercial Press in the 1930s, was actively commissioning leading intellectuals to compile and publish integrated series associated with particular fields of knowledge.[42] This period also saw the compilation of new dictionaries, such as a Shanghai slang dictionary, which strove to make sense of Shanghai's confusing vernacular.[43] In the 1930s, this commitment to ordering knowledge continued in the public library movement under the Nationalist regime, which resulted in the Shanghai publication of the *Encyclopedia Mini-collectanea* (*Baike xiao congshu* 百科小

---

41   See Dikotter 2006.

42   Culp 2011.

43   My thanks to Christopher Rea (personal communication, March 21, 2009) for pointing this out to me.

叢書). Such projects, moreover, came to shape the emerging popular interest in science. Unlike the earlier modern-science collectanea of the elite Foreign Affairs Movement of the latter part of the nineteenth century, the Republican era encyclopedia projects sought to render science accessible to a far broader urban audience.[44] Chen Diexian's collectanea were part of this larger movement to order, authenticate, and disseminate knowledge—particularly knowledge on science and industry—in Republican China.

Perhaps even more notable was how these collectanea addressed a more general social anxiety about industrialization, mass production, and global capitalism. In an era of heightened patriotism, they sought to promote both industrial and domestic science as the crucial foundation for China's commercial and political strength. In a time of uncertainty, these publications were also treatises on industrial invention and masterful copying. They sought to define authentic production know-how, not as original or native technological knowledge per se, but rather as dependent on hands-on and masterful ownership of knowledge—even if that knowledge originated from outside China— through application and practice (*shiyan*). Imitation of foreign technologies and ingenious use of indigenous knowledge, applied to industrial innovation and strengthening the nation, qualified as authentic and virtuous methods in manufacturing. Industrial innovation and the artful copying and adaptation of technologies were hardly at odds with each other.

Beyond articulating a discourse of masterful copying, these texts also reveal ways of making guarantees and verifying knowledge in the 1930s. In a period when the mass circulation of goods and words could easily render a commodity somehow inauthentic, and when patriotic movements treated foreign-made products as problematic, these collectanea functioned to authenticate knowledge of manufacturing. The methods used to authenticate knowledge in these collectanea were hardly unprecedented. For example, transmission of knowledge through compilation and editing, rather than original authorship or singular creation, had a long and valued history. Confucius fashioned himself as only one of a series of authorities, stating that virtue lay in "transmitting [the Dao] and not in creating" (*shu er bu zuo* 述而不作). Imperial China's

---

44   Culp (2011) argues compellingly that Wang Yunwu's *congshu* projects stemmed from his dedication to providing the general reader with access to a range of modern knowledge such as the sciences, as well as the essentials of Chinese learning. Another example might be Ye Shengtao's 葉聖陶 *Xin shaonian* 新少年 (*The New Youth*), a pamphlet that made science and technology available to apprentices in print workshops. According to Christopher Reed (personal communication, March 21, 2009), such publications constituted a layman's culture of technology.

book history saw the institutionalization of this classical idea and the ordering and effective transmission of virtuous knowledge in literary habits, state-sponsored book projects, and a vibrant and diverse printing industry. Hardly anachronistic, such strategies of transmitting through editorial and compilation endeavors, rather than "inventing" or "authoring" knowledge, continued to have a prominent place in these 1930s texts and were seen as crucial in imbuing scientific and technical knowledge with virtue and authority in the modern age.

Bibliography

Alford, William P. 1995. *To Steal a Book Is an Elegant Offense: Intellectual Property Law in Chinese Civilisation.* Stanford: Stanford University Press.

Chen Diexian 陳蝶仙 [Tianxuwosheng 天虛我生, pseudonym] 1933. *Shiye zhi fu congshu* 實業致富叢書 [Collectanea on how to acquire wealth in industry]. 5 vols. Shanghai: Xinhua Press.

___ 1935–1941. *Jiating changshi huibian* 家庭常識彙編 [A collection of household knowledge]. 8 vols. Shanghai: Jiating gongye she.

___ 1999. *The Money Demon: An Autobiographical Romance*, translated by Patrick Hanan. Honolulu: University of Hawai'i Press.

China Industrial Handbooks Kiangsu 1933. Shanghai: Bureau of Foreign Trade, Ministry of Industry.

Clunas, Craig 2004. *Superfluous Things: Material Culture and Social Status in Early Modern China.* Honolulu: University of Hawai'i Press.

___ 2007. *Empire of Great Brightness: Visual and Material Cultures of Ming China 1368–1644.* Honolulu: University of Hawai'i Press.

Cochran, Sherman 2006. *Chinese Medicine Men: Consumer Culture in China and Southeast Asia.* Cambridge, MA: Harvard University Press.

Culp, Robert 2011. "A World of Knowledge for the Circle of Common Readers: Commercial Press' Partnership with China's Academic Elite." Manuscript.

Dikotter, Frank 2006. *Exotic Commodities: Modern Objects and Everyday Life in China.* New York: Columbia University Press.

Elman, Benjamin 1984. *From Philosophy to Philology: Intellectual and Social Aspects of Change in Late Imperial China.* Cambridge, MA: Council on East Asian Studies, Harvard University.

___ 2004. *On Their Own Terms: Science in China 1550–1900.* Cambridge, MA: Harvard University Press.

Genette, Gérard 1997. *Paratext: Thresholds of Interpretation.* Cambridge: Cambridge University Press.

Gerth, Karl 2003. *China Made: Consumer Culture and the Creation of the Nation.* Cambridge, MA: Harvard University Asia Center.

Harrison, Henrietta 2005. *The Man Awakened from Dreams: One Man's Life in a North China Village 1857–1942.* Stanford, CA: Stanford University Press.

Lean, Eugenia 1995. "The Modern Elixir: Medicine as a Consumer Item in the Early Twentieth-Century Press," *UCLA Historical Journal* 15: 65–92.

____ 2013. "The Chinese Copycat on the World Stage: What Was New in Trademark Infringement Cases in Early Twentieth-Century China." Manuscript.

Meng, Yue 1994. "A Playful Discourse, Its Site, and Its Subject: 'Free Chat' on the *Shen Daily* 1911–1918." M.A. thesis, University of California–Los Angeles.

____ 1999. "Hybrid Science versus Modernity: The Practice of the Jiangnan Arsenal 1864–1897," *East Asian Science, Technology, and Medicine* 16: 13–52.

Mittler, Barbara 2004. *A Newspaper for China? Power, Identity, and Change in Shanghai's News Media 1872–1912.* Cambridge, MA: Harvard University Asia Center.

Narramore, Terry 1989. "Making the News in Shanghai: *Shen Bao* and the Politics of Newspaper Journalism 1912–1937." Ph.D. dissertation, Australian National University.

Shanghai shi nianjian 上海市年鑒 [Shanghai city yearbook] 1935. Shanghai: Zhonghua shuju.

Tsin, Michael 1999. *Nation, Governance, and Modernity in China: Canton 1900–1927.* Stanford, CA: Stanford University Press.

Waara, Carrie 1999. "Invention, Industry, Art: The Commercialization of Culture in Republican Art Magazines," in Sherman Cochran, ed., *Inventing Nanjing Road: Commercial Culture in Shanghai 1900–1945.* Ithaca, NY: Cornell University Press 61–89.

Zanasi, Margherita 2006. *Saving the Nation: Economic Modernity in Republican China.* Chicago: University of Chicago Press.

Zhang Yi'ou 張軼歐, ed. 1921. "Shanghai jiating she Wudipai camian yanfen jiamao yinggai gai gan weibian" 無敵牌擦面牙粉假冒影戱改[*sic*]干未便 [Shanghai Household Industry Peerless brand face and toothpowder trademark cannot be copied]. *Jiangsu shiyeyuekan* 江蘇實業月刊 [Jiangsu industrial monthly] 29: 44–45.

Zhongguo huaxue gongye she 中國化學工業社 1931. *Zhongguo huaxue gongye she er'shi zhounian jinian kan* 中國化學工業社二十 周年紀念刊 [The twentieth-year anniversary publication of China chemical industries]. Shanghai: Hanwen zhengkai yinshuju.

The Controversy over Spontaneous Generation in Republican China

Science, Authority, and the Public

Fa-ti Fan

Abstract

How did science, especially the scientific experiment, gain and maintain its epistemological and cultural authority in modern China? How and why did scientists form a group identity and guard their institutional and intellectual status? How did science become part of public discourse? To explore these questions, this chapter examines an important but largely forgotten scientific debate in Republican China—the controversy over "spontaneous generation" in the 1930s. The controversy began with a pamphlet written by a doctor based in Guangzhou reporting his "successful" experiments on spontaneous generation. Living organisms apparently arose from nonliving material. Several faculty members at the Zhongshan University were incensed by his claim and responded vigorously. Soon both sides were engaged in a series of public debates, exhibits, and experiments. Emotions ran high. Local newspapers and national magazines were mobilized. Government authorities were involved. In describing and analyzing this episode, I focus on several issues regarding science and its context in Republican China, especially science and the public, science as cultural authority, and science and national politics.

On June 27, 1933, shortly after 9 a.m., in the assembly hall of Zhongshan University in Guangzhou, an audience of about one thousand gathered. A dozen or so distinguished-looking men entered the hall and sat in the front. The hall was buzzing with anticipation and excitement. A quarter before ten, a gentleman with his entourage arrived. After the routine ceremony, the chairman announced the purpose and procedure of this public meeting. He then invited the gentleman who had just arrived to take the stage and report the process by which he had made his discovery, which the latter did. Then, one by one, the men seated in the front stood up and challenged him with questions peppered with scientific terminology; some questions were long, others short, and many sharply worded. At one point, one of the speakers showed a large image of a pear-shaped arthropod and clusters of pealike objects. Throughout the process, the gentleman who was being questioned either remained silent or replied curtly in an attitude of defiance and suppressed indignation. Most of the audience looked on with intense interest; from time to time, they cheered

© KONINKLIJKE BRILL NV, LEIDEN, 2014 | DOI 10.1163/9789004268784_010 or roared in laughter. The questions and answers went on for several hours. Not until 1 p.m. was the meeting brought to a close.[1]

This showdown highlighted a scientific controversy in Republican China that has largely been forgotten. The controversy over spontaneous generation in the early 1930s should be considered one of the most significant controversies over science in the Republican period. It shared some similarities with but was also distinct from other major public debates over science at the time. The best known of them, the debate over science and the philosophy of life in the 1920s, involved many prominent intellectuals, and the exchanges continued for a couple of years in newspapers and magazines.[2] That debate clustered around prominent problems in the search for the cultural identity of China, and for this reason, it received a good deal of attention at the time and has been a noted episode in modern Chinese intellectual history. However, it was really a debate among intellectuals about the ideology and cultural authority of science. Few trained scientists, other than Ding Wenjiang (丁文江), took part in the debate; it is not surprising that the many studies of the debate have revealed little about the scientific community and scientific practice in Republican China.[3]

In comparison, the controversy over spontaneous generation combined in-depth scientific discussions and broad public participation. Ostensibly, the debate was about particular scientific theories and scientific experiments, and most of the key participants were highly trained scientists. And yet the controversy was not limited to the laboratories or tucked away in specialized science journals. On the contrary, it involved a range of historical actors and took place in various public forums. Moreover, it engaged with important issues in the intellectual life of Republican China—the search for China's intellectual and cultural identity, the formation of scientific institutions and communities, the intersections of national and international intellectual currents, the regional and national politics permeating the intellectual communities, and the epistemic and cultural authority of science.[4] Therefore, the controversy provides a wide-angle view of certain prominent aspects in the cultural history of science in Republican China. First, the practice of science in a particular social and

---

1 Dong 1933, 99–115.

2 Kwok 1965; Lin Yusheng 1989; Luo Zhitian 1999.

3 A few other scientists, such as the psychologists Tang Yue and Lu Zhiwei, also participated in the debate. In Fan 2008 and Fan 2013, I look at two other important controversies that occurred in the intersections of science and culture/politics in Republican China.

4 Fan 2007 sketches out some of these themes in science and modern China during the long twentieth century.

cultural context often requires the historical actors to draw and negotiate the boundaries of science.[5] As we shall see, the controversy over spontaneous generation involved the drawing of boundaries between science and nonscience, between the scientific and the nonscientific community, and between the disciplinary boundaries of particular scientific expertise.[6] The reason for demarcating and guarding the boundaries lay in the perceived necessity of protecting the authenticity of science, expertise, and modernity against the contamination of amateurism, fraudulence, politics, commerce, and traditional culture. Participants in the controversy agreed on these principles but fought over how to characterize them and draw the corresponding boundaries.

Second, this episode enables us to examine an important but much neglected subject—the uses of experiment in scientific practice and in establishing the authority of science in China at the time. In what ways did the experiment as a scientific practice and a new way of acquiring knowledge play a role in fashioning the image of science and in determining the legitimacy of particular kinds of knowledge?[7] Since the late nineteenth century, Chinese intellectuals had come into contact with modern scientific experiments, either at Western-style schools or in books and magazines, and by the early 1930s, many of them had accepted the scientific experiment as a useful and even privileged way of acquiring factual and reliable knowledge.[8] Because in its technical aspects the controversy over spontaneous generation came down to disagreements over scientific experiments, a close examination of the episode will help us better understand the role and uses of the experiment in science and how experimental science was perceived among educated Chinese.

This leads directly to our next and third issue, science and the public.[9] The controversy over spontaneous generation took place across a gamut of public

---

5   Gieryn 1999.
6   If the episode is still remembered at all, it is cited to illustrate the necessity and urgency to attack pseudoscience (*wei kexue*). See Xue 2002; Zhong 2005. The problem with this approach is that it reduces a complex and multifaceted historical event to a battle between good and bad, right and wrong, science and pseudoscience (as understood in the terms of contemporary China). Instead of using preconceived categories, we should examine how the boundaries between these categories were contested and drawn.
7   E.g., Shapin 1988; Shapin and Schaffer 1989; Gooding, Pinch, and Schaffer 1989; Collins and Pinch 1993; Golinski 1998, esp. chaps. 3–5.
8   On scientific demonstrations/experiments in late Qing China, see Wright 2000, chap 5. See also chap 6, which deals with science and the popular press.
9   Science and the public is a broad and loosely defined topic. For my purposes here, some of the more relevant works are Golinski 1992; Morus 1998; Stewart 2007; Nyhart and Broman 2002; Z. Wang 2002.

venues or sites. There were public debates in front of large and sometimes rowdy audiences, exchanges in newspapers and magazines, and experiments on display in a university building as well as at a commercial exhibition. Presenting science to the public involved various strategies and techniques of representation—for example, rhetoric, performance, and public spectacles. All of them were relevant to the reception and perception of science in a particular society and culture.[10] As we shall see, a range of historical actors, not all of them scientists, played a role in the unfolding of the controversy. Various institutions—universities, hospitals, popular media, and the government— were involved. The scope of the public's participation in the controversy encompassed students, readers of major magazines and local newspapers, and visitors to the meetings, among others; most of them were educated young Chinese interested in science and education—who were seen as the future of the nation. This episode therefore provides an excellent opportunity for the historian to investigate the topic of science and the public and science popularization in Republican China.[11]

The Coming Storm

The controversy over spontaneous generation broke out over the claims made by Dr. Luo Guangting (羅廣庭) that his experiments had demonstrated the existence of spontaneous generation. In the winter of 1932, Luo self-printed a small pamphlet on spontaneous generation, called *Ziran fasheng zhi faming* (*The Discovery of Spontaneous Generation*), in which he summarized his experiments and declared that they proved spontaneous generation. A few months later, in the spring of 1933, the magazine of the Science Society, *Kexue zazhi* (*Science Magazine*), surely the most important general science magazine in Republican China, published the paper in the form of correspondence (*lai jian*, that is, less a formal article than a letter to the editor).[12] Certain students had

---

10   For a relevant historical context, see Lean 2007. See also Wright 2000, chap 5; Claypool 2005.

11   More than sixty years later, a biologist remembered well this scientific controversy. According to him, the event attracted much attention in Guangzhou and beyond at the time. He was then a high school student in Guangzhou and read about it in the newspapers. In fact, this event spurred his interest in biology so much that he entered the biology department of Zhongshan University and eventually became a biologist. See Liang 1997. On the topic of science popularization in the European and American contexts, see Topham 2009.

12   Luo 1933b.

brought Luo's pamphlet to the attention of their professors in the biology department at Zhongshan University, who maintained that earlier in 1932, Luo had submitted a paper on spontaneous generation to *Ziran kexue* (*Natural Science*), a science journal published by the university, and was rejected. The biologists at Zhongshan University also learned that Luo had showcased his experiments at the well-attended Guangzhou City Exhibit in February 1933 (some of them had actually gone to see Luo's exhibit).[13] In the spring of 1933, Luo launched another attack on the established view in science—this time, on evolutionary theory. In April, the venerable *Dongfang zazhi* (*Eastern Miscellany*) printed his paper disputing evolutionary theory, which incorporated his opinions on spontaneous generation.[14] The paper had previously appeared in pamphlet form and in a major local newspaper, *Minguo ribao*, and now gained a national audience.[15]

In the eyes of the biologists at Zhongshan University, Luo had gone too far. It was one thing to encounter Luo's claims in local venues, but it was another to see them in magazines of national reputation and circulation. Perhaps they also felt that the attack on Darwinian evolutionary theory had raised the stakes. However unorthodox spontaneous generation was in early twentieth-century science, it did not have the same broad cultural implications as an antievolutionary pronouncement. Darwin and evolutionary theory had become major signifiers in a range of political and cultural discourses in China.[16] By challenging evolutionary theory, Luo had transgressed. At this point, the biologists of Zhongshan University decided that they must take him to task.

Luo was a bone fide scientist and medical doctor. He was born in 1901 to a poor family in Beihai, Guangxi Province. His mother was employed by a Catholic missionary establishment there, doing laundry and other chores. Because of this connection, Luo received a French education and was sent to study in Vietnam, then a French colony. He obtained a medical degree and

13   Dong 1933, 314–322. On the history and importance of exhibitions in twentieth-century
      China, see Fernsebner 2006; Gerth 2004. Many exhibitions in Republican China were plat-
      forms for consumerism, modernization, and regional/national pride. The Guangzhou
      City Exhibit was no exception. Luo's critics had a point when they characterized the
      venue as commercial. However, the Guanghua Hospital, being a notable establishment in
      the city, could hardly avoid being involved in the exhibition. In some ways, a scientific
      discovery seemed to fit the themes of modernization and regional/national pride of such
      exhibitions.

14   Luo 1933c.

15   For the exchanges in the *Minguo ribao*, see Dong 1933, 1–36.

16   On the reception of Darwinism, see, e.g., Elman 2005, 345–351; Pusey 1983. On the cultural
      importance of developmental and evolutionary thought, see Jones 2011.

began medical practice there. About this time, he encountered pressure either to join the Catholic Church or to become a French national. (Accounts differ. The former seems to be the more likely case.) He decided to leave Vietnam for France. Luo studied in the medical school of the University of Paris and received a doctoral degree in medicine. He also spent a year at the Pasteur Institute in Paris, probably researching tropical medicine. According to his own accounts, it was while he was at the institute that he became interested in the topic of spontaneous generation. In 1930, he returned to Beihai and worked in the Catholic hospital there until 1932, when he moved to the Guanghua Hospital, a major private hospital, in Guangzhou. Luo was primarily a surgeon and gynecologist, and his duties as a doctor at the hospitals were heavy. However, in his spare time, he pursued, seemingly obsessively, his research on spontaneous generation.[17]

The biologists at Zhongshan University who spearheaded the attack on Luo were Zhu Xi (朱洗) and Dong Shuangqiu (董爽秋), and they enjoyed strong support from their colleagues and students in the process. Dong Shuangqiu was born in 1896 and had received a doctorate in botany from the University of Berlin in 1926. He was the chairperson of the biology department at the time.[18] Born in 1900, Zhu Xi was younger and had only recently joined the department. He had gone to France through the "Qingong jianxue" work-study program in 1920. After five years of working in various factories, he entered the University of Montpellier and in 1927 became a graduate student there. He obtained his doctorate in 1931 with a dissertation on embryology and cytology under the supervision of the distinguished experimental biologist Eugène Bataillon. Zhu returned to China in the winter of 1932 and took up a position at Zhongshan University shortly after.[19] Both scholars would go on to have distinguished careers in science.

In order to rein in Luo, a group of biologists, joined by some other faculty members, at Zhongshan University submitted a petition to the president of the university, Zou Lu (鄒魯), a Nationalist (GMD) politician.[20] Since the biologists at Zhongshan University played a crucial role in this controversy, it is necessary to say a few words about the Zhongshan biology department.[21] The

---

17   The most detailed biographical account of Luo is Luo Jizhou and Huang Shujin 2008.

18   Guoli Zhongshan daxue jiaowu bu 1933, 368.

19   Chen 2000. For a brief account of experimental biology in France during the time when Zhu Xi received his education, see Fischer 1990.

20   Dong 1933, 94–98.

21   On the biology department of Zhongshan University, see Zhang Yi 1934, 102–103; Guoli Zhongshan daxue jiaowu bu 1933, which includes lists of the faculty, students, and other university personnel, curricula, facilities and equipment, etc.

department had come into existence only recently, and it was small. In mid-1932, there were only four professors, and Zhu Xi would become the fifth.[22] Fei Hongnian (費鴻年) was an ichthyologist trained in Japan. Zhang Zuoren (張作人), who had only recently received his doctorate from the University of Brussels, worked on protozoans. In 1932, Jing Libin (經利彬), with both an M.D. and a Ph.D. degree from the University of Lyon, served as a visiting professor. At the time of the controversy, the department was staffed with a young, active, and well-educated faculty who worked closely with a small number (sixteen in total) of students. Together, they were prepared to do what they believed was right.[23]

In the petition, the professors pleaded with President Zou to forward their request to the Jiaoyu Gaige Weiyuanhui (Board of Education Reform) under the Xinan Zhengwu Weiyuanhui (Southwestern Political Affairs Committee) to call a public meeting in which Luo would demonstrate and explain his experiments to a panel of selected academics, experts, and respected doctors in Guangzhou. They made a series of arguments for its urgency. According to the professors, the university is the highest educational institution and it has responsibilities to society. In addition to educating and advising the young, professors have the duty to do research, search for the truth, and judge and expose frauds and falsehoods. This is the way it is in advanced countries, where intellectual activity flourishes through the discussions and debates among experts and scholars. In fact, those who commit fraud should face moral and even legal consequences. Moreover, scholarship is related to the honor of a nation. Luo sent his immature research results to international scientific institutions, an act that undoubtedly damaged the reputation of China's scientific community. At universities and middle schools, the faculty teach evolutionary theory and Louis Pasteur's scientific discoveries, and now Luo spreads his false theories to students and the public, which will only lead them astray. Hence, the petitioners demanded that Luo perform a supervised open experiment. To ensure its accuracy, the procedure of the experiment had to pass muster with the panel of experts. The petitioners saw themselves as guardians of scientific truths and were willing to put their jobs on the line. Citing an example of a Western academic in China who had voluntarily resigned over his mistake, the petitioners promised that they would follow this course of action if they were proved wrong. In the meantime, they implored the government to take up its

---

22   Guoli Zhongshan daxue jiaowu bu 1933, 368–369.

23   We do not know if the professors had privately tried to persuade Luo to give up his theories. They probably thought that since Luo had presented his views to a broad public at both local and national levels, it would require more than personal conversations to correct the wrongs already done.

responsibility of censuring pernicious discourses. They argued that the government should summon Luo to demonstrate his claims openly to others, and if he refused to comply or if he failed, then he should not be allowed to deceive the public further. If, however, he continued spreading his false claims, he ought to face moral and legal consequences.[24]

The petition reveals certain fundamental conceptions about the identity and community of elite scientists in Republican China. These were critical years for the professionalization and institutionalization of science in China; the scientists had made impressive progress in claiming respect, identity, and professional niches for themselves.[25] Having earned advanced degrees from major universities in Europe, Japan, or the United States, they took up positions at universities and research institutes in China. These scientists also organized learned societies and specialized associations along disciplinary lines. The Science Society was founded by overseas Chinese students in 1914 and moved back to China in 1918. The Geological Society of China, one of the earliest scientific associations devoted to a particular subject or discipline, was founded in 1922. The Physical Society of China came into existence in 1932, the Botanical Society in 1933, and the Zoological Society in 1934. As for research institutes, the Academia Sinica was established in Nanjing in 1928; in the same year, the Fan Memorial Institute of Biology was inaugurated in Beijing.

Chinese scientists (and other intellectuals) who had received education in Europe or the United States often held high social and intellectual status in China (more so than those who studied in Japan). They had taken academic pilgrimages to seek science and modernity and presumably returned with the best of what the West had to offer. To be sure, there were questions among Chinese intellectuals about the limitations of science and Western modernity, as reflected in the debate over science and the philosophy of life. But in fact most of the major participants in that debate were intellectuals who had studied in the West, and they often presented themselves as seekers who had found the best and newest wisdom from the West (e.g., Ding Wenjiang's Karl Pearson and Ernst Mach; Zhang Junmai's Henri Bergson and Hans Driesch). In China, scientists began to acquire a scholarly and professional identity. They demarcated themselves both from the traditional literati and from the nonscientist intellectuals. They regarded and presented themselves as professionals and experts in an area of activity essential to a modern state and society. This identity did not exclude them from being part of the intelligentsia of Republican

---

24   Dong 1933, 94–98.
25   On scientific societies and institutions in Republican China, see, e.g., Zhang Jian 2005; Wang Shiping 2008; Zhang Jiuchen 2005; Hu 2005.

China, and many of them were active in politics, administration, and education. Believing in the pivotal role of science in society at large, many of them
were also practitioners of science popularization. For instance, of the Zhongshan biologists, Zhu Xi, Zhang Zuoren, and Fei Hongnian would go on to write
many science books for students and general audiences. This intellectual and
institutional environment significantly influenced the controversy over spontaneous generation.[26]

Spontaneous Generation and Evolutionary Theory

The idea of spontaneous generation was nothing new. Luo and his critics listed
many examples from both Western and Chinese traditions. Both sides also
agreed that the long-running debate between advocates and opponents of
spontaneous generation in Europe came to an end with Pasteur's famous
experiments in the 1850s. It was generally accepted that Pasteur had demonstrated that what had previously been thought to be instances of spontaneous
generation actually resulted from the actions of microbes. Historians of European science have questioned this simple history and pointed out that even
after Pasteur's famous experiments, certain scientists continued research on
and claimed discoveries of spontaneous generation.[27] Similarly, Luo's experiments and claims about spontaneous generation were a continuation of and
response to Pasteur's experiments (albeit decades later), and both he and his
opponents in China pursued their contentions along these lines.

In his pamphlet, Luo claimed that he discovered cases of spontaneous generation in a series of experiments. He maintained that he observed living
organisms taking form in well-sterilized and well-sealed test tubes containing

---

26   It might be worth noting that Luo and his nemesis Zhu Xi were both educated in France
      about the same time. There is no evidence to suggest that they belonged to contending
      schools of thought in French biology, yet it is possible that Zhu was appalled by the fact
      that Luo had received his education in France. He might have feared that Luo's claims
      would reflect badly on people who studied in France like himself and that Luo would give
      a bad name to Chinese scientists when he reported his research to French scientists. (This
      latter concern was not limited to the Zhongshan biologists. See, e.g., Xian 1933, 23.) It
      should also be mentioned that neither Luo nor his opponents brought overt political or
      ideological beliefs to the fore in the debate. Zhu held anarchist beliefs at the time. It is not
      clear if Luo had strong political opinions. This doesn't mean, however, that the contro-
      versy was not "political," as will be shown later.
27   On the debates over spontaneous generation in Europe, see, e.g., Farley 1977; Roe 1981;
      Strick 2002; Secord 1989.

various kinds of culture medium, such as agar, glucose, egg yolk, and so on. The process unfolded as follows. Evaporation from the culture medium accumulated on the wall of the test tube and formed dew droplets, which gradually turned into spots of yellow and, later, a darker color. Under a microscope, one observed that these spots consisted of little pealike organisms, which would form clusters. Next, these little organisms would grow into a variety of creatures; some were animals and others plants. They might develop legs, forming first the upper or lower half of the body, and grow into arthropods, or they might metamorphose into maggots. Or they might form endospores and eventually become plants. The whole process could take weeks or months, and Luo declared that he had observed about ten different kinds of creatures that had emerged this way.[28] Luo reported that he had carefully and diligently performed the experiments for almost three years, and the outcome left no room for doubt.[29]

His claims flew in the face of the scientific orthodoxy that all life comes from seeds, eggs, parents, and the like (i.e., not from spontaneous generation). This notion of biogenesis was rendered into Chinese as *zhongzi shuo* 種子說 or *zhongsheng shuo* 種生說. Now, if Luo had simply claimed that some extremely basic life-forms could arise from nonliving matter, it would have been bold enough. (The experiments done by Pasteur and his opponents were about the spontaneous generation of microbes from organic matter, such as beef broth.) Yet he had gone much farther and declared that he observed highly complex organisms such as arthropods originating from water drops (not even from the growth medium). No wonder it didn't go down well with the biologists at Zhongshan University. Even so, his critics might have let the matter slide if Luo had not thrown down another gauntlet to the scientific establishment. In April 1933, he published a paper in *Dongfang zazhi* criticizing evolutionary theory.[30] Founded in 1904 and based in Shanghai, *Dongfang zazhi* was one of the oldest and most influential general intellectual magazines in Republican China.[31] Although the essay did not examine the details or mechanism of Darwin's theory, it highlighted the name Darwin in its criticisms of evolutionary theory.

Luo challenged Darwin's evolutionary theory primarily on two grounds. First, Darwin's theory does not explain the origin of life. This opinion wasn't anything new. It was a point that had not been lost to readers of Darwin, and in fact, Darwin himself would not have disagreed. His theory simply purported to

---

28   On the description of Luo's experiments, see Luo 1933b   841–855, esp. 842–846.

29   Dong 1933, 244, 332–333. See also p. 328 for another, similar kind of criticism.

30   Luo 1933c.

31   Huters 2008.

explain how and why life-forms changed over long periods of time—that is, the origin of species, not the origin of life. Unlike modern creationists who also clamor about the inability of Darwin's theory to address the origin of life, Luo was not religious, despite his early connections to Catholic institutions. He dismissed religious explanations of the origin of life, and his theory of spontaneous generation was purely physicochemical: molecules gathered together and became organic substances, which in turn congregated and formed organisms. We know from Luo's criticism of this "deficiency" in Darwin's theory that Luo was deeply interested in the problem: Whence comes life? Ultimately, his view was not too different from Lamarck's or those of many other advocates of spontaneous generation in early modern and modern Europe. As long as the conditions are right, life-forms can arise from nonliving matter. And the occurrences are ongoing. The origin of life is not something that happened long ago and has now stopped. It still happens around us.

The second front of Luo's criticisms of Darwin's evolutionary theory was on evidence. He argued that many classes of evidence, including anatomical and paleontological, posed problems for evolutionary theory. For instance, if humans and other primates are related, then why do only human females have the hymen, but not other primates? And why do only human males not have the penis bone when all other primates have it? Evolutionary theory, Luo asserted, cannot explain these phenomena. What about the purported embryological evidence? According to Luo, supporters of evolutionary theory claimed that human fetuses have gills in their early stage of development, which shows that humans have evolved from gilled animals. (Apparently, Luo was referring to Ernest Haeckle's theory and his famous illustrations of ontogeny recapitulating phylogeny.)[32] Luo did not dispute the anatomy. However, he asserted that although human fetuses may display gills, it cannot prove that humans have evolved from gilled animals. Rather, it simply shows the developmental process of humans in spontaneous generation. Then, there is also paleontological evidence. There are obvious gaps in fossil evidence, which does not support graduated evolution over time. It's more reasonable to assume that in a particular period there was a particular environment in which certain types of life-forms emerged. When the environment changed, old life-forms died out and new ones appeared—hence the abrupt changes in successions of life-forms. Moreover, there is also the evidence of blood types. If humans have evolved from the same ancestors, why are there four different blood types among humans? Isn't it likely that they might have actually originated from different environments? Evolution suggests that humans evolved from mon-

---

32   Zhu Xi disputed Luo on this point in Dong 1933, 137. See also Hopwood 2006.

keys. If so, why don't monkeys evolve into humans anymore? Who has seen that?[33]

Luo's general conclusion was that life-forms come into existence from spontaneous generation. Some of these life-forms, once in existence and when faced with unfavorable circumstances, may develop means of producing progeny. This does not mean that these organisms begin to follow the course of evolution, however, because Luo believed in the immutability of species.[34] Here Luo parted ways with Lamarck. He thought that both Lamarck and Darwin were wrong in their beliefs in the evolution of life-forms. In Luo's view, life-forms and species are fixed. Particular life-forms emerge in a particular environment through spontaneous generation, and they don't really change.

Criticisms of Darwinism in themselves were nothing unusual in Europe as late as the first decades of the twentieth century.[35] While most scientists agreed on the phenomenon of evolution, they had diverse opinions on its primary mechanism. Modern evolutionary synthesis was still years away. In France, neo-Lamarckism remained strong, and some of Luo's critics, including Zhu Xi, were not free of its influence.[36] (They were equally familiar with neo-Darwinism, such as the work of August Weissmann.)[37] Luo's position was extraordinary in that he went so far as to criticize the notion of biological evolution itself (and not simply the details or specific mechanisms of any one particular evolutionary theory). Luo's opponents clearly found his dismissive treatment of evolutionary theory outrageous. "Which evolutionary theorist said that humans had evolved from monkeys?"[38] "Evolution is the phenomenon of the unfolding of the biological world over a vast period of time. How can humans [in their brief existence] see the changes?"[39] Different blood types can be found in the same group of people, even in the same family or clan. How could they be products of different environments of the spontaneous generation of humans?[40] They saw themselves as guardians of science no less than their colleagues did ten years before in the debate over science and the philosophy of life, when Darwin's evolutionary theory was also a point of contention. Zhang Junmai (張君勱), who unwittingly stirred up the whole controversy, was a fol-

---

33   Luo 1933c   36–38.
34   Luo 1933c   35.
35   Bowler 1992. From the perspective of scientific controversy, the famous case of Paul Krammerer might be of interest to some readers (Gliboff 2006).
36   Farley 1974; Glick 1988; Corsi and Weindling 1988.
37   Dong 1933, 11.
38   Dong 1933, 140, 176.
39   Dong 1933, 20.
40   Dong 1933, 17–18, 139–140.

lower of vitalism *à la* Hans Driesch and Henri Bergson. His skeptical remarks on Darwinism irritated his critics, some of whom considered the philosophy of Driesch and Bergson mere shibboleth. Luo went much farther than Zhang in challenging not only Darwinism but also biological evolution in general. He also expressed no sympathy for vitalism.[41]

There is no denying that Luo could be cavalier in venturing opinions and offering evidence. This might have had something to do with the fact that his essays were written for a general audience, not specialists (and his critics readily acknowledged this possibility). Nevertheless, his statements about humans and monkeys, blood types, and the like were so extraordinary that one can hardly see how a doctor trained at prestigious institutions could have made them. Similarly, Luo's opinion about spontaneous generation was a little startling. It is true that many conjectured about how life first arose on earth. Both Luo and his foes agreed that unless one accepted a theistic explanation or the extraterrestrial origin of life, one must come to the conclusion that life first originated from nonliving matter. However, few accepted spontaneous generation of the kind that Luo championed. Luo's critics vehemently opposed the idea that organisms—indeed complex organisms like arthropods—came into being in test tubes in a matter of weeks.[42] The question here is not whether Luo was right or wrong, which is not crucial to our inquiry, but why did his opinions depart so far from the commonly accepted scientific views at the time and why was he so careless about building up a persuasive case (i.e., persuasive to his peers)? A person who knew Luo well in his later years told me that Luo had a stubborn and self-confident personality.[43] Apparently, he was not someone who yielded or gave up easily, although even his critics admitted that his manners were typically gentle and friendly.[44] In the early 1930s, Luo was convinced by what he saw in his experiments. And for some reason and perhaps quite independently, he came to the conclusion that evolutionary theory was mistaken. Whichever came first, both theories merged into an overarching and highly idiosyncratic biological and medical framework. Luo probably did not expect that his essays would ignite so much hostility and scrutiny. Otherwise, he would certainly have been more cautious about what he said in

---

41 Yadong tushuguan 1925 [1990], esp. the essays by Zhang Junmai and Ding Wenjiang. Earlier in the century, Li Wenyu, a Chinese Catholic, had tried to refute Darwin's evolutionary theory on scientific and theistic grounds, but Zhang's and Luo's critiques did not come from a religious perspective. Nevertheless, it is still possible that Luo's skepticism of Darwinism was a by-product of his Catholic school education. Wang Tiangen 2007.

42 Secord 1989.

43 Personal conversation with the writer Su Wei at Yale University, January 29, 2010.

44 Dong 1933, 123.

his initial publications. Once the attack on him started, he felt bullied and became defensive, resistant, and combative.

Battles of Words

Being a doctor at a private hospital, Luo was on the periphery of the scientific establishment. When the university professors went for him, Luo had to enlist his own sources of support in order to deflect his opponents. He found the local newspapers a good ally. Several newspapers in Guangzhou and Hong Kong printed positive reports and editorials on his work throughout the controversy.[45] Luo even brought a newspaper reporter with a camera to the official open experiment (more on the experiment later), much to the surprise of the other panelists.[46] For the committee had decided that the presence of the media was unwelcome and that no panelist should leak any information to newspapers before the official report and meeting two months later. The next day, however, local newspapers already reported favorably on Luo's conducting of the experiment. According to Luo's opponents, Luo admitted that he had spoken to reporters on the phone about the experiment. We cannot know what actually happened. But the fact that throughout the controversy, several local newspapers published articles sympathetic to Luo indicates that Luo had good relations with the local press. At one point, Luo even gave an interview

---

45   The report in a local newspaper (*Xianggang gongshang ribao*, July 1, 1933) a couple of days after the first public meeting may serve as an example. The article simply repeated what Luo had told the reporter: that he had solid scientific evidence, that his goal was to make a contribution to science, that he had done the experiments repeatedly and diligently ("getting out of bed every fifteen minutes to take observations"), and that he "loved truth and especially scientific truth." "I am a doctor. I could have simply focused on medical routines rather than devoting my heart and energy to scientific research and discoveries." All these points were highlighted as headings and in bold font. The article also reported that Luo felt strongly about moving China's science ahead and striving to catch up with the West. He dismissed the questioning and debating during the public meeting as meaningless, comparing it to village women squawking. He related with a sense of irony that in the public meeting, when he was baited by the so-called experts, he looked up and saw a banner on the wall of the assembly hall of Zhongshan University that announced, "We must catch up with the front rank of world science." The article ended with a description of Luo's laboratory and experiments. The reporter was invited into the laboratory. He looked through a microscope at certain test tubes and saw that they were teeming with life—numerous organisms were wiggling inside.

46   Dong 1933, 297.

with newspaper reporters at his hospital—a press conference!—in which he presented his side of the story and defended his experiments.[47]

Luo's opponents did not stand idly by and let Luo rule the press war. The explosion of magazines, newspapers, and other press media in the late Qing had fostered a realm of public opinion that could exert intellectual, cultural, and political influence.[48] Luo's critics reckoned that they could not let Luo control public opinion, for it might steer the course and outcome of the controversy. If that happened, their continual plead to the government for action might be ignored, and their criticisms of Luo's research might run into a wall of suspicion and hostility. They wrote to the editors of the magazines in which Luo published his pieces.[49] They sent articles to newspapers to counter the reports and editors favorable to Luo.[50] Zhu Xi and Dong Shuangqiu were particularly diligent in waging the press war; Zhu even published a line-by-line rebuttal to Luo's translation of the letters from French colleagues included in his pamphlet on spontaneous generation. Zhu and Dong didn't fight this battle alone.[51] A few others also challenged Luo in the popular press, notably the writer Ba Jin (巴金), who was gaining a literary reputation for his novels and short stories.

Ba Jin had known Zhu Xi for quite a while. Both were products of the Qinggong jianxue work-study program and both shared anarchist sympathies.[52] At the time of the scientific controversy, Ba Jin was visiting Zhu in Guangzhou. He frequented Zhu's laboratory, observed the latter's experiments on parthenogenesis in frogs, and listened to him dissect Luo's theory of spontaneous generation.[53] In fact, Ba Jin was in the audience of the public meeting

---

47 Dong 1933, 298.

48 E.g., Wagner 2001; Reed 2004; Mittler 2004.

49 Consequently, the editor of *Kexue*, Bing Zhi, a renowned zoologist, expressed regret over publishing Luo's paper at all. Since *Dongfang zazhi* was not a journal of the scientific community and had always been eclectic in its views, it was more resistant to the campaign waged by the Zhongshan University scientists. The magazine published a short essay by someone named Lu Xuanzhi that both criticized and commended Luo's article. Dong Shuangqiu and Zhu Xi blasted the editor of *Dongfang zazhi* for publishing the essay and reproached Lu and his piece (Dong 1933, 182–185, 309–310).

50 To their credit, the newspapers published them. However, it is fair to say that, overall, the local newspapers were sympathetic to Luo. As a successful doctor, Luo seemed to enjoy good social standing and esteem. His deeper local ties certainly made a difference.

51 Some of Dong's and Zhu's newspaper articles are collected in Dong 1933, 116–161.

52 Zhang Zhijie 2008.

53 Ba 1995, 176; Tang and Zhang 1989, 301, 324, 327; Ba 1989, 181–185.

in June 1933 described at the beginning of the present essay.[54] To assist Zhu's campaign against Luo, Ba Jin published a criticism of Luo's work in a popular magazine for middle school students, *Zhongxuesheng*. Ba Jin had first submitted his paper to *Dongfang zazhi*, but it was rejected on the ground that its tone was too strident. According to Ba Jin, even when he then sent the paper to *Zhongxuesheng*, the editor of *Dongfang zazhi* still wouldn't drop the matter and demanded that *Zhongxuesheng* prune the text. Years later, Ba Jin would call this experience his first brush with expurgation. Regarding biology, Ba Jin's essay didn't add much to what others had said about the topic; it was basically a simplified version of a long article by Zhu published in a science journal put out by Zhongshan University.[55] However, Ba Jin's piece highlighted a theme that would have been familiar to all Chinese intellectuals at the time—"Save the youngsters [from the bane of Luo's theory]!", a variation of Lu Xun's famous cri de coeur, "Save the children!", and a stab at Luo and his enablers (e.g., certain magazines and newspapers).[56]

Experiments

Exchanges in words could do only so much, however. Since the controversy began with disagreement over experiments, it seemed that it could be settled only in the laboratory. A newspaper article opined, "The present debate is important and it is different from the debate over science and the philosophy of life. This debate is not simply arguments on paper. It is about experiments."[57] As the result of the mobilization of Luo's opponents, a panel was formed to supervise an open experiment. The Committee on Supervision and Custody of the Open Experiment consisted of nine members; Luo, Zhangshan University (i.e., Luo's opponents), and the committee chair, Deng Zhiyi (鄧植儀), each recommended three members. Deng recommended Peng Li (彭利), a commissioner of the Laboratory of the City Bureau of Hygiene; Huo Qizhang (霍啟章), president of the Second Mental Hospital of the city; and Li Guochang (李國昌), a school inspector in the Department of Education. Deng Zhiyi was an education official and a respected agriculturalist. Peng Li, also known as Peng Huali (彭華利), obtained a master's degree in microbiology from the

---

54   Ba Jin claimed that he had personally heard Luo's reply to a question posed by Dong
       Shuangqiu in the meeting (Dong 1933, 173).

55   Dong 1933, 37–72.

56   Dong 1933, 162–181.

57   Luo 1933a   41.

University of California at Berkeley and was known for inventing a new method to produce smallpox vaccine. Huo Qizhang had a medical degree from the University of Lyon. Li Guochang was selected because of his official position as a school inspector. Zhongshan University recommended Dong Daoyun (董道蘊), a physician and editor of a local medical magazine, *Dazhong yikan*; Li Qifang (李其芳), who was a military doctor and a professor in the university's Medical School; and Zhu Xi. Besides himself, Luo recommended He Chichang (何熾昌), director of the City Bureau of Hygiene; and William Cadbury, superintendent of the Canton Hospital, a venerable establishment founded by American missionaries in the 1830s.[58]

The main task of the committee was to supervise an experiment. The experiment was to take place in the Laboratory of the Bureau of Hygiene, and the duration of the experiment would be two months. The panelists would agree on the procedure of the experiment. It must be noted that this "open experiment" (*gongkai shiyan*) was not open to the public; it was open only to the selected experts—who presumably possessed the knowledge and personal qualities essential for qualified expert witnesses—although the final experimental result would be announced to the public. After the experiment was set up, the committee would write its first report, detailing the purpose, process, and arrangement of the experiment. The object of the experiment was the arthropod referred to in Luo's pamphlet on spontaneous generation.[59] (Luo did not identify the creature, but based on Luo's illustration, Zhu believed that it was the cheese mite, a common mite that often infests grain, flour, cheese, and other foodstuffs.)[60]

Now, we come to the part over which the two sides would lock horns: did the committee have the official power to deliver a verdict on Luo's theory and experiments? If the committee supervised the arrangement and execution of the open experiment, and after two months there was no evidence of spontaneous generation, could it then determine that the experiment disproved Luo's theory? Luo's foes insisted that the committee should have the power, but Luo strenuously opposed it. They also disagreed over how long the experiment should be—two months (Luo's critics) or six months (Luo). It seems that both sides eventually made compromises. The committee would not have the authority to deliver a verdict on Luo's theory and recommend punishments, and the experiment would last for two months. After the committee examined

---

58 Cadbury and Jones 1935.

59 For accounts of the arrangement, purpose, etc. of the open experiment, see, e.g., Dong 1933, 5–9, 337–344; Luo 1933a   6–20.

60 Dong 1933, 107–109.

the result of the experiment, there would be a second report and a second public meeting, in which the result would be announced.[61] With the procedure and authority clarified, the panel convened in the Laboratory of the Bureau of Hygiene on July 23, 1933, to set up the experiment.[62]

There was another public experiment on spontaneous generation going on in the summer of 1933. When the faculty and students of the Zhongshan biology department reckoned that they should dispute Luo's claims, one of their actions was to start an experiment of their own and invite the public to see. The experiment served two main purposes. First, it aimed to counter Luo's theory of spontaneous generation; that is, it would prove the point that if an experiment was properly designed and conducted, then there shouldn't be any signs of spontaneous generation, even in tropical China. Second, the experiment also served the pedagogical purpose of repeating Pasteur's experiments, through which the students would learn how to design and conduct a biological experiment. It must have been very exciting for them to participate in a public controversy while doing something "relevant." They designed and arranged dozens of different glass flasks and setups from April till June, and the experiment would last for more than three months and end on October 20. The experiment was on display in three large cabinets in the hallway of the biology department and was open to the public. "We welcome anyone who is interested in the problem to come and observe the experiments."[63]

The Face-Off

Let us return to the official open experiment. Luo insisted that the experiment had to follow his design and procedure; otherwise, he would not be willing to take part in the experiment because it would not necessarily work. This appeared to be a reasonable request: how else could one evaluate his experiment if he wasn't allowed to replicate it? Luo prepared about one hundred test tubes (with agar, beef broth, and egg white and yolk as culture mediums) and submitted them to sterilization an hour a day, for three days. He then raised the

---

61   Luo suggested that the committee focus on three aspects of the open experiment: (1) the
      sterilization procedure; (2) the methods used to protect the contents from contamina-
      tion; and (3) the length of the experiment, which he argued should be six months. But the
      Zhongshan professors felt that six months was too long and wanted the experimental
      period to be two months. See *Xianggang gongshan ribao*, July 1, 1933.

62   Luo 1933a   6–20; Dong 1933, 5–9, 337–352.

63   A detailed report on the experiment is included in Dong 1933, 73–93. The quotation is on
      p. 77. Liang (1997) claims that the experiments remained on display until 1937.

room temperature to 34 degrees Celsius (by adding light bulbs to the room). After six days, he retrieved the test tubes with the panel present. He then kept the test tubes at regular room temperature for five days. On August 5, the panel convened again and under their eyes Luo proceeded to seal the test tubes. According to his account, he carefully sealed thirty-two test tubes with cotton swabs, filter paper, and wax, but since he was rushed and couldn't treat the remaining tubes in the same thorough manner, he simply corked them with cotton swabs (not an unusual method for similar experiments). He insisted that even these were well sealed (which his critics disputed). Luo then arranged the test tubes in various spots and positions—on the table, on the floor, and so on. The windows of the room were closed, but a water hole was left open. He also added three basins of water to the room. For comparison, five unsterilized test tubes and one with a small crack in the opening were placed outside the laboratory for observation. These were checked every few days. Luo also took several agar tubes with him for his own personal experiment—in these he would claim that he observed arthropods after three weeks or so.[64]

On the afternoon of October 4, the committee and two guests, Ernest Hartman, a professor of biology at Lingnan University, and Chen Yanfen (陳衍棻), president of the Guanghua Hospital, gathered in the laboratory to unseal and examine the test tubes. They were greeted by a chaotic scene. Many test tubes (especially those placed on the floor) were broken or otherwise damaged. The seals were punctured—probably chewed open by mice. But there were two or three dozen that looked undisturbed. The experts in the room proceeded to examine these. Hartman would observe the contents of one tube under the microscope and then Luo would do the same. On that day, the panel couldn't finish examining all the test tubes due to time constraints. So, the committee decided to meet again the next morning, which they did.[65]

Luo and his opponents would probably agree on the bare-bones account above of what happened that day. After that, there were many sharp disagreements. According to Luo, on examination, more than ten test tubes contained "plants and animals of the lower order" (e.g., mold, clusters of eggs, and even mites).[66] But the other panelists dismissed the claim. Luo alleged that the whole process of examination was hasty, perfunctory, and incomplete. First of all, Hartman and the others didn't look carefully at the test tubes. There were eggs and even mites, but Hartman and certain panelists insisted that those

---

64　On the process of the experiment, see Luo 1933a 　6–20; Dong 1933, 337–345. On the test tubes Luo took with him, see Luo 1933a 　12.

65　Dong 1933, 8–9, 347–352; Luo 1933a 　12–17.

66　Luo 1933a 　1.

were bodies of organisms that had entered the test tubes from the outside. The same object that appeared to Luo to be an arthropod in the process of formation appeared to Hartman to be no more than a decomposed body of a long-dead mite. Due to lack of time, some test tubes were set aside for the next morning, "including [in Luo's words] one that was best sealed and now contained moving organisms."[67] However, on the next day, the panel glanced over those and quickly retreated to another room to draft the report. The test tubes were then packed up and transferred to the second public meeting that afternoon. Luo protested that the process of examination had not even been completed.

Not surprisingly, Luo's opponents disagreed with Luo on what had actually happened in the experiment. To begin with, Luo asserted that all the test tubes and their contents were well sterilized and well sealed, a point his critics took pains to dispute. They maintained that the experiment was already ruined before it even started. The agar tubes were prepared on July 23–25, but they weren't sealed until August 4, after the high-temperature/room-temperature observation that Luo insisted on doing. During the ten days or so, there were plenty of opportunities for the test tubes to become contaminated by organisms of various kinds. The whole experiment, as designed and conducted by Luo, was therefore meaningless.[68] What Luo's critics wanted to establish was that Luo was an ignorant and incompetent lab scientist. To rub it in, one of Luo's critics gave a detailed account of how Luo had attempted to extract egg white from an egg.[69] The author broke down the process into steps, commenting on the possibilities for contamination at each step. For example, Luo used his ungloved fingers to pick up an egg and rotate it over the flame of an alcohol lamp. Luo's critic argued that this procedure not only could introduce contamination (from hands) but also did not ensure adequate sterilization of the surface of the egg. He concluded from his observations that Luo obviously did not know how to properly conduct an experiment and that all Luo's experiments were therefore questionable and the results unreliable. Along these lines, Luo's opponents also attacked him for ignoring standard laboratory practices. He had not kept proper laboratory journals, recording the time, place, and other crucial data of the experiments he had done over the years.[70]

In addition, Zhu Xi attacked Luo's skill as a microscopist. Trained as an experimental biologist, Zhu routinely used the microscope in his research. As

---

67   Luo 1933a   3.
68   Dong 1933, 342. See Luo's counterargument in Luo 1933a   20–25, 29–34.
69   Dong 1933, 214–217.
70   Dong 1933, 336, 339–340, 345.

a doctor who had some background in tropical medicine, Luo must have been familiar with microscopy, too. But Zhu pounced on two problems he saw in Luo's microscopy. First, Luo used only low-power microscopes (capable of a magnification of a couple hundred times), which, according to Zhu, were deficient for accurate and precise observation for this kind of research. Zhu himself used microscopes with a power of more than one thousand times.[71] Second, Luo's drawings of microscopic observations were inadequate and even wrong. Luo did not provide the magnification scale. How could the viewer know the actual size of the grossly enlarged microscopic creature?[72] Moreover, Luo's depiction of the bug-like creature was poor and wrong in places. The legs were not divided into sections as arthropods' legs were. Also, how the legs connected to the body was drawn incorrectly. It was not easy to operate a microscope and draw a picture of a moving creature at the same time. Without proper skill, knowledge, and training, one could easily miss small but important details. By pointing out the problems, Zhu was claiming his credibility as a lab biologist and undermining Luo's. He also reproved Luo for not having identified the creature and studied its life cycle, as should be done in proper entomological research. Zhu declared that he had identified the creature from Luo's picture, which he believed was a cheese mite.[73] Moreover, he charged that Luo dismally misunderstood what he observed. He treated eggs, bacterial filaments, and mites, all of which were represented together in the illustrations, as though they were different stages of development of the same organism.[74]

Thus, microscopy and visual representation joined the other laboratory apparatus and practices, such as sterilization, in producing "evidence" for scientific observations and theory. Indeed, evidence could not be separated from the ensemble of objects, operations, and images.[75] Luo certainly believed

---

71 Dong 1933, 387. Zhu's microscopic illustrations included one of 1600× magnification (Dong 1933, 108), while Luo's illustration in *Gongkai* (1933a, inserted between p. 4 and p. 5) was noted as 180× magnification. To explain the difference, Zhu said that with his higher-power microscope, he could distinguish individual spores in the large spore balls, which Luo, using a low-power microscope, mistook to be objects formed from water.

72 Dong 1933, 106, 129.

73 Dong 1933, 107–111.

74 Dong 1933, 111. Zhu Jieping, a microbiologist and public health engineer whose time at the University of Paris and the Pasteur Institute must have overlapped with Luo's, but who appears not to have known him, also dismissed Luo's experiments. Interestingly—or perhaps not surprisingly, considering that he was a microbiologist—he also didn't give much stock to medical doctors' training in microbiological techniques. See Zhu 1934.

75 Dong 1933, 385–386. The introduction of microscopy to China opened up new empirical, epistemological, and conceptual possibilities for Chinese intellectuals. Chinese Buddhists,

that images would help the audience understand his observations and theory, yet he must have also felt that the images would be strong evidence. That was why he included in his pamphlets illustrations of the experimental setups and the organisms he observed. His critics similarly took the power of images seriously. They provided many drawings of the experiments—of the shapes of the test tubes and flasks and how they were hooked together—created by students in the biology department. In the first public meeting, Zhu Xi went on the stage armed with an illustration that he himself had drawn, showed it to the audience, and interrogated Luo on his microscopic images.[76]

The controversy over the experiment went deeper than skills and observations, however. It also had to do with how to design and interpret the overall purpose and meaning of an experiment. In explaining spontaneous generation and the outcome of an experiment, Luo cited "the environment" (*huanjing*) as the most important and general factor.[77] "The environment" was used to refer to the different conditions in which spontaneous generation occurred or did not occur, and if it occurred, the environment determined how fast the process was. For example, Luo concluded that it was easier for spontaneous generation to occur in tropical China than in Europe because the former was warmer and more humid; in other words, the environment of tropical China was more favorable to spontaneous generation.[78] In the open experiment, Luo placed the test tubes in various places and positions in the room, insisting that these differences would contribute to differences in the environment (and therefore the outcome of the experiment). Luo put some of the test tubes on the table, some on the floor, some upright, some tilted, and so on, all in the name of the environment. He also kept three basins of water and a light bulb in the room. His critics were not impressed. Although they had to let Luo set up the experiment in the way he desired, they scoffed at his elaborate arrangement and thought it was meaningless. All the test tubes were in the same room, in a space of a few meters, so how could the environment be so different from spot to spot? How did the environment differ when a test tube was placed on the table rather than on the floor? How did the environment differ depending

---

for example, turned microscopic observations into evidence for the compatibility of science and Buddhism. See Hammerstrom 2012. The literature on microscopy, observation, and/or visual representation in science is sizable. See, e.g., Wise et al. 2006; Daston and Galison 1992; Breidbach 2002; Gooday 1991; Hopwood 1999; Ratcliff 2009, esp. chap. 9.

76   Dong 1933, 107–109. Howard Hsueh-Hao Chiang examines the illustrations in Zhu's science books in Chiang 2008.

77   E.g., Luo 1933b   842.

78   Luo 1933b   841–842.

on whether a test tube was placed flat or tilted? Could the differences be specified?[79]

Basically, his critics complained that Luo never precisely defined what he meant by "the environment." What is the environment? It seemed to include temperature, humidity, and different kinds of mediums. What else? Light? The pH level? How to break down the environmental factors? Is it possible to measure or predict how certain changes in the environment affect the outcome of the experiment? Is it possible to have certain environmental factors under control and manipulate the others? Here Luo's critics were asking him for his operational definition of "the environment." In their view, Luo wielded "the environment" as a fool-proof explanation for why and how spontaneous generation takes place—when the environment is right, it happens; when it fails to happen, the environment is not right—yet "the environment" remained a vague, elusive, and all-embracing category that was never properly defined. It was, in the words of one of his critics, "a mysterious entity ... devoid of scientific meaning."[80]

From Luo's perspective, this demand for precision was premature and counterproductive. For "the environment" was simply too sensitive, subtle, and complicated to be easily defined. It was still more difficult to measure. Nevertheless, this problem could not negate the validity of his experiments. The purpose of his experiments was to prove spontaneous generation. If spontaneous generation was observed, then it was a fact. If it was thus proved, then his experiment was successful. Since the purpose of his experiments was not to determine how different factors affect the process or phenomenon of spontaneous generation, it mattered little if such factors were isolated and measured. The important thing was to make sure that the test tubes, growth mediums, and other objects used in the experiments were properly sterilized and sealed, and that suitable environments (difficult to define exactly but knowable from experience) were provided.[81] That was what his experiments were designed to do, and he believed that they succeeded. To his critics, however, these kinds of experiments were simply too crude, primitive, and simple-minded. They only revealed Luo's lack of understanding of rigorous lab experimentation.

---

79   Dong 1933, 209, 353.
80   Dong 1933, 144, 205–211, 339.
81   Luo 1933a   21–22.

What Was at Stake?

In the end, the committee decided against Luo. After checking the results of the experiment, they retreated to another room to compose the second committee report while Luo stayed in the laboratory and studied the rest of the test tubes under a microscope. Sitting alone in the laboratory, if not earlier, Luo must have felt that he was being ganged up on. In any case, when the panel finished their task, they packed and sealed the test tubes and transferred them to the second public meeting. Later, Luo would protest that the panel looked at only a small number of the test tubes and did so in haste. The panel, however, would deny this and rebut his accusation by affirming that they had examined all those undisturbed by mice; the rest of the test tubes were deemed contaminated and set aside. If Luo chose to examine them, as he did by himself, that was his own business and should not affect the panel's decision.[82] I shall not try to solve the discrepancy here.

In any case, at 1 p.m., the members of the panel moved to the assembly hall of Zhongshan University. There, a large crowd had already gathered, eagerly waiting. The panel sat at a table in the front on which were placed specimens, illustrations, and three microscopes. The second public meeting began. Dozens of Luo's supporters—identified by Luo's critics as his students at the Guanghua Hospital—attended in eagerness and solidarity. They wanted to speak. The chair insisted that the purpose of the public meeting was for the experts—that is, the panelists—to announce the results of the official experiment. It was not an open discussion for all. Thus, only the panelists were permitted to speak at the meeting. According to Luo, he had only ten minutes to make his case, while Zhu Xi and Dong Shuangqiu delivered an hour-long tirade. One of Luo's supporters stood up and protested that "I have done research on spontaneous generation. I also have expertise on this topic. Why am I not allowed to speak?" Grumbles and hecklings created confusion. Finally, Luo walked to the podium, picked up a microscope, and set it down heavily on the table. "If it has to be like this, what's the point of doing experiments?" At this point, Luo and his supporters, a couple dozen of them, walked out of the room in a huff.[83]

---

82   Dong 1933, 230–231, 329–330, 370–372.

83   Dong 1933, 306–308, 335; Lin Youying 1933. Lin Youying is most likely a pseudonym. In the article, the author assumes the persona of a student attending Zhongshan University who happened to pass by the public meeting. I am not convinced that this was the case. The point and tone of the article are so similar to those of Luo's supporters in the local papers that it might actually have been written by one of them and sent to *Libailiu*, a popular magazine based in Shanghai, for publication.

Luo's frustration was due in part to his perception of the injustice done to him. He felt that the open experiment had departed from the realm of scientific research and had become a tool to attack him personally. In his opinion, the committee and the whole process were controlled by a few stubborn, prejudiced individuals. He also disputed the notion that the committee had the authority to determine the validity of his experiments over the past few years. Science was not a court trial, and the committee report was not a verdict. Luo argued that this kind of attempt to silence or punish him infringed on his freedom of scientific research, a political move that would only damage the progress of science. Science is about facts and truth. No subjective view, nor the power of any group, can be the self-appointed judge of science.[84]

But his opponents also believed in the objectivity and purity of science. They also believed that science must transcend politics, and they accused Luo of violating this principle. In his pamphlet on spontaneous generation, Luo secured the endorsement of major intellectual and political figures, such as Cai Yuanpei (蔡元培), Yu Youren (于右任), and Chen Jitang (陳濟棠).[85] Requesting endorsements by major politicians or other famous people for one's books was not unusual in China at the time. Moreover, Luo was a medical doctor. In China, a doctor often received plaques praising his skills in stock phrases (such as "Huatuo reincarnated" and "Skillful hands bringing back the spring"). Therefore, Luo might not have felt that including blurbs from dignitaries violated the code of conduct in science or at least in medicine, his profession. All the same, Luo's critics argued that what he had done was distasteful. A modern scholar, a true scientist, should not and would not seek these kinds of traditional trappings. A scientist pursues truth. Political dignitaries are not scientists. If they are not scientists, what they say is irrelevant to the value of a scientific work.[86] What Luo did only confirmed that he was not a true scientist. His critics thus portrayed him as someone fishing for gain and for fame. In the eyes of his critics, Luo committed another scholarly crime. He published his research results, not in standard scientific venues, but in newspapers, self-

---

84   Luo 1933a   20–21, 24–25.

85   Luo claimed that he had met Cai Yuanpei on one of the latter's visits to Guangzhou and that Cai took an interest in his work and encouraged him to publish his research results. Although I cannot find any mention of Luo in Cai's collected works and chronological biography (*nianpu*), I see no reason to doubt the truthfulness of Luo's claim. Cai's schedule was always crowded. He met many people every day. As the great promoter of science and education in China, he was invariably generous, eclectic, and encouraging. It's entirely plausible that he met Luo on a public or social occasion, heard the young doctor give a brief account of his research, and responded positively.

86   Dong 1933, 1, 36.

printed pamphlets, and general magazines. Worse, he even set up a display at the Guangzhou City Exhibit, a commercial event, which they considered to be blatant commercialization of science.[87]

Both Luo and his critics also concurred that science and modernity in China called for a new breed of intellectuals. In their petition and other writings, Luo's critics argued that traditional Chinese intellectuals, the literati, were muddle-headed and lacked public spirit. They lived by such mottoes as "sweeping only the snow in front of one's own door" and connived at the wrongs and evils in society.[88] That would not do. Luo's opponents asserted that professors have responsibilities to society and to the nation. They should expose what is wrong in society. Similarly, the government has the responsibility to correct wrongs and punish the perpetrators. According to his critics, a case like Luo's was particularly dangerous in the China of their day. They believed that Chinese society was backward, unenlightened, and that most Chinese were credulous about the strange and the mysterious. What Luo did was no different from a magic show—a conjurer performing the goldfish trick with a rice bowl.[89] It confused the common people, who do not know science and lack discernment. Luo was a charlatan and his tactics of publicizing his experiments were the same as common quack advertisements for aphrodisiacs.[90] His critics also had in mind the popularity of spiritualism in China some years before, a phenomenon that had similarly caused alarm among many modern-minded intellectuals; moreover, their rhetoric chimed with the GMD modernization project in Guangdong, which included a campaign against superstition.[91]

Their view of science was so lofty that they even took a swipe at another group of scientifically minded Chinese intellectuals. They remarked that of course "the ghost of metaphysics" (*xuanxue gui*) should be banished, as many had advocated. But science was much more than that. In China, however, scientific development was still at such a rudimentary stage that many of those who claimed that they were doing science were actually imposters. How could *zhengli guogu* (reorganizing the national heritage) and the like count as science?[92] It was simply rummaging in old books. This view helps explain a main concern of Luo's critics. Compared with Hu Shi and other similar cham-

---

87   E.g., Dong 1933, 33–34.

88   Dong 1933, 301–305, esp. 304.

89   Dong 1933, 103.

90   Dong 1933, 1; see also 395–396.

91   E.g., Dong 1933, 195–196; Huang 2007.

92   Dong 1933, 248–249.

pions or ideologues of science, Luo had a far better claim to be a scientist. He was a medical doctor educated at highly respected scientific institutions in Europe. Perhaps it was precisely for this reason that his critics felt that he posed a real threat to the development of science in China. Luo had glittering credentials, and what he advertised as his scientific discovery had resulted from laboratory work, complete with test tubes, microscopes, illustrations, and scientific terminology only specialists could pronounce. Interestingly, both Luo and his supporters also spoke from science and modernity; in fact, they used some of the same language as that of Luo's opponents. To them, this controversy displayed some of the historical ills of the Chinese intellectual and scholarly community—rampant factionalism, personal jealousy and envy, and the ghost of Confucianism (i.e., the tradition that a particular school of thought dominated and stifled intellectual life).[93] This implied that the opponents of Dr. Luo hounded him because they mindlessly followed the scientific doctrine and because they still had the pettiness of traditional Chinese intellectuals. They lacked creativity but were quick to criticize other people's work. Chinese intellectuals, Luo and his supporters argued, don't use their hands, don't use their eyes. They don't do experiments. Instead, they simply accept whatever the books or their teachers tell them. Luo himself certainly thought so and thus impugned his peers: many Chinese scientists received education abroad but accomplished nothing. They came back, regurgitated whatever they learned in a foreign country, fed that to their students at Chinese universities, and never did any real scientific research. They had no ambition and were already looking for a burial place for the coming day.[94]

Here he distinguished between two types of modern Chinese scientists: the few who were worthy of the name and the many who were fakes. An authentic modern scientist was one who did real experiments, experiments that aimed to make discoveries and advance science, and was one who used his hands and his eyes. In Luo's view, a whole class of fraudulent modern scientists—embellished with a foreign patina but stuffed inside with old, rotten straw—sleepwalked through the professorial lives. Luo spoke from the margin and from the high ground at the same time. His critics portrayed him as an irresponsible charlatan. He retaliated and called them shams. Both sides erected an ideal—namely, the authentic modern scientist, pure and uncontaminated, a genuine departure from the traditional Chinese scholar, an intrepid seeker of empirical truth, a new intellectual who must help shoulder the future of the new nation.

---

93   Luo 1933a   45; see also 39–40, 42.

94   Dong 1933, 392–394.

Several editorials and articles in local newspapers echoed this sentiment.[95] Science is not politics. The scientific truth is not determined by a majority vote, and the majority opinion is not necessarily right.[96] In fact (one newspaper article opined), history tells us that the progress of science—and, indeed, the progress of society—has been made by revolutionaries, people whose ideas were considered absurd and people who were misunderstood, ignored, criticized, or ridiculed by the majority. If there hadn't been Sun Yat-sen, where would China be today? Would we have our Republic? Wasn't Sun Yat-sen slandered and attacked for his revolutionary ideas and actions? (This analogy must have been very potent in Guangdong, where Dr. Sun was widely revered.) Scientific theories, including the theory of evolution, are accepted in China as truths without critical examination. In fact, evolutionary theory has its problems. Hasn't the notion of survival of the fittest provided a cover for imperialism?[97]

In a few sentences, the article wove a heroic narrative of science together with a heroic narrative of nationalism and anti-imperialism. Luo's experiments and theory could be right or wrong, but in his unwillingness to follow the accepted opinion, he was a true scientist. He might well be a scientific visionary just as Dr. Sun had been a political visionary. But there was more. "People of Guangdong are not united. That's why we Guangdong people are dominated by people from outside."[98] The article admonished its readers to defend Luo, whose family had originated from Guangdong, against the attacks from the outsiders (members of the biology department at Zhongshan University). This statement had immediate political references. Explicitly, it was referring to the attack of the Guangxi clique (a warlord coalition) on Guangdong a couple years before, but to many readers of the newspapers, it must also have pointed to the current conflicts between Chiang Kai-shek and the political coalition led by Chen Jitang, a native leader.[99] Regionalism was strong in Guangdong and was intertwined with political factions within the GMD at the time. The local political power distanced itself from and even challenged Chiang Kai-shek's

---

95 Some of the pieces are included Luo 1933a   38–45. It should be noted that we do not know who actually wrote these newspaper editorials and articles, for the bylines were either noms de plume or anonymous. Regardless, once an editorial or article was published in a newspaper, it became a voice in the public domain. See also Lin Youying 1933. Lin's piece, sent as a news story from Guangzhou, was published in a major popular magazine based in Shanghai.

96 Luo 1933a   39.

97 Luo 1933a   43.

98 Luo 1933a   44.

99 Fitzgerald 1997; A. Lin 2002.

authority. Therefore, the editorial essentially compared the controversy over spontaneous generation to the political struggle in Guangdong: the outsiders who staffed the faculty of the biology department of Zhongshan University were just like the outsiders who seized power in the GMD and elbowed out the party elders, many of whom were from Guangdong.

Conclusion

Hence, both sides held notions of what authentic scientists and science should be like, though they defined them somewhat differently. Luo and his supporters projected the idea that any scientists worth their salt should fiercely pursue original research regardless of what the scientific establishment said. It was a trope of the scientist as a lone hero. This modern hero stood apart from traditional Chinese scholars, who listened unimaginatively to their teachers when young and dictated verbosely when old. A genuine scientist loved truths and facts, valued originality, and used his or her hands and eyes to investigate nature. Thus, the modern Chinese scientist was defined as a new type of intellectual who not only assumed a new cultural identity but also had a new body (privileging eyes and hands over ears and mouth), a body that was crucial to the acquisition of scientific knowledge. The scientific experiment was grounded in this *embodied* epistemology.[100]

Luo's opponents shared many of these ideas. They also believed that they were a new type of intellectuals and experts. They similarly dismissed traditional Chinese scholars as old fogies who were trapped in thoughts and ideas of bygone eras and who lacked social consciousness. To ensure its place and authority in society, science must be demarcated from other realms of activity. Science had to keep its purity, shielded from contamination by politics, commerce, personal greed, and uninformed public opinions and emotions (whether they were newspaper editorials or outbursts from an audience). Consequently, the epistemological authority of science—which was perceived to be vital to the progress of society and nation—also depended on the moral characters of its practitioners. The scientific community must uphold the standards by which scientific claims and individual scientists were judged. Adulterated science was deceptive and dangerous and had to be exposed. If a gentleman of science in seventeenth-century England subtly invoked his class and virtues, elite scientists of Republican China similarly presented themselves in a particular persona—one that embodied modernity, expertise and

---

100   Dong 1933, 247–248.

credentials, national hopes, public spirit, and moral rectitude.[101] Perhaps in a period of political and cultural instability, confusion, and rapid change—when the ground seemed to be shifting all the time—it became all the more necessary to cling to a sense of authenticity. There ought to be ways to determine what was good and valuable from what was bogus or merely faddish.

In science, the experiment was an established method for determining what was factual, reliable, and true. Both Luo and his critics accepted this principle. Yet, as science studies have taught us, experiments are complex processes whose outcomes are socially embedded. On the surface, the core of the controversy over spontaneous generation was about whether Luo's experiments could be replicated or not. However, just as in many other controversies over scientific experiments, the two sides were unable to reach a straightforward conclusion. They wrangled over a range and diversity of factors—not only skill, knowledge, and instrumentation but also cultural values and personal characters. Both sides also mobilized political and social resources against each other. Indeed, Luo's opponents wanted not only to settle the differences over scientific theory and experiment but also to bring the whole affair to a close by punishing or silencing Luo. Ultimately, they did not see this simply as a "scientific" matter (i.e., a matter within the scientific community), for, in their view, Luo had already transgressed the boundaries of science; he had debased science with commerce, politics, and personal avarice and vanity.[102]

However, it would be self-defeating to limit science to the select few. In order to introduce science into China and establish themselves in Chinese culture and society, the scientists believed that they had to recruit an audience and raise public interest in science. That was why many scientists regarded the popularization of science—in addition to institutional education and research—as an important part of their work. To reach out to the public and to bring it to the world of science—through various popular media and other venues—should, nevertheless, not be allowed to threaten the authority and claims of expertise of professional scientists. In this view, scientists needed to appeal to the public and to maintain control over it at the same time. Therefore, in the case of the controversy over spontaneous generation, the scientists felt that limits had to be set as to who could come to the open experiment and who could express their opinions at the meetings. Boundaries had to be drawn around proper participants and procedures of science.

---

101   Shapin 1995. On the issue of scientific personae, see Daston and Sibum 2003. In the context of early twentieth-century China, see, e.g., Fan 2004 (with regard to the study of *bowu xue*, or natural history); Lam 2011; Shen 2009.

102   Dong 1933, 34–35.

The controversy must have caused Luo much trouble and distress, but it did not change his mind. He never gave up his theory of spontaneous generation.[103] Decades later, he still talked about it and conducted experiments in his house, where test tubes and bottles attracted the curious eyes of his neighbors' children.[104] Apparently, organisms still emerged from his agar tubes in the warm climate of Guangzhou.

Bibliography

Ba Jin 1989. *Ba Jin quanji* [Collected works of Ba Jing], vol. 12. Beijing: Renmin wenxue chubanshe.

___1995. *Ba Jin zizhuan* [The autobiography of Ba Jin]. Nanjing: Jiangsu wenyi chubanshe.

Bowler, Peter J. 1992. *The Eclipse of Darwinism: Anti-Darwinian Evolution Theories in the Decades around 1900*. Baltimore, MD: Johns Hopkins University Press.

Breidbach, Olaf 2002. "Representation of the Microcosm: The Claim for Objectivity in 19th Century Scientific Microphotography," *Journal of the History of Biology* 35: 221–250.

Cadbury, William Warder, and Mary Hoxie Jones 1935. *At the Point of a Lancet: One Hundred Years of the Canton Hospital   1835–1935*. Shanghai: Kelly & Walsh.

Chen Fu 2000. *Zhu Xi*. Shijiazhuang: Hebei jiaoyu chubanshe.

Chiang, Howard Hsueh-Hao 2008. "The Conceptual Contours of Sex in the Chinese Life

---

103   Luo's writings in the 1940s showed no signs of his backing down on the issue of spontaneous generation. See Luo 1946, 1947, 1948. He also continued his attacks on Darwinism and evolutionary theory and included among their awful ideological progenies imperialism, racism, Fascism, and Nazism.

104   Based on my conversation with the writer Su Wei at Yale University on January 29, 2010. Su's father and Luo were neighbors and close friends. Su remembers that when he and his siblings were children in the 1950s and early 1960s, they frequently went over to Luo's house. At the time, Luo still did experiments on spontaneous generation. Luo suffered during the Cultural Revolution. See Su 2009, 170. Xie Yan (2006, 105–106) states that Luo was against materialism. His writings of the 1930s, however, did not take such a position. He did say at one point in the 1940s that it would be counterproductive to separate the material and the spiritual, materialism and idealism, and mechanism and vitalism. They should all be brought together and subsumed under the science of physics and chemistry. It was a sweeping statement without substance made during an occasional speech. See Luo 1946. Zhong Shaohua (2005, 162) claims that Luo tried to rehabilitate his theory of spontaneous generation, together with his political status, after the fall of the Gang of Four.

Sciences: Zhu Xi (1899–1962), Hermaphroditism, and the Biological Discourse of Ci
    and Xiong   1920–1950," *East Asian Science, Technology and Society: An International
    Journal*   2: 401–430.
Claypool, Lisa 2005. "*Zhang Jian* and China's First Museum," *Journal of Asian Studies* 64,
    3: 567–604.
Collins, Harry, and Trevor Pinch 1993. *The Golem: What You Should Know about Science.*
    Cambridge: Cambridge University Press.
Corsi, Pietro, and Paul J. Weindling 1988. "Darwinism in Germany, France, and Italy," in
    David Kohn, ed., *The Darwinian Heritage*. Princeton, NJ: Princeton University
    Press   683–729.
Daston, Lorraine, and Peter Galison 1992. "The Image of *Objectivity*," *Representations*
    40: 81–128.
Daston, Lorraine, and H. Otto Sibum 2003. "Scientific Personae and Their Histories,"
    Science in Context 16, 1–2: 1–8.
Dong Shuangqiu, ed. 1933. *Xiancun shengwu ziran fasheng shuo zhi pinglun wenlu* [The
    collection of existing comments on spontaneous generation]. Guangzhou: [Biology
    department of] Zhongshan daxue.
Elman, Benjamin 2005. *On Their Own Terms: Science in China   1550–1900*. Cambridge,
    MA: Harvard University Press.
Fan, Fa-ti 2004. "Nature and Nation in Chinese Political Thought: The National Es-
    sence Circle in Early Twentieth-Century China," in Lorraine Daston and Fernando
    Vidal, eds., *The Moral Authority of Nature*. Chicago: University of Chicago Press
    409–437.
____ 2007. "Redrawing the Map: Science in Twentieth-Century China," *Isis* 98:
    524–538.
____ 2008. "How Did the Chinese Become Native? Science and the Search for National
    Origins in the May Fourth Era," in Kai-wing Chow et al., eds., *Beyond the May Fourth
    Paradigm: In Search of Chinese Modernity*. Lanham, MD: Lexington Books
    183–208.
____ 2013. "Circulating Material Objects: The International Controversy over Antiquities
    and Fossils in Republican China," in Bernard Lightman, Gordon McQuat, and Larry
    Stewart, eds., *The Circulation of Knowledge between Britain, India and China: The
    Early Modern World to the Twentieth Century*. Leiden: E. J. Brill   209–236.
Farley, John 1974. "The Initial Reactions of French Biologists to Darwin's *Origin of
    Species*," *Journal of the History of Biology* 7, 2: 275–300.
____ 1977. *The Spontaneous Generation Controversy from Descartes to Opurin*. Baltimore,
    MD: Johns Hopkins University Press.
Fernsebner, Susan 2006. "Objects, Spectacle, and a Nation on Display at the *Nanyang*
    Exposition of 1910. *Late Imperial China* 27, 2: 99–124.

Fischer, Jean-Louis 1990. "Experimental Embryology in France (1887–1936)," *International Journal of Developmental Biology* 34: 11–23.

Fitzgerald, John 1997. "Warlords, Bullies, and State Building in Nationalist China: The Guangdong Cooperative Movement 1932–1936," *Modern China* 23, 4: 420–458.

Gerth, Karl 2004. *China Made: Consumer Culture and the Creation of the Nation.* Cambridge, MA: Harvard University Asia Center.

Gieryn, Thomas 1999. *Cultural Boundaries of Science: Credibility on the Line.* Chicago: University of Chicago Press.

Gliboff, Sander 2006. "The Case of Paul Kammerer: Evolution and Experimentation in the Early 20th Century," *Journal of the History of Biology* 39: 525–563.

Glick, Thomas, ed. 1988. *The Comparative Reception of Darwinism.* Chicago: University of Chicago Press.

Golinski, Jan 1992. *Science as Public Culture: Chemistry and Enlightenment in Britain 1760–1820.* Cambridge: Cambridge University Press.

——— 1998. *Making Natural Knowledge: Constructivism and the History of Science.* Chicago: University of Chicago Press.

Gooday, Grame 1991. "'Nature' in the Laboratory: Domestication and Discipline with the Microscope in Victorian Life Science," *British Journal for the History of Science* 24, 3: 307–341.

Gooding, David, Trevor Pinch, and Simon Schaffer, eds. 1989. *The Uses of Experiment: Studies in the Natural Sciences.* Cambridge: Cambridge University Press.

Guoli Zhongshan daxue jiaowu bu 1933. *Guoli Zhongshan daxue ershiyi niandu gailan* [The 1933 yearbook of National Zhongshan University]. Guangzhou: Guoli Zhongshan daxue chubanbu.

Hammerstrom, Erik J. 2012. "Early Twentieth-Century Buddhist Microbiology and Shifts in Chinese Buddhism's 'Actual Canon,'" *Theology and Science* 10: 3–18.

Hopwood, Nick 1999. "'Giving Body' to Embryos: Modeling, Mechanism, and the Microtome in Late Nineteenth-Century Anatomy," *Isis* 90: 462–496.

——— 2006. "Pictures of Evolution and Charges of Fraud: Ernst Haeckel's Embryological Illustrations," *Isis* 97: 260–301.

Hu Zonggang 2005. *Jingsheng shengwu diaocha shuo shigao* [A draft history of the Jingsheng Institute of Biological Survey]. Jinan: Shangdong jiaoyu chubanshe.

Huang Kewu 2007. "Minguo chunian Shanghai de lingxue yanjiu: Yi 'Shanghai Lingxue Hui' wei li" [Research on spiritualism in Shanghai in the early years of Republican China: The example of the Shanghai Society of Spiritualism]. *Zhongyang yanjiuyuan jinduishi yanjiusuo jikan* 55: 99–136.

Huters, Ted 2008. "Culture, Capital, and the Temptation of the Imagined Market: The Case of the Commercial Press," in Kai-wing Chow et al., eds., *Beyond the May Fourth Paradigm: In Search of Chinese Modernity*. Lanham, MD: Lexington Books 27–50.

Jones, Andrew F. 2011. *Developmental Fairy Tales: Evolutionary Thinking and Modern Chinese Culture*. Cambridge, MA: Harvard University Press.

Kwok, D. W. Y. 1965. *Scientism in Chinese Thought   1900–1950*. New Haven, CT: Yale University Press.

Lam, Tong 2011. *A Passion for Facts: Social Surveys and the Construction of the Chinese Nation-State   1900–1949*. Berkeley: University of California Press.

Lean, Eugenia 2007. *Public Passions: The Trial of Shi Jianqiao and the Rise of Popular Sympathy in Republican China*. Berkeley: University of California Press.

Liang Qishen 1997. "Lishi de huigu: Guangdong sheng dongwu xuehui de chengli" [Looking back in history: The founding of the Zoological Society of Guangdong Province], *Zhongshan daxue xuebao luncong* 1: 3–4.

Lin, Alfred H. Y. 2002. "Building and Funding a Warlord Regime: The Experience of Chen Jitang in Guangdong   1929–1936," *Modern China* 28, 2: 177–212.

Lin Youying [pseud.] 1933. "Shengwu ziran fasheng de bianlun" [The debate over spontaneous generation], *Libailiu*   573–574, 690–691.

Lin Yusheng 1989. "Minchu kexue zhuyi de xingqi yu hanyi—dui Minguo shiernian 'kexue yu xuanxue' lunzheng de shengcha" [The rise and meaning of scientism in early Republican China—an examination of the debate over science and metaphysics in 1923], in Lin Yusheng, *Zhengzhi zhixu yu duoyuan shehui*. Taibei: Lianjing chuban gongshi   277–302.

Luo Guangting 1933a. *Gongkai shiyan shengwu ziran fasheng zhi jingguo*. Guangzhou: n.p.

_____ 1933b. "Shengwu ziran fasheng zhi faming" [The discovery of spontaneous generation], *Kexue* 16: 663–677, 841–855.

_____ 1933c. "Yong zhenping shiju lai dafu jinhualun xuezhe" [Reply based on concrete evidence to scholars of evolutionary theory], *Dongfang zazhi* 30, 8: 33–38.

_____ 1946. "Mingzhu, kexue, he xingyang" [Democracy, science, and belief], *Shidai gonglun* 8: 1–2.

_____ 1947. "Shengming qiyuan de yanjiu" [Research on the origins of life], *Dazhong yixue zazhi* 1: 17–21.

_____ 1948. "Shengwu youlai yu renlei mingyun" [The origins of life and the fate of the human race], *Keguan* 6: 19–24.

Luo Jizhou and Huang Shujin 2008. "Luo Guangting ji qi Shengwu ziran fasheng shuo jishu" [An account of Luo Guangting and his theory of spontaneous generation], in Guangdong sheng zhengxie xuexi he wenshi ziliao weiyuanhui, *Guangzhou wenshi ziliao cungao xuanbian*, vol. 7. Beijing: Zhongguo wenshi chubanshe. http://www.gzzxws.gov.cn/gzws/cg/cgml/cg7/200808/t20080826_4530.htm (last accessed August 25, 2010).

Luo Zhitian 1999. "Cong kexue yu renshengguan zhi zheng kan hou Wusi shiqi dui Wusi guannian de fansi" [Examining the reflections on the May Fourth concepts from the post–May Fourth era from the debate over science and the philosophy of life], *Lishi yanjiu* 3: 5–23.

Mittler, Barbara 2004. *A Newspaper for China? Power, Identity and Change in Shanghai's News Media   1872–1912*. Cambridge, MA: Harvard University Press.

Morus, Iwan R. 1998. *Frankenstein's Children: Electricity, Exhibition, and Experiment in Early-Nineteenth-Century London*. Princeton, NJ: Princeton University Press.

Nyhart, Lynn K., and Thomas Broman, eds. 2002. *Science and Civil Society*. Osiris 17. Chicago: University of Chicago Press.

Pusey, James R. 1983. *China and Charles Darwin*. Cambridge, MA: Harvard University Asia Center.

Ratcliff, Marc 2009. *The Quest for the Invisible: Microscopy in the Enlightenment*. Burlington, VT: Ashgate Press.

Reed, Christopher 2004. *Gutenberg in Shanghai: Chinese Print Capitalism   1876–1937*. Vancouver: University of British Columbia Press.

Roe, Shirley 1981. *Matter, Life, and Generation: Eighteenth-Century Embryology and the Haller-Wolff Debate*. Cambridge: Cambridge University Press.

Secord, James 1989. "Extraordinary Experiment: Electricity and the Creation of Life in Victorian England," in Gooding, Pinch, and Schaffer 1989, 337–383.

Shapin, Steven 1988. "The House of Experiment in Seventeenth-Century England," *Isis* 79: 373–404.

____ 1995. *A Social History of Truth: Civility and Science in Seventeenth-Century England*. Chicago: University of Chicago Press.

Shapin, Steven, and Simon Schaffer 1989. *Leviathan and the Air Pump: Hobbes, Boyle, and the Experimental Life*. Princeton, NJ: Princeton University Press.

Shen, Grace 2009. "Taking to the Field: Geological Fieldwork and National Identity in Republican China," in Carol Harrison and Ann Johnson, eds., *National Identity: The Role of Science and Technology*, Osiris 24. Chicago: University of Chicago Press   231–252.

Stewart, Larry 2007. "Experimental Spaces and the Knowledge Economy," *History of Science* 45: 1–23.

Strick, James E. 2002. *Sparks of Life: Darwinism and the Victorian Debates over Spontaneous Generation*. Cambridge, MA: Harvard University Press.

Su Wei 2009. *Zoujin Yelu* [Walking into Yale]. Beijing: Fenghuang chubanshe.

Tang Jinghai and Zhang Xiaoyun, eds. 1989. *Ba Jin nianpu* [The chronicle of Ba Jin's life]. Chengdu: Sichuan wenyi chubanshe.

Topham, Jonathan T., ed. 2009. "Historicizing 'Popular Science,'" *Isis* 100: 310–368.

Wagner, Rudolf 2001. "The Early Chinese Newspapers and the Chinese Public Sphere," *European Journal of East Asian Studies* 1, 1: 1–33.

Wang Shiping 2008. *Wuli xuehui shi* [The history of the Chinese Society of Physics]. Shanghai: Shanghai Jiaotong daxue chubanshe.

Wang Tiangen 2007. "'Tianyanlun boyi': Kexue yu zongjiao shiye zhong de jinhualun pipan" [Broad discussions on the evolutionary theory: Criticisms of the evolutionary theory between science and religion], *Shixue yuekan* 7: 92–101.

Wang, Zuoyue 2002. "Saving China through Science: The Science Society of China, Scientific Nationalism, and Civil Society in Republican China," in Nyhart and Broman 2002, 291–322.

Wise, Norton, et al. 2006. "Science and Visual Culture," *Isis* 97: 75–132.

Wright, David 2000. *Translating Science: The Transmission of Western Chemistry into Late Imperial China 1840–1900*. Leiden: Brill Academic Publishers.

Xian Rongxi 1933. "Ziran fasheng yu fei ziran fasheng zhi zhengbian" [The debate over spontaneous generation], *Shidai gonglun* 70: 23–26.

Xie Yan 2006. *Luohong huhua: Su Bo de gushi* [Fallen red blossoms protecting the flower: The story of Su Bo]. Beijing: Qunyan chubanshe.

Xue Pangao 2002. "1932 nian shengwu ziran fasheng shuo zai Zhongguo chenzha fanqi—yichang kexue tong fankexue de douzheng" [Flashback of the theory of spontaneous generation in 1932: The struggle between science and antiscience], *Zhongguo keji shiliao* 23: 1, 9–17.

Yadong tushuguan, ed. 1925 [1990]. *Kexue yu renshengguan* [Science and the philosophy of life], in Mingguo congshu bianjiweiyuanhui, ed., *Mingguo congshu*, vol. 1, pt. 3. Shanghai: Shanghai shudian.

Zhang Jian 2005. *Kexue shetuan zai jindai Zhongguo de mingyun: Yi Zhongguo kexueshe wei zhongxin* [The fate of science societies in modern China: Using the China Science Society as the primary example]. Jinan Shi: Shandong jiao yu chu ban she.

Zhang Jiuchen 2005. *Dizhixue yu minguo shehui 1916–1950* [Geology and society in Republican China 1916–1950]. Jinan Shi: Shandong jiaoyu chubanshe.

Zhang Yi, ed. 1934. *Guoli Zhongshan daxue chengli shi zhounian xinxiao luocheng jiniance* [The commemorative album for the tenth anniversary of the founding of National Zhongshan University and for the completion of the new campus]. Guangzhou: Zhongshan daxue chubanbu.

Zhang Zhijie 2008. "Zhu Xi yu wuzhengfu zhuyi—wei shengwu xue jia Zhu Xi zhuanji buyi" [Zhu Xi and anarchism—an addendum to the biography of biologist Zhu Xi]. *Kexue wenhua pinglun* 5, 3: 21–34.

Zhong Shaohua 2005. *Youyou lu* [Records of wanderings (in thought)]. Beijing: Xueyuan chubanshe.

Zhu Jieping 1934. "'Shengwu ziran fasheng' zhi jieshi chongti" [Revisiting the explanations of "spontaneous generation"], *Jiaoda tangyuan zhoukan* 64: 12–14.

Bridging East and West through Physics

William Band at Yenching University

Danian Hu[1]

Abstract

Trained in Liverpool, England as a professional physicist, William Band volunteered to teach and serve at Yenching University, an American-funded Christian mission college in Peking, between 1929 and 1941. Instead of evangelizing, however, Band came to China "to establish and promote cultural relationships between East and West" through modern physics. As a theoretical physicist at a leading university in China during the 1930s, Band mentored a large group of top native physicists (including several distinguished women), fostered the study of theoretical physics, and participated in the indigenous Rural Reconstruction Movement. William Band's story illustrates some previously overlooked aspects in the development of modern science in Republican China, highlighting transnational contributions by individual Western scientists and the special role played by mission colleges.

Introduction

Having landed in China in the aftermath of the notorious Opium Wars, the early evangelists discovered that they had little influence on the Chinese until some of them began to turn to educational work and founded mission schools.[2] These schools, particularly well known for their scientific curricula, developed quickly during the Self-Strengthening period (1860–1895), when Western technical and scientific learning and knowledge of foreign languages were in great demand. As a result, not only was the number of mission schools increased and their quality enhanced, but their social status was also greatly advanced in

---

1   I gratefully acknowledge the vital support of multiple small research grants from both PSC-CUNY Research Awards and CCNY President's Fund, which made my trips to various archives in the United States, United Kingdom, and China possible. I am very grateful to the support from the Institute for the History of Natural Science at the Chinese Academy of Sciences and especially the assistance provided by Professor Guangbi Dong, Professor Baichun Zhang, and Dr. Lie Sun. I thank Benjamin A. Elman and Jing Y. Tsu for inviting me to participate in the workshop held at Yale in January 2010 and to contribute to this volume. Thanks also go to Fa-ti Fan and all other participants of the workshop for their helpful comments and critiques.
2   Lutz 1971, 12–17; He and Shi 1996, 32–33.

China.[3] Moreover, their educational methods and academic programs often served as models for the newly established governmental schools.

In 1905, the Chinese imperial government finally decided to abolish the traditional civil service examination system completely in order to promote Western-style school education nationwide. This new policy was a double-edged sword for the mission schools; it cleared many barriers hindering their development but at the same time pressured them to enhance their competence.[4] As a pragmatic and attractive alternative for many to pursue Western learning, mission schools' enrollment soared from 16,836 in the early 1890s to 57,638 in 1906.[5] To maintain their superiority, especially in higher education, and prepare for the upcoming competitions with the new public schools, mission schools moved to concentrate their resources, merging, upgrading, or expanding smaller schools to form better colleges. By the early 1920s, Christian colleges in China had reached the pinnacle of their development: the number of students attending Christian colleges amounted to up to 80 percent of national enrollment in higher education; and of the twenty-four colleges existing then in China, sixteen (or nearly 70 percent) were Christian colleges. A key to the early success of these Christian colleges was their increasingly professionalized faculty: Christian institutions not only actively recruited missionaries with professional training in higher education but also eagerly sought professional scientists who had no desire to be missionaries, as demonstrated in this essay.[6] The recruitment of nonmissionaries was also likely a response to the secularist policy imposed by the Nationalist government.

Until the mid-1920s, the Christian schools in China were completely independent from the Chinese national educational system. With burgeoning nationalism and patriotism in post–May Fourth China, autonomous Christian schools, which openly promoted Western religions and were under the protection of extraterritoriality, became chief targets of public attacks. In the 1920s, the anti-Christian and "Restore Educational Rights" movements emerged first in Shanghai and Peking and quickly spread to other major cities. By the end of 1925, both the central and some provincial governments had officially demanded that all schools founded by foreigners be registered with the Chinese authority, requiring them to comply with "minimal prerequisites" such as "Chinese participation in the administration of the schools, elimination of requirements concerning religious worship and study, equality in the

---

3   He and Shi 1996, 43–50.
4   He and Shi 1996, 59.
5   He and Shi 1996, 59.
6   He and Shi 1996, 60–63.

treatment of Chinese and Western faculty members, and an acknowledgment that the purpose of the school was educational rather than evangelistic." [7] These terms, especially those concerning religion, of course met with protests from missionaries and mission college administrators. Nevertheless, most mission schools eventually accepted the government's demands despite their initial resistance. By the early 1930s, all but one Christian college had completed registration with the government.[8]

Not everyone accepted these changes passively. In fact, a few evangelists had long entertained "radical" ideas such as establishing mission colleges "in Chinese life independent of treaties with western countries or any other extraneous factors, with only such protection as the Chinese people themselves possessed and wanted to share with [the missionaries]." A leading representative of this group was John Leighton Stuart (1876–1962), the founding president of Yenching University (Yanjing daxue 燕京大学), who believed that "foreigners and Chinese were to take part on equal terms in every aspect of University affairs."[9]

Yenching University was born from the union of four mission schools around Peking between 1915 and 1920.[10] As early as 1921, "Yenching's Western faculty had voted to register with the Peking Board of Education." The university also promptly registered with the Nationalist government in Nanking after its establishment in April 1927. The registration with the government in fact removed from Stuart and his colleagues the "burden" to make China Christian. From then on, they could "devote all energies to making Yenching academically more respectable and 'more Chinese.'"[11]

Under the liberal leadership of John Stuart, Yenching University developed steadily, both in size and in academic quality.[12] In 1926, Yenching established "the first fully organized graduate school program in China," which "presented opportunities and stimulus for the training of a large number of brilliant Chinese students up to the Master degree level."[13] Since 1925, Yenching had already been ranked among the top ten universities, both public and private, in China.[14] By the mid-1930s, the Yenching faculty proudly found their students to

---

7   He and Shi 1996, 64–71; Lutz 1971, 252.

8   He and Shi 1996, 69, 71.

9   Stuart 1954, 71.

10   West 1976, 34–35.

11   West 1976, 95, 96; Stuart 1954, 71. Yenching applied for the registration with the Nanking government in November 1927; see Zhang 2000, 9.

12   Zhang, Wang, and Qian 2000, 1162, 1278.

13   Band 1959, 7.

14   Kuno 1928, 56; Lutz 1971, 202; Zhang, Wang, and Qian 2000, 16.

be "consistently superior" after comparing them with the contemporary American national standard for college-level courses.[15] As a privately funded institution with an international faculty, Yenching's contributions to China were transnational.

Likewise, Christian mission colleges generally made many transnational contributions to scientific development in China during the first half of the twentieth century. They consciously recruited "[Western] men trained in scientific specialties for college grade work" as early as the first decade of the 1900s; by the early 1920s, "the number of western scientists with post graduate degrees, in the Christian Colleges in China, had increased to nearly fifty." These foreign scientists quickly and successfully fostered a large group of indigenous scientists. By 1925, "there were more Chinese with [Western] Ph.D. degrees in science on the faculties than westerners, and in at least a few of the colleges, major responsibility was in the hands of highly competent Chinese scientists." The mission colleges pioneered postgraduate scientific training programs in China, awarding master's degrees as early as 1917. During the 1930s and 1940s, mission college scientists continued high-quality teaching and research despite various difficulties caused by the Japanese invasion. After the closure of all mission schools in mainland China in late 1951, William Band, a longtime British faculty member at Yenching University, summarized their accomplishments in natural sciences: "the colleges have anticipated the needs of the Chinese people, intensified their scientific work and raised the academic standards as the country became prepared for further progress"; "the Christian Colleges remained in the lead in the excellent quality of their graduates, if not in their numbers."[16]

Band joined the Yenching faculty as a young theoretical physicist in the autumn of 1929, when the university was undergoing its rapid expansion and the Department of Physics was desperately looking for new blood. At Yenching, Band established his career in physics, rising from an instructor to a full professor and the department chair. He not only published many scientific papers but also nurtured many distinguished Chinese physicists, making significant contributions to physics education and research in Republican China. This essay examines Band's career up to late 1941 with a focus on his most productive time in Peking.

Band's experience at Yenching University is an enlightening case. It calls our attention to two significant but largely overlooked themes in Chinese scientific

15   Band 1959, 12.
16   Band 1959, 4, 5, 7, 13–14, 15. For the closure times of all twenty-one mission colleges, see He and Shi 1996, 366.

developments during the Republican period. First, dozens of Western scientists serving in China for years became a significant transnational force advancing Chinese science and its professionalization. Unlike their counterparts in the seventeenth and late nineteenth centuries, these foreign scientists "were not particularly interested in propagating any church organization"; instead, they wished to put "their own skills and training at the service of [Chinese] people."[17] Until recently, most historical studies concerning Republican Chinese science have focused on native-Chinese scientists' work and overlooked, if not ignored, the contributions of foreign professionals who devoted the best years of their lives to Chinese scientific education and research. Second, mission colleges served as prominent bases for scientific studies in Republican China, especially after their compulsory registration with the Chinese government. This essay demonstrates that it was Yenching University that provided a stage for Band to play such an active and efficacious role in the development of modern physics in China.

The Making of a "Mathematical and Practical" Physicist

William Band (1906–1993) was born in Wallasey, Cheshire, England, on August 27, 1906. Since his great-grandfather, the Bands had been in the tailoring business.[18] After receiving his "Higher School Certificate" in July 1923, Band enrolled at the University of Liverpool (UOL) that autumn, becoming the first college student in his family.[19]

The UOL is a few miles to the east of Band's home, crossing the Mersey River. Founded in 1881, the UOL immediately appointed Oliver J. Lodge (1851–1940) as the first chair of the physics department. A pioneer of radio telegraphy and one of the best-known scientists of his day, Lodge led the department for nineteen years.[20] Lionel Robert Wilberforce (1861–1944), a Cambridge graduate with first-class honors in the Natural Science Tripos, succeeded Lodge and chaired that department for the next thirty-five years. Wilberforce was "a brilliant lecturer and an originator in the field of lecture demonstration." According to a pupil, "He had a thorough understanding of all types of students ... He took great pains to make physics attractive to his students." In the early years of the twentieth century, the department maintained a strong and fruitful research

---

17 He and Shi 1996, 5.
18 Band 1981, i, chap. I, chap. II: 1.
19 Band 1981, chap. IV: 8.
20 Rowlands 2006, 3–4; Millar 2002, 229–230.

program, especially in experimental study of X-rays, as evidenced by the work of Charles Barkla, a UOL graduate and faculty member, who received the 1917 Nobel Prize in physics.[21]

The high time of 1917 was quickly followed by a heavy personnel loss in the physics department during the Great War. Several important faculty members, including Barkla, left Liverpool. Fortunately, James Rice (1874–1936), a theorist trained in Belfast, joined the department in 1914 and remained in Liverpool for the rest of his life. Rice was a famed relativist, who gave lectures as early as 1917 "on the 'startling new theory' of relativity." He also served as "the main host" when Albert Einstein visited Liverpool and toured the department in June 1921. In the post–World War I years, Rice continued his studies on relativity and made an "outstanding contribution to physics" by publishing his *Relativity; a Systematic Treatment of Einstein's Theory*, a monograph many considered "one of the most thorough early works on the subject, and, for years, 'the most popular university textbook' on relativity 'in the English language.'"[22]

The appearance of Rice's book on relativity might well have coincided with Band's enrollment at UOL in the autumn of 1923. The British Association for the Advancement of Science meeting held in Liverpool that September must also have acquainted this college freshman with Niels Bohr and his work. Bohr received an honorary degree at UOL and discussed the correspondence principle at the meeting.[23] As a result, Band began his physics study in a milieu of active discussions concerning both revolutionary theories: the theory of relativity and quantum theory.

Stimulated by his boyhood reading at home, Band chose to major in physics at UOL. In his freshman year, he took the first-year physics course with Wilberforce and enrolled in Applied Mathematics, a course that was usually for second-year students. In his second year, Band attended all the available courses in undergraduate physics. He easily passed the required final examinations and received "the ordinary B.Sc. degree" in June 1925. During his third year, Band "took the recommended graduate courses and passed the examina-

---

21   Rowlands 2006, 8–9; Band 1981, chap. V: 1–2; Whitaker 1988, 38–39. L. R. Wilberforce was
        better known as the great-grandson of William Wilberforce, a famous abolitionist.

22   Band 1981, chap. V: 1–2; Rowlands 2006, 10–11; Rice 1923. For a summary of Rice's life see
        "Obituary: Prof. James Rice" 1936.

23   Rice had apparently completed his manuscript by May 1923, which is when his preface is
        dated, and it probably took a few months before the book was available to readers. Band
        entered UOL in the autumn of 1923. See Rice 1923, viii; Band 1981, chap. IV: 8. For informa-
        tion on the 1923 British Association meeting and Bohr's trip to Liverpool, see Rowlands
        2006, 12; Bohr 1976, 44–45.

tions with the highest score in class, receiving in summer 1926 the B.Sc. 1st class Honours degree" and the Oliver Lodge Prize.[24]

Band began work on a master of science degree while teaching as an "ungraded lecturer" in the autumn of 1926. An Honours graduate usually undertook some experimental research for the master of science degree in his first year of postgraduate study, but Band chose to explore theoretical issues and submitted his thesis at the end of 1927. This thesis, titled "An Examination of Professor Whitehead's Theory of Relativity," examined and compared Alfred Whitehead's and Einstein's treatments of the general theory of relativity. A. S. Eddington, the Cambridge astrophysicist and world-renowned advocate of Einstein's theory of relativity, served as the external examiner of Band's thesis. Eddington appraised Band's thesis "in very favourable terms, stating that it fell only [a] little short of the standard on which one would ordinarily award a Ph.D. degree in this country." It is remarkable that in the thesis Band referred not merely to many works about the theory of relativity but also to brand-new works by Heisenberg, Born, and Dirac concerning quantum mechanics, demonstrating his early attention to and up-to-date knowledge of the latest developments in theoretical physics.[25]

After earning his master's degree in December 1927, Band continued to teach as an ungraded lecturer at UOL. Despite his "active interest in theoretical work," Band decided "to enrich [his] experience in experimental physics before planning a possible transfer to Cambridge University for Ph.D. research." Collaborating with another graduate student, Band "very ably" carried out an experimental X-ray project to determine the crystal structure of titanium dioxide (TiO_2). For that purpose, they designed and built from scratch an X-ray tube of "a novel type." Although Band later claimed that they "never did succeed in unraveling the crystal structure," R. W. Roberts, the lecturer in physics who supervised their project, considered their work "invaluable to the development of this subject by subsequent workers in this department."[26] By the time he left for China, the twenty-three-year-old Band "was exceptionally well equipped for his age on the mathematical and practical sides of physical science."[27]

24   Band 1981, chap. V: 1–2.
25   Band 1981, chap. V: 3; Band 1927, 100; Rice 1934.
26   Roberts 1934; Band 1981, chap. V: 6–7.
27   Rice 1934.

A New Physics Lecturer at Yenching (1929–1931)

Having turned down a job offer from the Imperial Chemical Industries and given up a possible opportunity to study at the Cavendish Laboratory as a doctoral candidate under J. J. Thomson, Band accepted a decent offer from Yenching University early in 1929.[28] At Yenching's earnest request, he arrived in Peking on September 30, just in time "to take part on October 1, 1929 in the formal opening ceremonies on the new campus near Empress Dowager's Summer Palace northwest of Peking."[29]

Physics teaching at Yenching began no later than 1920 when its College of Natural Sciences was founded, and physics instruction was first carried out in collaboration with Peking Union Medical College. Yenching University produced its first physics major graduate as early as 1922, one year before its physics department was founded. The first department chair (acting   1923–1925) was an American missionary, Charles Hodge Corbett 郭查理 (1881–1963).[30] Corbett was born in China to a Presbyterian missionary family from Pennsylvania and had a good command of both spoken and written Chinese. After graduating from Union Theological Seminary in New York with a bachelor of divinity degree in 1907, he spent another year taking courses at Columbia University and the University of Chicago before returning to China to join the faculty of the North China Union College (NCUC), one of Yenching's four constituent mission schools. At NCUC, Corbett taught science and was especially interested in teaching physics through practical examples, a practice he continued at Yenching after its foundation. In 1921, Corbett recruited Y. M. Hsieh, a former student of his at NCUC, as an instructor of physics experiments, and the two soon coauthored a general physics textbook, *Principles of Physics and Their Modern Applications*. The book, which illustrated physics principles with everyday objects in Chinese life, was an exemplary volume implementing John Dewey's pragmatist philosophy in scientific education. In fact, Dewey offered suggestions to the authors and, in the foreword he wrote for the book, deemed it "a notable advance in the presentation of the subject of physical science."[31]

---

28   Band 1981, chap. VII: 1; Wilberforce 1934.
29   Band 1981, chap. VII: 4. The date of Band's arrival in Peking can be inferred from a letter he wrote; see Band 1929c. The schedule for the opening ceremony can be found in Yanjing daxue Beijing xiaoyouhui 1999, 9.
30   Zhang, Wang, and Qian 2000, 201, 1417; Smith and Corbett 1965, 75.
31   Smith and Corbett 1965, 74–75; Wu 1996, 77; Pan, Wu, and Fan 1993, 494–495; Corbett and Hsieh 1924, preface; Corbett and Hsieh 1926, foreword.

The physics department at Yenching had multiple missions, which included training premedical and preengineering students, introducing scientific methods and the significance of modern physics to students from other departments, equipping physics teachers with the updated knowledge of theoretical achievements, and nurturing research physicists. The last two missions were more likely added by Corbett's successors. When Corbett resigned from the university in January 1926, Paul A. Anderson (1898–1990), who came to Yenching in the previous year, was appointed as the acting department chair. Anderson was an experimental physicist who had earned his doctorate at Harvard University and had taught in Berlin, Germany. He became the full chair in 1927, and during his tenure, he helped create China's first master of science program in physics.[32]

At first, enrollment in the physics department was very small. Between 1925 and 1929, only six students received their bachelor of science degrees and two earned their master of science degrees in physics. The two master of science physics degrees, earned in 1929, were the very first awarded by a university in China. The department encouraged its faculty to find "opportunity for research work," intending to maintain the quality of its graduate program. Anderson, for instance, "did some experiments in the purification and distillation of barium which required a very high degree of manipulative skill and considerable apparatus." He personally made many experimental apparatuses and trained a couple of technicians for the department. Due to the sudden death of his wife, however, Anderson left the university in 1928.[33]

Anderson's successor was Chinese physicist Y. M. Hsieh (Xie Yuming 谢玉铭, 1893–1986), who began his career at Yenching in 1921. Sponsored by the Rockefeller Foundation, Hsieh went to the United States in 1923, where he earned his master's degree at Columbia University in 1924 and his doctorate under A. A. Michelson at the University of Chicago in 1926. Hsieh returned to Yenching after receiving his doctorate. Like Anderson, Hsieh was also a skillful experimentalist who stressed lab work in teaching and believed that a fine laboratory ought to have a good machine shop at its disposal. He set up a well-equipped workshop where the faculty, technicians, and students at Yenching made their own instruments for class instruction or research.[34]

---

32   Zhang, Wang, and Qian 2000, 201–202, 1417–1418; Dong 2009, 5.

33   Zhang, Wang, and Qian 2000, 201–202, 1418; Dong 2009, 5; Tayler 1929a. On Anderson's background, I also referred to the webpage at http://www.supportscience.wsu.edu/endowment.html (accessed on September 12, 2006). The quotations appeared in Tayler's letter.

34   Wu 1996; Zhang, Wang, and Qian 2000, 202–204; Pan, Wu, and Fan 1993, 494–495. For Hsieh's opinion on workshops, see Qian 1987.

In the 1920s, experimentalists monopolized most physics departments in China, including that at Yenching. The Chinese interest in theoretical physics seems to have emerged after the relativity craze in the early 1920s. By the end of the 1920s, leading physics departments in China had begun to seek theorists. The fact that the first two Chinese theoretical physicists, S. C. Wang (Wang Shoujing 王守竞, 1904–1984) and P. Y. Chou (Zhou Peiyuan 周培源 1902–1993), both returned to China in 1929, the very same year when Band arrived in Peking, was not merely a coincidence but rather a clear indication of significant Chinese development in physics.[35]

As the department head, Hsieh must have realized this, which explains why he eagerly recruited Band. By adding the first theoretical physicist at Yenching, Hsieh evidently attempted to enhance the department's competence and cope with the growing enrollment. Between 1929 and 1932, undergraduate enrollment began to increase significantly. The department admitted twenty freshmen in 1931, and it maintained a similar annual admission rate in the following few years. During this three-year period, eleven students earned their bachelor's degrees and six graduate students received their master's degrees (excluding the two awarded in 1929). The growth in enrollment, especially in the number of graduate students, naturally demanded a more diversified and advanced curriculum.[36] The university and the department certainly expected that Band's academic background and strengths could help them meet the rising demands and challenges in the years to come.

Band's initial appointment was as an "Instructor in Theoretical Physics,"[37] and his teaching indeed helped cultivate theoretical studies in physics. During his first academic year (1929–1930), he taught three courses: Mathematical Physics, Vector Analysis, and the Principle of Relativity.[38] His "mathematical physics" was new at Yenching both in title and in contents. In fact, it represented a significant new trend in China's college physics education: to strengthen students' theoretical training. Such emphasis was perhaps more perceptible in the course's corresponding Chinese title, Lilun wulixue dagang 理论物理学大纲, or "an outline of theoretical physics."[39] In this endeavor, Yenching was in the vanguard in China because even several years later "math-

---

35   For Chou's background, see Hu 2005, 116–121. For Wang, see Dong 2009, 22, 175–176.

36   Before Band's arrival, the faculty members who could direct graduate students were Paul Anderson, Y. M. Hsieh, and D. K. Yang (Yang Jinqing 杨荩卿) (Pan, Wu, and Fan 1993, 494–495), but none of these professors was a theorist or able to offer advanced courses in theoretical physics.

37   Band 1934, 4.

38   Yenching University 1929a   29–30.

39   Yenching University 1929b   60.

ematical physics" still rarely appeared in the physics curricula on other campuses. According to a survey of the Chinese Ministry of Education in 1933
(more than three years after Band began to teach it at Yenching), only seven
out of twenty-three (30 percent) participating universities had "mathematical
physics" listed in their curricula; among these seven schools, five designated it
as a required course for physics majors and two as elective.[40]

Both Vector Analysis and the Principle of Relativity were also brand-new
courses at Yenching University. The former taught students "[e]lements of vector algebra and calculus and the linear vector function in three dimensions,"
while the latter covered not only "[t]he restricted theory as formulated by
Einstein and Minkowski, with applications to dynamics, gravitation and electricity" but also "[t]he general theory of relativity," "its application to gravitation and optical phenomena," and "its extension to continuous media."[41] The
course Vector Analysis was truly unique in China: none of the twenty-three
universities in the above-mentioned survey had it in their physics curricula.
On the other hand, eight of the twenty-three universities offered "theory of
relativity," usually as an elective course for advanced students, demonstrating
the popularity of relativity.[42]

Band apparently performed rather well in his first year because Yenching
promoted him to "Lecturer" before August 1930. In the following academic year
(1930–1931), Band worked even harder, offering six different courses. Besides
repeating Vector Analysis, he offered two sequent courses: Introduction to
Mathematical Physics, I and II. Mathematical Physics I used vectors and vecto-
rial operators throughout the course and covered topics such as kinematics,
particle dynamics, general principles applied to planetary motion, statistical
theory of thermodynamics, and electrostatic and magnetostatic field. Mathematical Physics II dealt with the theory of wave functions, Maxwell's theory of
electromagnetic fields, solution of field equations in particular cases, radiation
pressure, energy and mass, restricted relativity and Lorentz transformation,
Minkowski's 4-vectors, general solution of field equations, motion of electrons,
tensors, and invariant relativity form to field equations.[43] These two courses in

---

40   National Institute for Compilation and Translation (NICT hereafter) 1933, 327, 343 (here I
      include a course titled Lilun wulixue fangfa 理论物理学方法 [Methods of theoretical
      physics])   407, 436, 447, 469, 490.
41   Yenching University 1929a   30.
42   NICT 1933, 299, 327, 334, 343, 402, 436, 457, 485; six of these eight universities offered a
      course on the theory of relativity as an elective.
43   Yenching University 1930a   37.

mathematical physics certainly covered more diversified and advanced topics than what he taught in the previous year.

Band also began to teach Thermodynamics and Quantum Theory in 1930–1931. The former covered "the principles of thermodynamics and their applications to physical and chemical processes," while the latter discussed "statistical theory of heat, quantum statistics, quantum theory of radiation, statistical mechanics and wave mechanics of the atom."[44] This was the first time "quantum theory" appeared in Yenching's curriculum. Despite its title, this new course actually covered both "old quantum theory" and "quantum mechanics" in present terms. Chinese physicists had begun to learn about "old quantum theory" in the early 1920s.[45] Although quantum theory never had the same publicity as the theory of relativity did in the following decade, the former appeared more popular than the latter in college teaching before the mid-1930s. Still, most Chinese universities' curricula did not cover such topics. In the 1933 survey of college curricula, only ten of the twenty-three (43 percent) physics departments in China included quantum theory or quantum mechanics or both in their curricula.[46] As a result, Yenching became one of a few leading institutions in China where one could learn recent advances in theoretical physics.

The sixth course that Band offered was certainly unusual for such a young theoretical physicist: a seminar titled the Natural Philosophy of Modern Physics, which began with "a summary of the theories of Relativity, Wave Mechanics of the Atom, and Statistical mechanics" and explored "the Natural Philosophy of Whitehead, Broad, Russell and Eddington." Band intended to give students in various disciplines "a grasp of the significance of [modern physics]." By design, it was not "exactly elementary in nature, but the physics included [would] be as non-technical as possible." Due to the "sufficient difficulty" "in understanding the philosophical part of the subject," Band adver-

---

44   Yenching University 1930a   38.

45   Xu Chongqing (1888–1969) seems to be the Chinese scholar who, in 1917, was the first to use the term *liangzilun* 量子论, or "quantum theory," in Chinese literature, but it was Zhou Changshou (1888–1950) who was the first to introduce China to the scientific contents of the theory. Both Xu and Zhou graduated from Tokyo Imperial University (Hu 2007, 1, 4). For the detailed story about the introduction of relativity into China, see Hu 2005, chap. 2.

46   NICT 1933, 299, 327, 334, 343, 349, 402–403, 407, 436, 457, 485. In these Chinese curricula, "quantum theory" was rendered as *liangzi lun* 量子论 or *yuanliang lun* 原量论, and "quantum mechanics" as *liangzi lixue* 量子力学 or *yuanliang lixue* 原量力学; sometimes quantum mechanics was taught under the name "wave mechanics" (*bodong lixue* 波动力学 or *bo lixue* 波力学).

tised the seminar as more suitable for "mature" or graduate students, and participants had to obtain his permission before taking it.[47] In fact, this seminar attracted not only a small group of students but also a few faculty members, including Dr. Randolph C. Sailer of the psychology department and Dr. Lechung Tsetung Hwang (Huang Zitong 黄子通) of the philosophy department, both of whom offered "many helpful suggestions" and had "thought-provoking discussions" with Band.[48] Hwang was a full professor and the department chair at the time.[49]

This seminar was apparently successful and popular, as Band appears to have offered it repeatedly until his sabbatical leave in 1936–1937.[50] Encouraged by the seminar's success, Band prepared a manuscript in the early 1930s titled "The Philosophy of Modern Physics" and intended to publish it in London.[51] In preparing this manuscript Band had two objectives: besides helping professional physicists grasp "the significance of their subject in its broad relations with the rest of life," he mainly intended "to provide the interested layman with an intelligible discussion of how the modern developments of physics affect our attitude to the world we live in."[52] His seminar and his manuscript represented Band's serious attempts to explore the profound philosophical nature of modern physics theories, focusing mainly on causality and statistical features. Band was probably the first person in China to undertake such a philosophical survey, especially on the new quantum mechanics.[53]

Expanding Scientific Work under "Militant Nationalism"

As Band correctly observed, in the 1930s "the major concerns of the native Chinese students were centered on the Japanese invasion of Manchuria and the apparent inability of the National Chinese regime under Chiang Kai-shek to resist the advance of the Japanese into China." As a result, Yenching students joined those from "all the national universities" in the Peking area and staged

---

47 Yenching University 1930a 39.
48 Band 1931, foreword. Band spelled Hwang's name as "L. T. Huang."
49 Yenching University 1930b 5.
50 Yenching University 1930a 1931a 1931b 1932, 1934, 1936.
51 Band 1931, front cover.
52 Band 1931, "Introductory Remarks" 2–3.
53 In 1926, Zhou Changshou (1888–1950), a Chinese physicist trained in Japan, translated *Modern Natural Science* by Japanese philosopher Tanabe Hajime (Tianbian Yuan 田边元, 1885–1962), which discussed the theory of relativity and quantum theory but not quantum mechanics.

"frequent anti-Japanese demonstrations," "protesting Chiang Kai-shek's inaction in the face of the Japanese threat."[54]

In Band's opinion, the period between 1929 and 1931 was a "utopian" time for Yenching and other Christian colleges in China. During these years, the "vigorous Christian College system," which was "recognized by the firmly established National Government, held in high popular esteem in all parts of China and in academic circles throughout the world," was "full of confidence and enthusiasm [and played] an important and steadily growing part in the regeneration of the country." This "utopian situation," however, largely ended with the Japanese invasion of Manchuria on September 18, 1931. "From then on, there was no more peace on the campuses of the Christian Colleges in China. Everything that was done had to be rationalized in terms of national salvation." One may tend to think that, under such conditions, teaching and research in science at these colleges would be neglected. It turned out that "[a]n even stronger urge set in towards expanding work in the natural sciences, including pre-medicine and pre-nursing."[55]

Having taken over the physics department chair in 1932, Band and his colleagues attempted to "channel [their] students' energies in constructive directions." The department, for example, not only offered courses such as Radio Telegraphy and Telephony and Radio but also built a "ham" station that was "operated almost entirely by physics students, maintaining daily schedules with other stations in China, the Philippines and the United States, incidentally handling over 3,000 amateur messages between Peking and California during one year."[56] Both Band and his wife actively participated in the station's operation, and the department trained students "in handling messages and servicing the equipment." Students even "built portable sets suitable for field work in time of war."[57]

Indeed, there was "no let-up in academic standards" at Yenching University during these years of "militant nationalism," when the physics department was mostly under Band's leadership.[58] Band chaired the department until the Japanese shut down the campus in December 1941. It was during Band's tenure that the department's enrollment increased substantially, especially after 1938,

---

54   Band 1981, chap. IX: 3–4.

55   Band 1959, 10.

56   Band 1981, chap. IX: 4; Band 1959, 10. D. K. Yang (or Yang Chin-ching) taught Radio Telegraphy and Telephony during 1931–1933; Pi Te-hsien (Bi Dexian 毕德显) and Chang Wen-yu (Zhang Wenyu 张文裕) taught Radio in 1934–1935. See Yenching University 1931b   72; 1932, 66; 1934, 52.

57   Band 1981, chap. IX: 4.

58   Band 1959, 10.

when all national universities in North China had fled south. The increase in enrollment was a university-wide phenomenon (see table 1). Between 1937 and 1941, Yenching University was a powerful magnet for Chinese students in occupied North China because it was the only "oasis of academic freedom in a desert of military control."[59]

TABLE 1   *Wartime Enrollment for Individual Schools at Yenching University*

Year	1936–37	1937–38	1938–39	1939–40	1940–41
School of Arts 文学院	284	187	269	306	380
School of Sciences 理学院	284	216	307	346	369
Law School 法学院	234	177	303	314	310

SOURCE: LIU AND LIU 2003, 128.

The department awarded sixty bachelor's and twenty-five master's degrees during 1932–1941. Band directed most of the master of science theses, many of which were published in scientific journals in Great Britain and America, promoting young Chinese physicists and their works in the international scientific community.[60] Band was a beloved teacher at Yenching University, and his courses were very popular among students. His teaching was broad in scope and rich in content: Band taught at least twenty-one different courses at Yenching, covering both theoretical and experimental physics and ranging from freshman College Physics to advanced relativity and quantum theories, from the specialized Thermo-magnetic Effect to the comprehensive Natural Philosophy of Modern Physics.[61]

All graduates from the physics department received solid scientific training, which was demonstrated by the many scholarships they won in nationwide competitions and by the achievements of those studying abroad in doctoral programs. Between 1933 and 1937, "more than ten students with M.S. degrees"

---

59   Zhang, Wang, and Qian 2000, 205. Y. M. Hsieh served as the acting chair while Band was on leave during the 1936–1937 academic year (Yenching University 1936, 69). The quotation is from Edwards 1959, 347.

60   Zhang, Wang, and Qian 2000, 205.

61   See Yenching University Bulletin in respective years; Band 1934, 4.

from the physics department went on to pursue their doctorates in the United States or England and often at top schools such as Caltech, MIT, Harvard University, University of California–Berkeley, and Cambridge University.[62] Upon graduation, they returned to China and many went on to become leaders in their own professional field. Among these gifted pupils of Band's were Chang Wen-Yu 张文裕 (1910–1992, class of 1931), who became a leader in China's high-energy physics and successfully designed and constructed the world-renowned Beijing Electron-Positron Collider;[63] Wang Cheng-shu 王承书 (1912–1994, class of 1934), who was one of several distinguished female physics graduates from Yenching and later pioneered Chinese studies in controlled thermonuclear fusion and plasma physics;[64] and Huang Kun 黄昆 (1919–2005, class of 1941), who founded semiconductor physics research in China, initiated phonon physics studies, and won the 2001 State Preeminent Science and Technology Award, the highest national prize for scientists and engineers in contemporary China. [65] The achievements of these Yenching-educated physicists represent in part Band's contributions to Chinese physics.

In addition to his success as an effective teacher, Band was also a prolific researcher. Despite his heavy teaching and administrative duties, Band produced at least forty-eight articles between the autumn of 1929 and December 1941. Of these forty-eight papers, sixteen were coauthored with his graduate students and dealt with both experimental and theoretical problems. Band's broad investigations concerned general relativity, quantum mechanics, unified field theory, statistical and surface physics, thermoelectric and thermomagnetic phenomena, X-rays, and low-temperature superconductivity. The papers he published were often the first of their kind in China. Band's works typically reflected his profound philosophical thinking even though some of his topics apparently did not conform to the mainstream interests of contemporary physicists.[66]

Band did not bury himself in academic work and ignore the problems in the society he lived in. He actively supported the Rural Reconstruction Movement, which was spreading in China during the 1930s. In 1935, Band published an essay titled "The Place of Physics in the Rural Reconstruction of China," in

---

62   Band 1959, 10.

63   Zhongguo kexue jishu xiehui 1996, 474–490.

64   Zhongguo kexue jishu xiehui 1996, 516–526; Zhongguo kexueyuan xuebu lianhe bangong-shi 1995b   3.

65   Zhu 2002, 37–47, 144; Zhongguo kexueyuan xuebu lianhe bangongshi 1995a   28.

66   I thank Professor Dong Guangbi at the Chinese Academy of Sciences for sharing with me his insight on Band's research work.

which he pointed out multiple important services that physicists could offer to the movement, such as training the technical personnel needed in rural areas, improving rural radio broadcasting and communication, conducting tests and research for rural industries, initiating and enhancing agricultural physics and geophysics, and finding solutions to indigenous problems. He not only called on university physicists in China to help the movement with their knowledge and facilities but also let his students put their knowledge to use in support of the movement.[67]

Conclusion

William Band's distinguished service at Yenching University ended dramatically in the early morning of December 8, 1941. Upon hearing of the Japanese attack on Pearl Harbor over the radio, Band, his wife, and two other colleagues managed to flee from the campus just in time. Only minutes after their car dashed out Yenching's east gate, the Japanese army rushed in from the west gate and rounded up all the faculty and students on campus. The Bands were among only a few who escaped this Japanese raid and successfully reached their refuge in Communist guerrilla bases in the mountains to the west of Peking. After a two-year and 1,000-mile trek to Yan'an, the Bands went on to Chongqing, China's wartime capital, where they served in the Sino-British Science Cooperation Bureau headed by Joseph Needham. Band and his wife left Chongqing for England at the end of 1944, expecting to return soon to their beloved Yenching campus in Peking after the anticipated victory over Japan—a plan they never fulfilled.

Since the late sixteenth century and especially since the 1840s, numerous Westerners had come to China to preach, to adventure, to explore, or to teach. Many of them served as "advisers" of various kinds for the Chinese, among whom Jonathan Spence has examined sixteen representative individuals in his fascinating book *To Change China: Western Advisers in China*. William Band can also be considered an adviser, not only to his Chinese students but also to the reformers in the Rural Reconstruction Movement. It is therefore instructive to compare his career in China with those of earlier "advisers."

Band went to China both by invitation and of his own volition, but apparently due more to the latter. He certainly shared other advisers' desire to "help to bring ... improvement to China," which, in his case, meant to advance

---

67   Band 1935.

Chinese studies in modern physics.[68] James Rice, Band's mentor at UOL, was probably not completely mistaken in thinking that "Band had an ardent desire to undertake missionary work in China."[69] Band, however, was not a traditional Christian missionary who lectured on modern physics "as the wrapping" around the gospel or any other "ideological package," as Jesuit and Protestant missionaries had done in the seventeenth and nineteenth centuries, respectively.[70] Band was nevertheless a special kind of missionary who propagated modern physics instead of preaching Christian doctrines.

At this point it is still difficult to pin down all Band's motives for serving in China, but we can gauge his dominating ambition from a statement he made after living in the country for fifteen years and experiencing many volatile changes. In the spring of 1944, Band visited the campus of Christian West China Union University in Chengdu, where four other sister mission colleges in exile had been temporarily relocated from East and North China. Witnessing the striving of these colleges' faculty and students in their scientific studies under extremely difficult conditions, Band could not help making the following emotional remarks:

> The essential core of these centres of higher learning is constituted by their purely academic pursuits. This common purpose has always been and still is to establish and promote cultural relationships between East and West. It is a human, essentially religious, purpose, which is becoming more and more important, and a continually stronger binding force between otherwise distinct systems. *It was the impulse of this purpose which excited the present writer out of his normally stable state at home, stimulated his transition to Yenching University.*[71]

Apparently what motivated Band to come to China was mainly his "ardent desire" to bridge East and West through cultural, especially scientific, exchanges.

Indeed, Band was willing to sacrifice for his belief. For example, while most Western faculty at Yenching were apparently paid in US currency, either directly by the university or through their missionary societies, Band chose to accept "the same salary scale as the Chinese faculty" and was paid in local currency despite his knowledge of its rapid devaluation in the 1930s. As a result,

---

68   The quotation appears in Spence 1980, 291.
69   Rice 1934, 1–2.
70   The quotations are from Spence 1980, 290.
71   Band 1944, 3 (emphasis added).

Band's basic pay was Chinese $360 per month, which was equivalent to a monthly payment of US$95.65, or about US$1,150 annually. Although he couldn't help lamenting his meager "hand-to-mouth" income, he was proud of his conviction: "a fully effective cultural exchange between the Chinese and [the Westerners] was possible only in the absence even of the semblance of artificial material privilege based on race."[72]

To put his belief into practice, Band also made conscious efforts to help develop theoretical physics in China. After five years teaching at Yenching, Band realized that many Chinese students had "a considerable natural aptitude for theoretical work." However, he found that there were "practically no centres in China where training in Theoretical Physics receive[d] adequate attention." Consequently, he planned to spend his upcoming sabbatical leave specializing in theoretical physics at Harvard University "in order to bring back to Yenching a better stimulus for more complete and proportionate development therein."[73] Band also actively maintained professional exchanges with colleagues in the Chinese physics community. He participated in events organized by the Chinese Physical Society, which he joined soon after its foundation in 1932 and served as a member of its editorial board. At the society's eighth annual meeting convened in Kunming in September 1940, Band and his students at Yenching University contributed twelve papers, which amounted to 30 percent of all submitted papers; of these twelve papers, Band authored or coauthored five.[74]

William Band contributed much to the Chinese development in modern physics, especially the professionalization of Chinese studies in theoretical physics. It is also important to note that Band's accomplishments were based at Yenching University, which provided an effective stage for his outstanding performance. Band's story demonstrates the significant roles individual Western scientists and mission colleges have played in scientific development during Republican era China. Many similar stories still await historians to uncover and tell. Such stories are indispensable because no account of Republican Chinese science can be considered complete without an adequate evalu-

---

72   Band 1981, chap. IX: 2.

73   Band 1934, 2. Band did not end up going to Harvard University but spent his sabbatical at
        Cambridge University during 1936–1937. Because of the Japanese invasion, Band never
        fulfilled his wish to build a center for theoretical studies in physics at Yenching.

74   Band became a member of the Chinese Physical Society no later than 1934. See his CV in
        Band 1934, 3. For the society's editorial board and eighth annual meeting, see Department
        of Physics and Physics Club of Yenching University 1941, 9; 1940, 21.

264

ation of the contributions of these foreign professionals and their hosting institutions in China.

Bibliography

Band, William 1927. "An Examination of Professor Whitehead's Theory of Relativity." Master's thesis, University of Liverpool.

____ 1929a. "Dr. A. N. Whitehead's Theory of Absolute Acceleration," *Philosophical Magazine* 7: 434–440.

____ 1929b. "A Comparison of Whitehead's with Einstein's Law of Gravitation," *Philosophical Magazine* 7: 1183–1186.

____ 1929c. Letter to F. H. Hawkins, October 1, 1929. Records of the Council for World Mission, box 26, 2981, London Missionary Society Archives at School of Oriental and African Studies Library, University of London.

____ 1931. "The Philosophy of Modern Physics." Manuscripts, Archives, and Special Collections, cage 617, box 3, F 22, Washington State University Libraries, Pullman.

____ 1934. Application for Fellowship for Advanced Study in Physics at Harvard University. Archives of the United Board for Christian Higher Education in Asia, Record Group no. 11, 320-4877, Special Collections, Yale Divinity School Library, New Haven, CT.

____ 1935. "The Place of Physics in the Rural Reconstruction of China." Peiping: Department of Physics, Yenching University.

____ 1944. *Science in the Christian Universities at Chengtu, China*. New York: Associated Boards for Christian Colleges in China.

____ ca. 1959. "The China Colleges and Instruction in the Natural Sciences." Archives of the United Board for Christian Higher Education in Asia, New File, RG 11, 64A-841, Special Collections, Yale Divinity School Library, New Haven, CT.

____ 1981. *Autobiography:* 班威廉 *William Band.* Manuscripts, Archives, and Special Collections, cage 617, box 2, F 12, 13, and 14, Washington State University Libraries, Pullman.

Bohr, Niels 1976. *Collected Works: Correspondence Principle 1918–23.* Amsterdam: North-Holland.

Corbett, Charles Hodge 1926. Letter to the Trustees of Peking University, January 8, 1926. Archives of the United Board for Christian Higher Education in Asia, RG 11, 40-1016, Special Collections, Yale Divinity School Library, New Haven, CT.

Corbett, Charles H., and Yu-Ming Hsieh 1924. *Principles of Physics and Their Modern Applications*, pt. 1. Peking: Commercial Press, Branch Works.

____ 1926. *Principles of Physics and Their Modern Applications*. Shanghai: Commercial Press.

Department of Physics and Physics Club of Yenching University 1938. "Wuli xuexun" 物理学讯 [Yenching physics news]. Peiping: Department of Physics and Physics Club of Yenching University.

____ 1940. "Yenching Physics News." Peiping: Department of Physics and Physics Club of Yenching University.

____ 1941. "Yenching Physics News." Peiping: Department of Physics and Physics Club of Yenching University.

Dong Guangbi 2009. *Zhongguo xiandai wulixue shi* 中国现代物理学史 [A history of physics in modern China]. Ji'nan: Shandong jiaoyu chubanshe.

Edwards, Dwight Woodbridge 1959. *Yenching University*. New York: United Board for Christian Higher Education in Asia.

Gaunt, Reverend L. H. 1906. "The Chronicle of the London Missionary Society." London: London Missionary Society.

Guoji liuti lixue he lilun wuli kexue taolunhui zuzhi weiyuanhui 国际流体力学和理论物理科学讨论会组织委员会 1992. *Kexue jujiang shibiao liufang* 科学巨匠 师表流芳 [Peiyuan Chou: Great scientist and reputable teacher]. Beijing: Zhongguo kexue jishu chubanshe.

Hawkins, F. H. 1914. *Through Lands That Were Dark: Being a Record of a Year's Missionary Journey in Africa and Madagascar*. London: London Missionary Society.

He Xiaoxia 何晓夏 and Shi Jinghuan 史静寰 1996. *Jiaohui xuexiao yu zhongguo jiaoyu jindaihua* 教会学校与中国教育近代化 [Mission schools and the modernization of Chinese education]. Guangzhou: Guangdong jiaoyu chubanshe.

Hu, Danian 2005. *China and Albert Einstein: The Reception of the Physicist and His Theory in China 1917–1979*. Cambridge, MA: Harvard University Press.

____ 2007. "Zhou Changshou and the Introduction of Quantum Theory in China," paper presented at HQ-1 Conference on the History of Quantum Physics, Max Planck Institute for the History of Science, Berlin, Germany.

Kuno, Yoshi S. 1928. *Educational Institutions in the Orient, with Special Reference to Colleges and Universities in the United States*. Berkeley: University of California.

Liu Jiafeng 刘家峰 and Liu Tianlu 刘天路 2003. *Kangri zhanzheng shiqi de jidujiao daxue* 抗日战争时期的基督教大学 [Christian universities during the Anti-Japanese War]. Fuzhou: Fujian Educational Press.

London Missionary Society and James Sibree 1923. *A Register of Missionaries, Deputations, etc., from 1796 to 1923*. London: The Society.

Lutz, Jessie Gregory 1971. *China and the Christian Colleges 1850–1950*. Ithaca, NY: Cornell University Press.

Millar, David 2002. *The Cambridge Dictionary of Scientists*. Cambridge: Cambridge University Press.

National Institute for Compilation and Translation (NICT), ed. 1933. *Jiaoyubu tianwen shuxue wuli taolunhui zhuankan* 教育部天文数学物理讨论会专刊 [Proceedings of the symposium at the Chinese Ministry of Education on astronomy, mathematics, and physics]. Nanjing: Ministry of Education.

"Obituary: Prof. James Rice" 1936. *Nature* 137: 807–808.

Pan Yongxiang 潘永祥, Wu Ziqin 吴自勤, and Fan Shulan 范淑兰 1993. "Yanjing daxue wulixuexi shigao" 燕京大学物理学系史稿 [A draft history of the physics department at Yenching University], *Wuli* 物理 [Physics] 22: 493–500.

Qian Linzhao 钱临照 1987. "Daonian woguo wulixuejie qianbei Xie Yuming jiaoshou" 悼念我国物理学界前辈谢玉铭教授 [In remembrance of Professor Y. M. Hsieh, a forefather in Chinese physics], *Wuli* 物理 [Physics] 16, 3: 184.

Rice, James 1923. *Relativity; a Systematic Treatment of Einstein's Theory*. London: Longmans, Green.

_____ 1934. Letter to the Rockefeller Foundation, August 14, 1934. Archives of the United Board for Christian Higher Education in Asia, Record Group no. 11, 320-4877, Special Collections, Yale Divinity School Library, New Haven, CT.

Roberts, R. W. 1934. Letter to the Rockefeller Foundation, July 10, 1934. Archives of the United Board for Christian Higher Education in Asia, Record Group no. 11, 320-4877, Special Collections, Yale Divinity School Library, New Haven, CT.

Rowlands, Peter 2006. *125 Years of Excellence: The University of Liverpool Physics Department 1881 to 2006*. Liverpool: Science Communication Unit / PD Publications, Oliver Lodge Laboratory.

Smith, Harold Frederick, and Charles Hodge Corbett 1965. *Hunter Corbett and His Family*. Claremont, CA: College Press.

Spence, Jonathan D. 1980. *To Change China: Western Advisers in China   1620–1960*. Harmondsworth, England: Penguin Books.

Stuart, John Leighton 1954. *Fifty Years in China: The Memoirs of John Leighton Stuart*. New York: Random House.

Tayler, J. B. 1929a. Letter to W. Band Jr., March 14, 1929. Records of the Council for World Mission, box 27, 2073, London Missionary Society Archives at School of Oriental and African Studies Library, University of London.

_____ 1929b. Letter to F. H. Hawkins, March 14, 1929. Records of the Council for World Mission, box 27, 2073, London Missionary Society Archives at School of Oriental and African Studies Library, University of London.

West, Philip 1976. *Yenching University and Sino-Western Relations   1916–1952*. Cambridge, MA: Harvard University Press.

Whitaker, Robert J. 1988. "L. R. Wilberforce and the Wilberforce Pendulum," *Physics Teacher* 26 (January): 37–39.

Wilberforce, L. R. 1934. Letter to the Rockefeller Foundation, July 16, 1934. Archives of the United Board for Christian Higher Education in Asia, Record Group no. 11, 320-4877, Special Collections, Yale Divinity School Library, New Haven, CT.

Wu Boxi 吴伯僖 1996. "Xie Yuming" 谢玉铭 [Y. M. Hsieh], in Zhongguo kexue jishu xiehui,中国科学技术协会编, ed., *Zhongguo kexue jishu zhuanjia zhuanlue, lixuepian, wulixue juan 1* 中国科学技术专家传略, 理学编, 物理学卷 1 [Short biographies of Chinese specialists in science and technology, Collections on scientists, vol. 1 on physicists]. Shijiazhuang: Hebei jiaoyu chubanshe 76–85.

Yanjing daxue Beijing xiaoyouhui 燕京大学北京校友会 1999. *Yanjing daxue: Jianxiao 80 zhounian jinian lishi yingji* 燕京大学：建校80周年纪念历史影集 [Yenching University 80th anniversary historical photo album 1919–1999]. Beijing: Renmin Zhongguo chubanshe.

Yanjing yanjiuyuan 燕京研究院 2001. *Yanjing daxue renwu zhi* 燕京大学人物志 [Who's who in Yenching University]. Beijing Shi: Beijing daxue chubanshe.

Yenching University 1927. "Announcement of Courses 1927–28," in Yenching University, ed., *Yenching University Bulletin*. Peking: Yenching University.

____ 1929a. "Announcement of Courses, Graduate Division 1929–30," in Yenching University, ed., *Yenching University Bulletin* (August). Pe[i]ping: Yenching University.

____ 1929b. "Ziran kexue yuan kecheng yilan" 自然科学院课程一览 [The College of Natural Sciences announcement of courses 1929–1930]. [Peiping:] Yenching University.

____ 1930a. "Announcement of Courses, Graduate Division 1930–31," in Yenching University, ed., *Yenching University Bulletin*, vol. 12, no. 20 (August). Peiping: Yenching University.

____ 1930b. "Yenching University Directory" 燕京大学教职员学生名录, 1930–31, in Yenching University, ed., *Yenching University Directory*, vol. 12, no. 14 (October). Peiping: Yenching University.

____ 1931a. "Yenching University Bulletin: Graduate Division 1931–1932," vol. 16, no. 20 (May). Peiping: Yenching University.

____ 1931b. "Yenching University Bulletin: College of Natural Sciences 1931–1932," vol. 16, no. 25 (June). Peiping: Yenching University.

____ 1932. "Yenching University Bulletin: College of Natural Sciences," vol. 17, no. 25 (August). Peiping: Yenching University.

____ 1934. "Yenching University Bulletin: College of Natural Sciences 1934–1935." Peiping: Yenching Universit

____ 1936. "Yenching University Bulletin: Announcement of Courses 1936–1937." December. Peiping: Yenching University.

Zhang Weiying 张玮瑛, Wang Baiqiang 王百强, and Qian Xinbo 钱辛波, eds. 2000. *Yanjing daxue shigao 1919–1952*燕京大学史稿 [A draft history of Yenching University 1919–1952]. Beijing: Renmin zhongguo chubanshe.

Zhongguo kexue jishu xiehui, ed. 中国科学技术协会编 1996. *Zhongguo kexue jishu zhuanjia zhuanlue, lixuepian, wulixue juan 1* 中国科学技术专家传略, 理学编, 物

理学卷 1 [Short biographies of Chinese specialists in science and technology, Collections on scientists, vol. 1 on physicists]. Shijiazhuang: Hebei jiaoyu chubanshe.

Zhongguo kexueyuan xuebu lianhe bangongshi 中国科学院学部联合办公室 [General Office of Academic Division of the Chinese Academy of Sciences] 1995a. *Zhongguo kexueyuan yuanshi huace* 中国科学院院士画册 [Album of members of the Chinese Academy of Sciences]: *1955 and 1957.*

____ 1995b. *Zhongguo kexueyuan yuanshi huace* 中国科学院院士画册 [Album of members of the Chinese Academy of Sciences]: *1980.*

Zhongguo wuli xuehui liushi zhounian bianxiezu, ed.《中国物理学会六十周年》编写组编 1992. *Zhongguo wuli xuehui liushi nian* 中国物理学会六十年 [The sixtieth anniversary of the Chinese Physical Society]. Changsha: Hunan Education Press.

Zhu Bangfen 朱邦芬 2002. *Huang Kun—shengzi wuli diyiren* 黄昆一声子物理第一人 [Huang Kun: The first man in phonon physics]. Shanghai: Shanghai kexue jishu chubanshe.

Periodical Space

Language and the Creation of Scientific Community in Republican China

Grace Shen[1]

Abstract

Despite the chaotic circumstances of the early Republican period, China's small community of field geologists had defied the odds and worked out the basic structure of large parts of northern and eastern China by the early 1920s. Their efforts to communicate these findings and engage the international geological community were undermined by the paucity of suitable outlets and the weight of low expectations. It took the establishment of the Geological Society of China and an ambitious program of multilingual scientific diplomacy to reach a global audience, but with the start of the War of Resistance against Japan, this outward-facing and self-assertive strategy no longer served the needs of China's diversifying geological community. New efforts were needed to create a protected space in which local researchers could exchange information without the pressure of international scrutiny. This essay traces local scientists' deployment of language, sociability, and publications first to enroll outsiders in their cosmopolitan vision of Chinese geology and then to carve out a "private" sphere for domestic debate. Both moves reflect Chinese understandings of the international scientific economy and Chinese scientists' shifting place within it.

For historians who study science in Republican China, it has often been possible to sidestep the "Needham Question" in favor of the many modernization campaigns that marked the early twentieth century. Still, tempting traps remain, and just as soon as we have addressed the impact thesis by demonstrating Chinese agency in the quest for modern science, we are faced with the argument that Chinese efforts were both tardy and derivative. Studies of the appropriation of science have been more successful at highlighting Chinese initiative and originality, but in doing so they have often suggested that Chinese interest in science was really about something else, such as nationalism, professional advancement, or self-fashioning. While these auxiliary ends have always been crucial to the practice and development of science, in the Chinese case they have drawn attention away from technical and conceptual accom-

---

1 I would like to thank Jing Tsu and Benjamin Elman for their insight and effort in organizing the Yale conference "Routes of Science" and editing this volume. Parts of this article first appeared in my recent book entitled *Unearthing the Nation: Modern Geology and Nationalism in Republican China*. Permission to publish those parts has been granted by the University of Chicago Press.

plishments and focused it on the instrumentalization of science instead. Ironically, the prominence of Republican era science promoters, who imagined science as an all-encompassing abstraction, has only reinforced the notion that Chinese science was full of bluster but short on substance.

Recent work on the activities of Chinese scientists has provided a welcome corrective, but by foregrounding the quality of scientific research, these studies can inadvertently play into the same narrative of mastery and progress that underlies older reception theories. For us to see that our actors are doing "good science," new practitioners must demonstrate proficiency or excellence by recognized standards. If, however, we know that science has been properly grasped (or appropriated) only when it looks and functions like science elsewhere, then as analysts we are allowing an asymmetry in which shifts in theory or practice at the metropoles signal innovation, while adaptations in newer communities flag inexperience or compromise. Though we have been at pains to show the contextual specificity of science in new settings and have carefully selected cases to circumvent mastery from becoming the dominant discourse, historians of the Republican era often deal with actors whose own positivism and boundary-maintenance activities invite this line of thinking.

As the chapters in this volume show, Chinese scientific activity took many different forms, but whether these involved translation, domestic industry, academic laboratories, or fieldwork, expertise and authenticity were common preoccupations, and mastery of science mattered. Still, these chapters also suggest that while the value of science was regularly discussed in terms of objectivity, progress, and truth, Chinese who were directly involved with *doing* science were well aware that science was conducted through complex systems of access, credit, and circulation. Their insistence on mastery was as much about the right to speak as it was about quality, and scientific authority claims were strongly shaped by concerns of communicability and legibility.

This insight opens up interesting possibilities for historians, who can use the Chinese experience to explore the ways in which processes of exchange rendered differences commensurable enough to have assignable scientific merit. It also allows us to separate "good science" from fixed standards of comparison, by reminding us of how contingent such judgments are on the traffic of information. Here the case of modern geology is particularly instructive because the Chinese community's accomplishments gave them reason to anticipate international endorsement long before it was actually achieved, and the story of their entrée into foreign scientific circles shows a canny understanding of both the opportunities and the dangers of performing mastery on the international stage.

Ready for Their Close-up

In the wake of the 1911 revolution that overthrew the imperial order, the seeds of modern geology that had been sown by text translations and mining ventures in the last years of the Qing dynasty began to sprout with unexpected vigor. Returned students who had studied geology abroad, such as Ding Wenjiang 丁文江 (1887–1936), Zhang Hongzhao 章鴻釗 (1877–1951), Weng Wenhao 翁文灝 (1889–1971), and later Li Siguang 李四光 (1889–1971), were committed to training a native corps of geologists on Chinese soil, and through their determined efforts first a Geological School (Dizhi yanjiusuo 地質研究所, 1913–1916) and later a Department of Geology (established 1917) at Peking University took shape.[2] The Geological School, which was a short-lived but tightly knit community of passionately committed students and mentors, employed modern classroom and field methods to produce the basic staff for a newly established Geological Survey of China (1916–1950).[3] The Peking University Department of Geology, which shared several key personnel with the Survey, then put geological education in China on a stable footing and furnished the model on which many other geology departments would develop in the 1920s and 1930s. The best and brightest of China's young geologists moved easily between department and Survey, putting local twists on foreign methods and concepts and expanding the native geological community through personal contacts.

Despite political disruptions and a limited budget, by 1919, the Geological Survey had amassed enough field research of broad geological interest to establish a *Geological Bulletin* (*Dizhi huibao* 地質彙報, also known as the *Bulletin of the Geological Survey of China*), and in the following year it began putting out longer, more focused studies in several new series of *Memoirs* (*Dizhi zhuanbao* 地質專報). The founding of the *Geological Bulletin* was a particular point of pride for Chinese geologists, and it was intended to make a splash in international geological circles. After all, the Chinese Survey was not a colonial enterprise founded and run by foreigners but a native-run institu-

---

2   I am translating 地质研究所 (lit. "geology research institute") as the "Geological School" to clarify its function and distinguish it from the later Research Institute of Geology in the Academia Sinica. This choice follows two official publications, Andersson 1921, which uses "Geological School," and National Geological Survey of China 1931, which uses "School of Geology."

3   The Survey was actually founded in 1913, but with only one member and without any real financial support from the government. The year 1916, which is traditionally recognized as the beginning of the Survey, marks the intake of Geological School graduates and the start of serious geological fieldwork.

tion in which locally trained recruits and foreign experts worked together in the interests of the young Chinese Republic. By the end of its first decade, Chinese geologists had defied the odds to clarify the basic structure of large parts of northern and eastern China, and the Survey counted such luminaries as former director of the Geological Survey of Sweden, Johan Gunnar Andersson, and distinguished American paleontologist and geologist Amadeus Grabau among its members.[4] It seemed only fitting that the *Geological Bulletin* should highlight both the local and the international nature of the Survey.

To accomplish this, Chinese Survey members wrote their findings up in Chinese, while foreign Survey members were encouraged to use English. Since Western languages were printed horizontally and bound on the left, and Chinese was traditionally printed vertically and bound on the right, each issue could seamlessly divide papers by language category, with every article appearing in a full original version in one half and either a Chinese or English abridgement in the other.[5] Each section had its own cover and table of contents, and the two halves of each issue extended from the covers inward, with pages numbered in ascending Chinese or Arabic numerals so that neither section took precedence over the other. In design, the *Geological Bulletin* seemed ideally balanced and egalitarian, but in practical terms, this linguistic symmetry—or, more properly, chirality—hid the research of Chinese geologists in plain view. Like a Rorschach blot, each issue of the *Geological Bulletin* revealed the unspoken ambitions and assumptions of the reader, and while Chinese readers who read both Chinese and English were pleased to find their native language taken seriously as a scientific medium, foreign scientists limited to the English section found little incentive to decode the Chinese findings at the "back" of the journal. Though visually neat, the Survey's Janus-faced bulletin was in fact a physical representation of the deeper failure of local and international geological circles to meaningfully communicate.

---

4 Andersson discusses his own involvement in several sources, but a good account can also be found in Fiskesjö and Chen 2004. For more on Grabau, who came to China as both professor at Peking University and chief paleontologist of the Geological Survey in 1920, see Mazur 2004, 224.

5 One might consider, for comparison, bilingual journals in Canada, which reproduce text in French and English either on each page, on facing pages, or in segregated halves that require the reader to flip the issue over to read each language properly. In each case, the reader finds visual or physical cues to indicate the parity of the two languages, unlike in the *Geological Bulletin*, where the shared binding makes this parity obvious to Chinese readers and opaque to foreign readers (unless those readers are looking for it).

While many arguments have been made about the suitability of Chinese as a scientific language,[6] this supposed failure was not the result of any fatal linguistic flaw but of the power dynamics of the international scientific community in the early twentieth century. The visual and haptic properties of the Survey's *Geological Bulletin* might have made it easier for readers to privilege knowledge presented in a familiar language, but this familiarity was itself symptomatic of ingrained scientific hierarchies and expectations, for Chinese geologists did not have the luxury of reading only their mother tongue. Everything between the two covers mattered for them, and all serious Chinese researchers were multilingual by necessity.

Language choice linked knowledge to politics and frequently entangled estimates of value in questions of access. But if language could separate, it could also connect, and as the history of geological journals in Republican China demonstrates, all these functions could be deployed selectively. For Chinese, geological reports in English and Chinese had their own personalities and prospects, and it was sometimes just as valuable to "hide" behind Chinese as it was to reach out through Western languages. By taking a closer look at the serial publications of the Geological Society of China from the 1920s to the 1940s (in contrast to the *Geological Bulletin* of the Chinese Survey and others), we can not only analyze the gatekeeping potential of language choice but trace the Chinese geological community's use of publication language to fashion a collective identity within the shifting currents of international science.

Struggling to Be Heard

Despite a boom in the periodical press during the late Qing and early Republican periods, the development of China's native geological research capacity quickly outgrew available resources for disseminating geological findings.[7] In 1910, a journal like the *Earth Sciences Journal* (*Dixue zazhi* 地學雜誌, established 1909) was able to reach an outward-looking, politically conscious readership, and Zhang Hongzhao used it as a platform to set the

---

6   As recently as 1999, in his retirement lecture at the Eidgenössische technische Hochschule Zürich, geologist Kenneth Hsü discussed the propensity for innovation in science in terms of the orality of European languages and the visuality of written Chinese. See Hsü 1999. For a very different approach to Chinese as a scientific language, see Halliday 1993.

7   There are many new sources on the boom in book and periodical printing, but two useful starting points are Goodman 2004 and Judge 2006.

wheels in motion for institutions like the Geological Survey.[8] Once the Survey was active, however, the *Earth Sciences Journal* was no longer a suitable outlet. It continued to publish articles on geology in the late 1910s, but these were largely introductory essays or digests of foreign works meant to familiarize the educated public with scientific concepts and prepare it for the demands of modern citizenship. Though it had once carried the first explicitly geological maps ever published in Chinese, the journal was predominantly concerned with physical and economic geography, and this focus became increasingly explicit over time.[9]

The most basic problem that Chinese geologists had in the first decades of the twentieth century was reaching an audience that was qualified to evaluate and utilize their work. Domestic Chinese newspapers, which previously carried the bulk of writing on topics such as earthquakes, topographical features, and mineral resources, were largely uninterested in more technical reports.[10] Several universities began to put out scientific journals, but these were limited to a small community and often pedagogical in nature. Regional administrative journals and company mining reports tended to focus exclusively on resources and applications.

Until 1919, the Geological Survey of China circulated its work-in-progress informally and then published results of practical significance in ministry gazette (*gongbao*).[11] However, this left little scope for elaborating on the more theoretical aspects of Survey research, and it did not do much to attract those scattered Chinese outside officialdom who were beginning to get involved in geology, physical geography, or mining.

After 1918, Chinese geologists who wanted to engage a more scientifically or technically aware readership could publish in *Science* (*Kexue* 科學), the jour-

---

8   For the first of these seminal articles, see Zhang 1910–1911.

9   These first maps were a series of geological and mining maps of Zhili (now Hebei Province) produced by Kuang Rongguang 鄺榮光 from his own observations as chief prospector of the Zhili Provincial Bureau of Mines. Since the *Earth Sciences Journal* was not oriented toward original research and the visual and conceptual idiom of the maps was unfamiliar to readers, they sparked a great deal of patriotic pride but little substantive discussion. See Kuang 1910a   1910b.

10   Most of these articles were in the vein of travel descriptions, native place reports, or event narratives. Student field trips were sometimes reported as news items with scientific interest. Local gazetteers still largely concentrated on physical, political, and economic geography rather than incorporating modern geology.

11   Because of repeated reorganizations in the new Republican government, the Geological Survey of China published reports first in the Ministry of Industry and Commerce *gongbao* and then in the Ministry of Agriculture and Commerce *gongbao*.

nal of the Science Society of China.[12] Still, *Science* catered to a general, rather than a specialized, scientific audience, and in its first decade it concentrated on introducing various new disciplines to China's reading public.[13] Though *Science* did aspire to be a showcase for Chinese scientific advances, most of the geological articles it published dealt with pedagogy, international institutions, metallurgy, or historical overviews of the earth sciences.[14] Several other journals, such as the *Journal of the Association of Chinese and American Engineers* (established 1920) and the *Magazine of Natural History* 博物雜誌 (established 1919), covered various aspects of mining and related fields in the early 1920s, but none of these targeted a specifically geological audience.[15]

Chinese who had studied geology or mining overseas were sometimes able to address knowledgeable readers outside China, but this was not always the case. The full scientific results of Zhang Hongzhao's undergraduate thesis, which marked the first extensive geological fieldwork conducted in China by a Chinese researcher, were never made public, despite Zhang's influential position.[16] Others were more fortunate, but publication abroad did not guarantee an audience. Wang Chongyou, who studied with Amadeus Grabau at Columbia University, put out *Bibliography of the Mineral Wealth and Geology of China* in England and America in 1912, but this was merely a compilation of existing literature, and its impact was so limited that two other Chinese students earned mining and geology degrees in the United States based on comparable work, and both were unaware of Wang's earlier efforts. Neither Parkin Wong's

---

12   *Science* was established in 1915, but because it was published overseas before 1918, authors in China found it difficult to publish in the Science Society's journal until after it relocated to Shanghai and later Nanjing.

13   *Science* dealt with the promotion of science and scientific spirit rather than technical material, though substantive articles increased after the late 1920s.

14   Typical of these contributions were introductory pieces, such as "Zhongguo dizhi gangyao" 中國地質綱要 (Outline of Chinese Geology), or informal discussions, such as "Mantan sichaun de dizhi" 漫談四川的地質 (Conversations on Sichuan Geology). Often these articles did highlight new insights into China's geology, but they summed up relatively solid advances rather than the cutting edge of active research.

15   There were several "natural history" magazines in China around this time, but the most active in Beijing from 1920 to 1925 was published by the Beijing Higher Normal Natural History Society (Beijing gaodeng shifan bowu xuehui 北京高等師範博物學會). This should not be confused with the *Peking Natural History Bulletin* (*Beijing bowu zazhi* 北平博物雜誌), which became far more influential after 1926 and which shared much of its membership with the Geological Society of China.

16   His thesis can be found among the Hangzhou geological reports at the library of the University of Tokyo and is a rebuttal of Ishii Yamajiro's 1909 theory that West Lake was a volcanic formation.

"The Mineral Resources of China" (M.S., Cornell University 1914) nor Wah Seyle Lee's "Bibliography of the Geology of China" (M.S., Stanford University 1916) ever saw the light of day, however, and later geologists in China had to compile their own bibliographies to meet the demands of research.[17]

With the exception of Ding Wenjiang, who published preliminary resource reports in his capacity as director of the Geological Survey, those Chinese who managed to publish in foreign periodicals were often limited to prosaic reporting on the state of Chinese mining or superficial overviews.[18] Wang Chongyou wrote several short reports for the American *Engineering and Mining Journal,* and Weng Wenhao contributed "The Geology of China" to Samuel Couling's *Encyclopedia Sinica* in 1917, but none of these articles pushed any boundaries.[19] When, in 1921, Li Siguang turned his 1918 master's thesis into "An Outline of Chinese Geology" in Britain's *Geological Magazine,* he made a challenging and original synthesis of existing knowledge on Chinese geology available to a Western audience, but forced his compatriots once again to turn to a foreign journal for insight into the geology of their own territory.[20]

Interestingly, while Chinese researchers struggled to find appropriate ways of circulating work on the geology of their homeland, outsiders had a wide variety of options. Some studies of Chinese geology, like those of American geologist Bailey Willis and Swedish explorer Sven Hedin, were backed by foreign institutions and therefore automatically reported by their sponsors. But there was no shortage of specialized journals for less prominent geologists to publish in. Like popular magazines that saw China as distant and exotic, geological and mining journals were eager for news and experiences of Chi-

---

17   Wang Chongyou published a continuation of his earlier work in Shanghai in 1917 but this also had limited circulation. More influential were Young 1935; and Chi 1936, 1942.

18   Ding Wenjiang (as V. K. Ting) had a few articles from the *North China Herald* reprinted in the *Far Eastern Review.* Some of these were in turn reprinted in specialist journals, such as Ting 1917, which did reach a significant audience and was noted for its challenge to the coal estimates of Noah Drake.

19   In contrast, Weng Wenhao's acclaimed doctoral dissertation for the University of Louvain was published in Belgium and widely cited, but this was a local study of Belgium rather than a contribution to the geology of China, and it was published in French as part of ongoing local debates within an active geological community (Wong 1913).

20   Li Siguang's (J. S. Lee's) master's thesis was published as "An Outline of Chinese Geology" in *Geological Magazine* (1921) with the assistance of William Savage Boulton. This also formed the basis of Li's structural geology class but could not reach beyond that small audience.

nese territory, which was still terra incognita from the standpoint of modern geology.

More to the point, Ferdinand von Richthofen and Raphael Pumpelly's late nineteenth-century explorations had intimated that China possessed vast stores of mineral wealth, especially coal and iron, and readers were eager for detailed information.[21] Tantalizing collections of "dragon bones" from Chinese apothecaries hinted at important paleontological mysteries in China's fossil record,[22] and the tectonics of Asia were ever more important for a structural understanding of Europe.[23] Problems of mountain building, isostasy, glaciation, and stratigraphic correlation created a ready-made place for China in existing geological debates, and Western audiences welcomed everything from tidbits of mining news to speculative papers presented at international geological congresses.

Foreign geological workers with only short-term experience in China could speak as experts in diplomatic and consular briefs, company mining reports, and geological society meetings all over America and Europe. Western residents in China also put out expatriate publications, like the *Journal of the Royal Asiatic Society–North China Branch*, to exchange information with each other and other scientific centers.[24] Japanese geologists in China even established their own research institute in Manchuria and had several publications devoted to the geology of China.[25]

Information on Chinese geology peppered the proceedings and journals of scientific societies outside China and was widely scattered through specialized

---

21   As Shellen Xiao Wu points out, even though Richthofen's work was the most widely cited study of Chinese geology and geography from the late nineteenth to early twentieth centuries, it was really only his English-language work that had any broad impact in China. This was partially due to the local availability of his 1870–1872 letters to the Shanghai Chamber of Commerce (published by the *North China Herald*) and partially due to the wider legibility of English in China compared with Richthofen's native German. China's leading geologists were all able to work with German sources, but even for them, English was often preferred. Language affected the ability of foreign work to travel in as well as Chinese work to travel out. See Wu 2010, 68.

22   See Black 1925 and Matthews 1915, which strongly influenced Black.

23   See, e.g., Emmanuel de Margerie's 1913 proposal for "The Geological Map of the World"; and Argand 1922.

24   A few other examples include the *Journal of the West China Border Research Society* (1922) and the *China Journal of Science and Arts* (1923), which served foreign communities in different regions of China. There is also a considerable transnational missionary literature on the geology of China.

25   See Nakano 1970; and Liang and Feng 2002.

journals in countless languages.[26] Because Chinese-language material was unintelligible to all but Japanese geologists, this network of publication and exchange functioned with almost no reference to the Chinese geological community or the Chinese government. Chinese workers could participate only if they were willing and able to operate exclusively through foreign journals, and even so, readers reacted more favorably to reports by their own countrymen or by scientific notables than to those by Chinese themselves.

Against this backdrop, it is not surprising that the Geological Survey of China had hoped to make a splash with its *Geological Bulletin* in 1919. Survey members could discuss geological problems with a small circle of colleagues, but they were still eager for outside stimuli and anxious for their work to be more widely known. Geology could do little to serve the economic needs of the nation unless local companies had access to survey results. Similarly, native geological capacity was unlikely to diversify and grow if universities and interested individuals were excluded from the Survey's most challenging geological discussions. Perhaps even more important than this narrowing of audience was the lingering possibility that China's lack of visibility in international circles would encourage further encroachment from foreigners eager to tap China's territorial resources.[27] In fact, foreign researchers began returning to China almost as soon as the Great War was over, and since the bilingual strategy of the *Geological Bulletin* had not worked as planned, a better solution was needed if local geologists were to reach the many constituencies with a stake in Chinese geology.

A Cosmopolitan Voice

Besides the dedicated corps of Chinese and foreign Survey members, geology was an expanding discipline in universities and technical schools, and both government officials and commercial interests looked to geology for insights into Chinese resources and development. Moreover, foreign activity remained high, and parts of Chinese territory that were difficult for the small group of native geologists to access were often within reach of geologically minded missionaries, educators, mining speculators, or adventurers from abroad.

---

26   See Young 1935 for a sample of this work.

27   Geology was deeply implicated in the later phases of the Great Game as well as in struggles over Manchuria and Mongolia, but the clearest links were between Ferdinand von Richthofen's work on coal and later German claims in the Jiaozhou Bay area. For more, see Steinmetz 2007.

With a push from Yuan Fuli 袁復禮 and Xie Jiarong 謝家榮, two younger members of the Survey who had recently returned from advanced study overseas, the leaders of the Chinese geological community began organizing the Geological Society of China in late 1921. By early 1922, the society was over seventy-five members strong and met regularly in Beijing to "strengthen contacts amongst domestic geological workers" and "initiate scholarly communication and exchange with foreign geological societies."[28]

These gatherings were polyglot affairs, where local students and businessmen, resident foreign geologists, and temporary scientific visitors were equally welcomed by their Chinese hosts. All official society business was conducted in English, but foreigners were encouraged to speak in their native tongues, with English translation as needed.[29] Because Chinese geologists were all trained in English or another foreign language, they were discouraged from using Chinese whenever possible, which Li Siguang noted was "not without a sacrifice to our own convenience." As he explained it years later, the Geological Society of China "adopted the leading Western languages, principally English, as the official means of communication" because Chinese geologists did not want foreign guests to feel "shy" or "less at home" among Chinese speakers.[30]

This bit of altruism turned out to be a savvy move, and it attracted a large number of foreigners from all over China to society meetings. It also made the *Bulletin of the Geological Society of China* (*Zhongguo dizhi xuehui zhi* 中國地質學會誌), which was established in December 1922, a viable forum for non-Chinese geologists. Unlike Western-sponsored journals in China, the *Bulletin* was not bound to any one foreign language because it was not aligned with either a foreign metropole or a particular expatriate community.[31] The majority of its papers were in English, but contributors could submit in any of the leading scientific languages, including French and German. If in the past foreign geologists had lacked motive or opportunity to publish in China, the *Bulletin* provided both.

The combination of China's vast field for exploration and Beijing's cultural attractions lured the most famous geologists working in Asia into the *Bulletin*'s orbit, and the journal's linguistic flexibility meant that, for the first time, geolo-

---

28   Quoted in Yang Zunyi 1993, 288; and cited as part of an interview with Yuan on November 9, 1980, in Xia and Wang 1982, 7n2.

29   See chap. 3 in Shen 2014.

30   Lee 1942, 23–24. Interestingly, even foreign geologists who lived in China for years, like Grabau, did not learn Chinese.

31   Li Siguang actually considered China's short history of modern science as an asset because it freed Chinese scientists from entrenched schools of thought and conservative tradition.

gists of almost any nationality could publish on the spot instead of scattering their results across foreign journals. No single geological journal in the Americas or Europe could boast original research in so many different languages, and the *Bulletin*'s accommodative policy allowed the widest range of geological workers to reach the greatest concentration of other experts on Chinese geology without loss of fidelity or impact.[32] It also guaranteed that the contributions of Chinese authors not only would be accessible to foreign readers but would in fact be indistinguishable from other reports. Essentially, by joining the crowd rather than standing apart, Chinese papers were able to share the limelight.

Indeed, much of the success of the *Bulletin of the Geological Society of China* hinged on the decision to publish in Western scientific languages. Within a few years the *Bulletin* became such an important source for Chinese geology and its bearing on geology at large that non-Chinese workers from comparatively marginal scientific communities also began using the *Bulletin* to advertise their own scientific activities. Swedish, Russian, and Japanese geologists were among the most active groups studying China, but linguistic challenges often made their national journals inaccessible to a wider readership, and Anglo-European geologists frequently overlooked their China studies. By publishing in the *Bulletin of the Geological Society of China*, Swedish, Russian, and Japanese geologists not only publicized their own work but also raised awareness of these local scientific communities and made bibliographies of their compatriots' research available in English, French, or German. Even English-speaking geologists from colonial areas, such as Hong Kong, saw the *Bulletin* as a means of broadening their audience.

Naturally, the benefits of exposure were greatest for Chinese geologists, for whom the *Bulletin* was a primary vehicle. Over half of the papers printed in the *Bulletin* in its first decade were by Chinese geologists (many but by no means all of whom were in the Geological Survey), and these Western-language studies reached almost two hundred libraries in over twenty countries and territories by 1929.[33] Though the Geological Survey and the Geological Society both exchanged their publications with foreign geological organizations, the Sur-

---

32   Even reports and proceedings of the International Geological Congress were not published in multiple languages. Each host country selected either French or English for its published proceedings, and all papers written or presented in another language were translated into this official language, with an abstract provided in the other.

33   These estimates are based on libraries presently holding the *Bulletin of the Geological Society*, archival material on library exchanges, and references to the *Bulletin* in international conference proceedings. Many more libraries hold individual volumes, especially regarding Peking Man. For detailed exchange statistics, see Ji 1942.

vey's *Geological Bulletin* did not raise international awareness of Chinese geologists the way that the *Bulletin of the Geological Society of China* did. The segregation of English and Chinese papers in the *Geological Bulletin* made it too easy for foreign readers accustomed to colonial surveys to assume that the only scientifically productive researchers in China were the foreigners whose publications were printed in full "up front."[34] Instead, the society's journal published all its papers in Western languages without consideration of national origin, and this enabled the papers of Chinese geologists to "travel" wherever the *Bulletin* was found.

With this advantage, the *Bulletin of the Geological Society of China* took care to showcase the breadth of Chinese geology. Because the Geological Society of China was organized as an alternative to the utilitarian strictures of Survey work, it challenged the stereotype that Chinese geologists were only interested in practical applications by encouraging theoretical speculation. The *Bulletin* went so far as to redistribute the more applied papers given at society meetings to other publications, like the *Oriental Engineer*. This gave Survey geologists a dedicated space to explore problems beyond the purview of government research and helped them cooperate with geologists at a growing number of academic and research institutions. The *Bulletin* also published the thesis work of Chinese studying abroad, bringing new ideas and techniques back to China and demonstrating that Chinese interest in geology did not stop at China's borders. At the Geological Society's twenty-year Jubilee in 1942, Li Siguang reflected that:

---

34   The assumptions left unchallenged by the linguistic separation of the *Geological Bulletin* are summed up in a 1913 letter to the "Discussions" section of the US *Mining and Scientific Press* regarding the proposal of a Chinese geological survey: "If the Chinese are well advised, their National Geological Survey or Bureau of Mines, whichever it may be termed, will be organized by experienced American or British mining engineers, and at first will be largely composed of well-trained men, having directly under them picked natives who have had some technical training. In the course of time, the Chinese would naturally replace the foreigner, just as in the Chinese custom service so efficiently organized many years ago by Sir Robert Hart, and which has proved an invaluable mainstay through the stormy and trying days which have overtaken Chinese during the last twenty years" (Garrison 1913, 736). This view survived the establishment of the actual Geological Survey of China and was frequently embodied in references to the Survey that either vouchsafed it on the grounds of its foreign members or disregarded its importance because of its Chinese leadership. While the view was corrected in the 1920s–1940s, it survives in nonscientific circles to this day. Note, e.g., King 2011, 374: "While the Chinese directed the Survey, many research programs were in fact directed by Western scientists, such as Amadeus W. Grabau, an American, and J. G. Anderson [*sic*], a Swede."

Our society, though [it] carries the usual banner of a national organization, is in fact international in character. You need only turn over the
pages on which the names of our members are printed. They come from
all lands, and actively participate here in a fraternal atmosphere that, as
far as I can gather, is seldom so widely and deeply felt in many of the similar organizations in *other* countries.[35]

In its range of topics, diversity of contributors, and linguistic flexibility, the
Bulletin mirrored the international geological community in China and projected a cosmopolitan image of Chinese geology to distant centers all over the
world.

Growing Cacophony

This cosmopolitan character of the Chinese geological community was built
around a core group of returned students and the tight-knit cadre of Geological
School and Peking University graduates produced in the first two decades of
native geological training. These men mingled freely with a growing roster of
scientific visitors in the hothouse of the Republican capital and established an
international reputation for both the Geological Survey and the Geological
Society of China. Those few who did not remain in the Survey taught at Beijing-
area schools and later founded departments of geology in other major cities.
Wherever these early members of the geological community worked, they
were vertically bound to Beijing by student-teacher and subordinate-leader
ties and horizontally linked as former schoolmates and colleagues. They all
honed their geological expertise in eastern China, engaged directly with foreign research through the Geological Society of China or advanced study overseas, and shared an identity as part of China's scientific elite. Once the
Academia Sinica's more theoretically oriented Research Institute of Geology
(established 1928) became active in South China and provincial surveys
cropped up in the wake of the Northern Expedition (1926–1928), the Chinese
geological community began to splinter, but the founding generation of geologists was still influential enough to manage this new diversity and sustain a
sense of camaraderie through personal relationships.

After the outbreak of hostilities in the War of Resistance against Japan in
1937, the geological community swelled in response to wartime needs and rapidly outgrew these informal bonds. It did not help that Chinese geological

---

35   Lee 1942, 23.

leaders, who had been concentrated first in Beijing and then Nanjing, were forced to disperse or take up other duties. Weng Wenhao became minister of economic affairs in 1938, and he tried to keep tabs on the Geological Survey through his chosen successors, Huang Jiqing 黃汲清 and Li Chunyu 李春昱, but his schedule was too demanding, and he could make only occasional appearances at major events. Li Siguang spent the first years of the conflict traveling through the southwest trying to further the national cause while protecting his Guilin-based Research Institute from Guomindang interference.[36] At the other end of the spectrum, Zhang Hongzhao was unable to escape when Japanese forces occupied Beijing (by then, Beiping), and he had to isolate himself in order to avoid collaborating with the enemy. Ding Wenjiang had already died of carbon monoxide poisoning in 1936, and though many wartime geological gatherings were dedicated to his memory, he was no longer around to hold the community together by force of personality, as in years past.

During the war, the Geological Survey of China followed the Guomindang government from Nanjing to the interior and moved as many of its books, equipment, and personnel as possible to Beibei, just north of the wartime capital at Chongqing. The Geological Society relocated to Beibei as well but reduced its meeting schedule to accommodate the changeable wartime context. The first generation of natively trained researchers, now leaders in their own right, were scattered across the interior as they tried to establish new geological field offices and spearhead critical lines of wartime research. For the likes of Xie Jiarong, who moved from the Survey to the Jianghua Mining Bureau in Hunan in 1937 before relocating to Yunnan to head the National Resources Commission's new Mineral Prospecting Office in 1940,[37] or China's foremost vertebrate paleontologist, Yang Zhongjian 楊鍾健, who escaped occupied Beiping to take up posts in Kunming, Xinjiang, and Gansu before being sent abroad to work with the American Museum of Natural History,[38] the society's annual meetings were a precious opportunity to gather with as many old colleagues as could make it to the capital.

Geologists in the independent provincial surveys that sprung up before the war had little opportunity to think about society meetings, however. Those

---

36   Li's relationship with the Guomindang was fraught with tension and too complex to be explained here, but during the War of Resistance, he had to go into hiding to avoid being arrested by his own government. For more, see Ma and Ma 1999, 143–144.

37   Wang 1991. The National Resources Commission established the Xukun Railway Prospecting Bureau in June 1940. This was soon renamed the Southwest Mineral Prospecting Office and, under Xie Jiarong's direction, became the national Mineral Prospecting Office in 1942.

38   Yang Zhongjian 1983, 89–129.

provincial surveys that moved inland to escape occupation, like the Hunan and Jiangxi Surveys, were largely cut off from local funding.[39] Others, such as the Guangdong and Guangxi Survey, were forced to terminate activity completely until the end of the war. A few new local surveys began operations, but mostly with support from the central Survey or new Survey field offices.

In addition to the surveys, wartime migration also displaced several academic departments. The most famous of these, the Department of Geology, Geography, and Meteorology of Lianda, was formed after the amalgamation of Peking University, Qinghua University, and Nankai University. These universities relocated to Changsha in 1937 before making the trek to Kunming in 1938. Laboratory and reference facilities were almost nonexistent, but the Yunnan countryside provided considerable scope for investigation, and new geological students participated in the work of the Kunming branch Survey, the Yunnan Provincial Survey (established 1942), and several interdisciplinary university projects on the natural history of southwest China. Economic hardships, military service, and Japanese bombings led to a high attrition rate at the new university, but over one hundred and fifty students graduated from Lianda's rigorous geology program, and twenty-one of them later became academicians of the Chinese Academy of Sciences.

Not all geology departments fared as well as the one at Lianda. National Central University, whose geology program was second only to Peking University's before the war, followed the government to Chongqing and boasted an impressive roster of professors from the Geological Survey and the Research Institute of Geology. However, wartime conditions in Chongqing limited student fieldwork, and graduates received only basic research training. National Zhongshan University, which was forced to relocate repeatedly during the war, was barely able to maintain a full course load in geology and did not graduate any students after 1941.

Even as training, funding, and supervision deteriorated during the war, mining companies, petroleum-prospecting groups, and large-scale construction projects sprung up in a desperate attempt to sustain China's besieged economy. The demand for basic geological expertise far exceeded the number of available college graduates or geologists, and many new ventures simply advertised for technically minded individuals and offered on-the-job training in required techniques.[40] This proliferation of geological institutions, private

---

39   For details, see "Xiangnan dizhi diaochasuo gaikuang," n.d.; *Jiangxi sheng dizhi diaochasuo gaikuang* 1948; Liu 1985, 532–533.

40   There were many advertisements and calls for new geological workers. See, e.g., "Jingjibu gongkuang tiaozhengchu zhengji jishu renyuan zhanxing banfa" 1939.

companies, and ad hoc educational settings diluted the former unity of the Chinese geological community, and the geographical separation of these organizations magnified the problems of coordination and information exchange that were already beginning to emerge before 1937. Moreover, individuals at the edges of the geological community, who had the least connection to core members, were also the most in need of guidance and support to sustain their activities.

The Geological Society of China was still the natural forum for bringing these diverse new elements together, but many of its features were out of step with the times. Regular meetings, for instance, were no longer practicable, and the *Bulletin of the Geological Society of China* was better suited to promoting Chinese geology internationally than building domestic ties between veterans and new recruits. As the War of Resistance intensified, the international presence declined rapidly, and once the Pacific War began in 1941, there were very few foreigners for the society to host.[41] This shift rendered many of the Geological Society's multilingual practices obsolete, and in the absence of a responsive foreign audience, the use of Western languages (in person and in print) limited Chinese participation to the highly trained old guard without serving any real cosmopolitan function.

Preaching to the Newly Converted

The dislocations and austerities of wartime research made the problems of accommodating a growing native geological community more acute, but cracks had started to show as soon as local surveys and academic departments began spreading beyond the established centers of Beijing/Beiping, Nanjing, and Guangzhou. At the Tenth Annual Meeting in 1933, Weng Wenhao repeated the society leadership's position:

> Our President J. S. Lee [Li Siguang] has already emphasized the importance of continuing the high standard of our *Bulletin* and its representative nature for the whole country. It is sincerely hoped that his appeal for more scientific papers from different places outside Peiping [Beiping] will meet with quick response from our members, and the board of edi-

---

41   Most active during the war was Peter Misch, of Lianda. Amadeus Grabau remained in Beiping, but like Zhang, his activities were restricted because of the Japanese occupation.

tors will receive and publish with great pleasure any communication of scientific signification.[42]

This call was repeated several times, with the editorial board always insisting that they "warmly welcomed" contributions "from all the institutions working on Chinese geology."[43]

These requests from the Chinese geological elite were quite sincere, and Weng and others puzzled over the lack of response. The problem was not that geologists outside Beiping were too modest or unambitious to submit their work. The problem was that the limited resources, overtaxed personnel, and explicitly practical mandates of China's local geological institutions kept "scientific signification" out of reach for the majority of new geological workers. To ask them to publish in Western scientific languages only added insult to injury.

As the Geological Society's sole organ, the *Bulletin* was caught between its established role of representing the best of Chinese geology to the world and its duty to accurately reflect the changing composition of the geological community at home. Most new geological personnel consulted English or German sources out of necessity, for textbooks and reference materials in Chinese were still sorely lacking, but only those few trained in China's elite departments did so with any ease, and linguistic facility marked a complex set of differences.

Once the War of Resistance began, stratification of opportunity, ability, and purpose all intensified in the native geological community. Geological staff hired by local agencies and enterprises to take measurements, collect samples, and identify locations for future study were rarely exposed to much theory. They had even fewer chances to practice writing in Western languages, and they left polishing China's scientific reputation through the *Bulletin* to their more accomplished colleagues.

Finally, at the Twelfth Annual Meeting in 1936, Xie Jiarong proposed that the society establish a Chinese-language bimonthly, called the *Geological Review* (*Dizhi lunping* 地質論評). The motion was carried unanimously, and the *Review*'s first issue appeared before the end of the year, with Xie as chief editor and a roster of China's younger stars heading nine subdisciplinary departments.[44] According to the society's description of its publications, the *Review*

---

42   "Proceedings of the 10th Annual Meeting of the Geological Society" 1933, 9.

43   "Proceedings of the 11th Annual Meeting of the Geological Society" 1935, 5–6.

44   "Proceedings of the 12th Annual Meeting of the Geological Society" 1936, 13. The board of editors was divided into departments for dynamic geology, stratigraphy, structural geology, petrology and mineralogy, vertebrate paleontology, invertebrate paleontology, paleobotany, economic geology, and physiography.

would "introduce the newest geological concepts," "summarize research results," and offer book reviews, news, and notes.[45] The formation of the *Review* enabled the *Bulletin* to remain "the only important journal for the study of Chinese geology" by once again freeing it to focus on advanced scientific contributions without worrying about questions of inclusiveness.[46] While the *Review* served the society's native majority, the reinvigorated *Bulletin* continued to publish the work of China's best and brightest in internationally accessible form, and it kept foreign work on China available to domestic audiences long after the Geological Society's hosting days had waned.[47]

In his introduction to the 1942 *Overview of the Geological Society of China*, Weng Wenhao explained that the Geological Society "published the *Bulletin* first in order to reach out to the world and put out the *Review* second in order to present [geology] to the people of the nation." He described the two journals as natural complements, grouping them both under the society's ideal of "not asking where a contribution comes from" in the interest of "spreading achievements to all colleagues everywhere."[48] This approach was indeed the keystone of China's cosmopolitan reputation, but Weng's retrospective depiction glosses over the *Bulletin*'s struggles for geographical and institutional diversity within China and typifies the pioneering generation's ongoing preoccupation with bringing China into the scientific community of nations.

For China's younger geological leaders, who discovered geology during the May Fourth era and received rigorous native training before going abroad, the story of the *Bulletin* and the *Review* was less about orderly progress than about unfortunate necessity. In his introduction to the *Review*'s inaugural issue, Xie Jiarong was apologetic that the society's first publication officially adopted Western languages only, explaining that, at the time, "there was no alternative" and hoping the choice was not mistaken for "a purposeful attempt to be unconventional." Xie had helped draft the society's constitution in 1922, and he admitted that it had "truly been a regret" that the society did not have a Chinese

---

45   "Zhongguo dizhi xuehui chubanpin" 1937. It also welcomed contributions from interested nonmembers ("Dizhi lunping tougao jianzhang" 1937).

46   "Zhongguo dizhi xuehui chubanpin" 1937.

47   Many foreign papers that were only announced and not actually presented at annual meetings were still included in the *Bulletin*, and over time foreign participation in the society became dominated by publication rather than attendance at meetings. The *Bulletin* thus kept China in touch with the outside world, even through times of physical inaccessibility such as the War of Resistance and the later civil war.

48   Weng 1942, 1.

journal. Despite the great success of the *Bulletin*, Xie wondered, "How can the academic publications of a country depend exclusively on foreign languages?"[49]

This was not merely a point of pride, and the question called attention to the native geological community's shift from an exclusive circle that could communicate directly to an increasingly anonymous and disparate assortment of workers in need of both coordination and coherence. By publishing in Chinese, the *Review* not only made geological knowledge accessible to new audiences and a wider pool of domestic contributors but also created a much-needed "private" space where Chinese geologists could communicate on a national scale without the pressures of international scrutiny. No serious Chinese geologist could avoid learning at least one foreign language, but foreign geologists had no need to learn Chinese to study Chinese geology, and, with the exception of certain Japanese researchers, none of those living and working in China ever learned more than a few conversational phrases. Even Amadeus Grabau, who died in Beijing after almost thirty years at the heart of the Chinese geological community, worked exclusively in Western languages and depended on Chinese collaborators for local data. In scientific circles, Chinese was a medium that carried knowledge inward but did not export it outward. Though every society member was entitled to a copy of both the *Bulletin* and the *Review*, the *Review* did not even have an English table of contents, and its substance was thus completely opaque to most foreigners.

Relieved of the need to either demonstrate China's scientific prowess to the outside world or hold foreign interest, the *Review* devoted itself to the practical problems of the native community, chief among which were isolation and scarcity. By publishing rougher, more preliminary research, it allowed geologists in the periphery to get feedback from distant colleagues and subdisciplinary specialists on scientific questions. It also put regional mineral surveys and applied research in a wider context, aggregating local data and facilitating collaboration. These articles, while less polished or sophisticated than those of the *Bulletin*, still represented serious geological inquiries, and they served the middle field of Chinese geologists as well as more senior scientists hoping to circulate preliminary findings.

However, the *Review* really came into its own during the War of Resistance, when everyone, from less trained recruits needing basic information to established researchers struggling to stay connected, all found something in its pages. For them, the *Review* circulated firsthand accounts of international conferences and foreign geological organizations, reports of new instruments and methods, and professional information, such as job announcements and fund-

---

49   Xie 1936, 1.

ing opportunities. Its book reviews were substantive digests of important articles and reports on all aspects of geology. They provided Chinese-language
summaries of research by Chinese and foreign geologists alike, on both Chinese
and non-Chinese topics. These reviews covered rare and recently acquired
materials as often as newly published work, making important references
available to geologists in remote and underfunded areas.[50] Similarly, students
studying overseas frequently contributed reviews of literature located abroad,
alerting compatriots to critical new citations that would otherwise be unavailable. Through its features and reviews, the *Review* concentrated and redistrib-
uted scarce resources, and by lowering the bar to include such prosaic items, it
actually raised the general standard of Chinese research.

The *Review* traded refinement for functionality in terms of production values as well. At forty cents, its cover price for nonmembers was five times lower
than that of the *Bulletin*, making it more affordable for students. During the
War of Resistance, when the *Bulletin* opted to print in Hong Kong and accept
delays rather than compromise quality, the *Review* switched to newsprint and
cheaper ink.[51] Though wartime stoppages were inevitable, the *Review*'s commitment to timeliness made it a vital source of community news.[52] Besides
summaries of research activities at China's major geological institutions, the
news section announced upcoming publications, scholarships and awards,
student field trips, and open events like talks and exhibitions. It also communicated important initiatives, like the national map project and donation
drives for money, maps, or instruments. Personal items such as marriages and
illnesses were also covered, as were updates on geologists thought to be missing or dead. As people and institutions were reassigned and relocated, the
Review helped geologists keep track of each other despite the rapid changes
and uncertainty of war. In each issue Chinese geologists read about who was in
the field, which colleagues were studying abroad, and what areas were safe for
research.

In addition to building community through mutual aid and reportage, the
Review also encouraged scientific discussion. The *Bulletin* did not shy away
from controversial theories, like eustasy or continental drift, but as a publication designed to accommodate a transient foreign population and represent

---

50   Xie 1936, 1.

51   The Geological Society and *Bulletin* followed the government to Changsha and then
       Chongqing but had difficulty finding local printers who could produce high-quality
       plates.

52   The *Bulletin* had a news and notes section at one time but canceled it because it was too
       "local."

the breadth of Chinese geology, it rarely had enough continuity to support extended debate. The *Bulletin* was, moreover, explicitly intended to provide a congenial atmosphere for scientific cooperation.[53] The *Review*, on the other hand, was not bound by the constraints of hosting, and it catered to an extremely invested readership with complex personal loyalties. Within the safety of the Chinese-language context, even rank-and-file native geologists felt comfortable directly engaging one another on the quality and interpretation of data, and the editors of the *Review* welcomed "scientific discussion by correspondence." A few of these exchanges, such as those regarding Quaternary glaciation, were indecorously ad hominem, but more often than not they were nuanced and constructive, establishing, for example, the basis for standardization of terminology, mapping practices, and periodization.

Many other things that could not easily be expressed to the outside world in the *Bulletin* fit quite naturally in the *Review*, and leaders of the Geological Society often used the Chinese-language journal to bring their personal influence to bear on difficult issues. The war with Japan, which dominated all aspects of Chinese life from 1937 to 1945, appeared in the *Bulletin* only as occasional references to "unpleasant conditions ... caused by the present day situation"[54] or "frequent bombardment."[55] But while the *Bulletin* maintained as much of a business-as-usual air as possible, the *Review* featured war-related articles quite often and did not mince words. The "private" space of the *Review* allowed Weng Wenhao to address the sensitive issue of loyalty and collaboration with the enemy head on, and in one of two open letters meant to provide a "compass," or guide to action, for the entire Chinese geological community, Weng wrote with great emotion:

> Scientific truths do not have national borders, but scientific personnel, scientific data, and scientific workspaces all do. We absolutely cannot use science as a pretext to abandon the nation and cannot use the excuse of protecting our institution's data to forget the Republic of China. Under present circumstances, we would rather sacrifice everything that our institutions have to win back our character and preserve our national dig-

---

53   Because the Geological Society framed itself as cosmopolitan, the *Bulletin* was in many ways quite sanitized. Unlike established foreign scientific journals, which often encouraged heated debate on matters of scientific importance, the *Bulletin* was designed to present Chinese geology in its best, most welcoming light. Though Chinese geologists had made great advances, they were still concerned about projecting an image of decorum and erudition to the international scientific community.

54   "Proceedings of the 14th Annual Meeting of the Geological Society" 1938, 1.

55   "Proceedings of the 17th Annual Meeting" 1941, 119.

nity. My saying this is not without cause; these are reactions to actual circumstances, so it grieves me even more to speak.[56]

Practical matters of scientific conduct and research ethics that were important for shoring up the uneven quality wartime geology but awkward to publicize in English were treated with similar candor, and Yang Zhongjian and others frequently offered heartfelt opinion pieces to Chinese geologists at large.

These personal messages from the geological elite to the native community helped them extend their influence over new geological workers who fell outside their direct pedagogical and professional networks. With this authority, leaders set research agendas and tried to discipline the increasingly disparate Chinese geological community on everything from the labeling of samples to the etiquette of scientific criticism. The *Review*'s mixture of original research and community features made coordination of wartime mobilization and the maintenance of scientific standards possible when direct contact was not. Though the *Review* was not the standard-bearer for Chinese geology in international circles, the familiarity and shelter of the Chinese language enabled the *Review* to realistically address the increasingly complex nature of the domestic geological community without compromising the hard-won reputation that the first generations of Chinese geologists had struggled to establish.

Conclusion

The *Bulletin of the Geological Society of China* defied both the ongoing threat of imperialism and the fractured nature of the Republican Chinese state by engaging the international geological community and promoting a cosmopolitan view of Chinese geologists. The cornerstone of this cosmopolitanism, which was foregrounded in the sounds and sociability of society meetings as well as in the text and type of the *Bulletin*, was multilingualism. Though dependence on foreign languages for access to science made mastery of geology challenging for Chinese practitioners, it also conferred a notable advantage upon those accomplished geologists who could speak and read English, French, or German (sometimes even Russian or Japanese) in addition to their native Chinese. By being tolerant of linguistic diversity, the Geological Society of China and the *Bulletin* became critical collection points for all major research on the geology of China. In so doing, they gained entry into global scientific networks and discourse, and what was once a linguistic liability

---

56   Weng 1938, 98.

became a casual display of virtuosity. The *Bulletin* was a showcase for Chinese geology that strengthened the local community by facing outward.

The *Geological Review*, in contrast, was established during a period of great scarcity and urgency to expand the scope of professionalization to a new demographic. Over time the linguistic positioning of the *Bulletin* began to exclude Chinese involved in geology in less elite settings, and so the national language was used to strengthen internal ties across domestic, rather than international, space. As war with Japan became a reality, the *Review*'s deployment of Chinese actually carved out a "private" space for the selective communication of insider concerns across vast distances. Far from being immured by the seeming insurmountability of the Chinese language, the Chinese geological community felt protected from unwanted displays of its inevitable growing pains, and it developed a journal culture that replaced the prewar vitality of Geological Society meetings with vigorous written exchanges. This process was by some measures a step back from the polish of the *Bulletin*, but it not only fostered growth through hard times but also pushed the Chinese geological community past its early focus on representing and staging expertise to the more banal but sustainable goal of diversifying native geological capacity.

Despite their very different character, the Geological Society's two organs and their complementary use of language clearly revealed anxieties about the aims and conduct of modern science. Was science powerful because of its theoretical claims or because of its empirical foundations? Did scientific authority stem from the open examination of problems and ambiguities or from the exclusion of doubt and dissent? Should science be cultivated at the top or at the grass roots? And what did the answers to these questions suggest about China's identity as a latecomer to modern science? These anxieties reflect the understudied diversity of motivations that Republican Chinese geologists had for pursuing science and highlight a central tension throughout much of China's scientific history in the twentieth century: was scientific internationalism China's pathway to self-assertion, or was self-assertion China's royal road to a cosmopolitan universal? Looking at the case of Chinese geologists, we might also ask whether legibility and transparency are the secret to science's efficacy, or whether global science actually requires impenetrable corners where new ideas and practitioners can test their wings.

Bibliography

Andersson, J. G. 1921. "The National Geological Survey of China," *Geografiska Annaler* 3: 305–310.

Argand, Émile 1922. "La tectonique de l'Asie," *International Geological Congress, XIII* 1, 5: 171–372.

Black, Davidson 1925. *Asia and the Dispersal of Primates: A Study in Ancient Geography of Asia and Its Bearing on the Ancestry of Man.* Peking: Geological Survey of China.

Chi, Y. S. 計榮森 1936. *Science Bibliography of China, V. Section of Geology* 中國科學著作目錄, 第五組: 地質學. Nanking: National Research Council, Academia Sinica.

——— 1942. *Bibliography of Chinese Geology for the Years 1936–1940* 中國地質文獻目錄民国二十五年至二十九年. Pehpe: National Geological Survey of China.

de Margerie, Emmanuel 1913. "The Geological Map of the World," in *Compte-rendu de la XIIe session du Congrès géologique international, Canada* 1913. Ottawa: Impr. du gouvernement 182–188.

"Dizhi lunping tougao jianzhang" 地質論評投稿簡章 [General regulations for draft submissions to the *Dizhi lunping*] 1937. *Geological Review* 地質論評 2, 2: 222.

Fiskesjö, Magnus, and Chen Xingcan 2004. *China before China: The Discovery of China's Prehistory.* Stockholm: Museum of Far Eastern Antiquities.

Garrison, F. Lynwood 1913. "A Geological Survey of China," *Mining and Scientific Press* 107: 735–737.

Goodman, Bryna 2004. "Networks of News: Power, Language and Transnational Dimensions of the Chinese Press 1850–1949," *China Review* 4, 1: 1–10.

Halliday, M. A. K. 1993. "The Analysis of Scientific Texts in English and Chinese," in M. A. K. Halliday and J. R. Martin, eds., *Writing Science: Literacy and Discursive Power.* Milton Park, England: Routledge 124–132.

Hsü, Kenneth J. 1999. "Why Isaac Newton Was Not a Chinese," http://www.searchand discovery.com/documents/Hsu/newton.htm (last accessed January 9, 2012).

Ji Rongsen 計榮森, ed. 1942. *Zhongguo dizhi xuehui gaikuang* 中國地質學會概況 [Overview of the Geological Society of China]. Chongqing: Zhongguo dizhixuehui.

Jiangxi sheng dizhi diaochasuo guikuang 江西省地質調查所概況 [Overview of the Jiangxi provincial geological survey] 1948. Nanchang: Jiangxi sheng zhengfu jiansheting.

"Jingjibu gongkuang tiaozhengchu zhengji jishu renyuan zhanxing banfa" 經濟部工礦調整处徵集技術人員暫行辦法 [Ministry of Economic Affairs industrial and mining adjustment administration provisional personnel recruitment measures] 1939. *Geological Review* 地質論評 4, 2: 151–152.

Judge, Joan 2006. "The Power of Print? Print Capitalism and the News Media in Late Qing and Republican China," *Harvard Journal of Asiatic Studies* 66, 1: 233–254.

King, Ursula 2011. *Teilhard de Chardin and Eastern Religions.* Mahwah, NJ: Paulist Press.

Kuang Rongguang 鄺榮光 1910a. "Zhili dizhi tu" 直隸地質圖 [Geological map of Zhili], *Earth Sciences Journal* 地學雜誌 1, 1: n.p.

＿＿ 1910b. "Zhili kuangchan tu" 直隸礦產圖 [Mining map of Zhili], *Earth Sciences Journal* 地學雜誌 1, 2: n.p.

Lee, J. S. 1918. "The Geology of China." M.S. thesis, University of Birmingham.

＿＿ 1921. "An Outline of Chinese Geology," *Geological Magazine* 58: 259–265, 324–329, 370–377, 409–420.

＿＿ 1942. "Reflections on Twenty Years' Experience," *Bulletin of the Geological Society of China* 22, 1–2: 21–47.

Liang Bo 梁波 and Feng Hui 冯炜 2002. "Mantie dizhi diaochasuo" 满铁地质调查所 [Geological Survey Institute of South Manchuria Railway Company], *Kexue yanjiu* 科学研究 20, 3.

Liu Zhaomin 劉昭民 1985. *Zhonghua dizhixue shi* 中華地質學史 [History of Chinese geology]. Taipei: Taiwan shangwu yinshuju.

Ma Shengyun 马胜云 and Ma Lan 马兰 1999. *Li Siguang nianpu* 李四光年谱 [Chronicle of Li Siguang]. Beijing: Dizhi chubanshe.

Matthews, W. D. 1915. *Climate and Evolution.* New York: New York Academy of Sciences.

Mazur, Allan 2004. *A Romance in Natural History: The Lives and Works of Amadeus Grabau and Mary Antin.* Syracuse, NY: Garret.

Nakano Takamasa 1970. "Some Prevailing Trends of the Historical Development of Geosciences in the Far East," *Geoforum* 1, 3: 63–80.

National Geological Survey of China 1931. *The National Geological Survey of China 1916–1931: A Summary of Its Work during the First Fifteen Years of Its Establishment.* Beiping: Geological Survey of China.

"Proceedings of the 10th Annual Meeting of the Geological Society" 1933. *Bulletin of the Geological Society of China* 13, 1: 1–14.

"Proceedings of the 11th Annual Meeting of the Geological Society" 1935. *Bulletin of the Geological Society of China* 14, 1: 1–22.

"Proceedings of the 12th Annual Meeting of the Geological Society" 1936. *Bulletin of the Geological Society of China* 15, 1: 1–16.

"Proceedings of the 14th Annual Meeting of the Geological Society" 1938. *Bulletin of the Geological Society of China* 18, 1: 1–8.

"Proceedings of the 17th Annual Meeting" 1941. *Bulletin of the Geological Society of China* 21, 2–4: 111–129.

Shen, Grace Yen 2014. *Unearthing the Nation: Modern Geology and Nationalism in Republican China.* Chicago: University of Chicago Press.

Steinmetz, George 2007. *The Devil's Handwriting: Precoloniality and the German Colonial State in Qingdao, Samoa, and Southwest Africa.* Chicago: University of Chicago Press.

Ting, V. K. 1917. "Mineral Resources of China," *Mining Magazine* (London) 17: 188–190.

Wang Yangzhi 王仰之 1991. "Xie Jiarong yi kancha zhongguo shiyou dizhi de xianzhu" 謝家榮一勘查中國石油地質的先驅 [Xie Jiarong: Pioneer in the exploration of China's petroleum geology], *Zhongguo keji shiliao* 中国科技史料 12, 3: 54–55, 59.

Weng Wenhao 翁文灝 1938. "Weng Wenhao xiansheng zaizhi dizhi diaochasuo tongren shu" 翁文灝先生再致地質調查所同人書 [Second letter from Weng Wenhao to his colleagues at the Geological Survey], *Geological Review* 地質論評 3, 1: 97–100.

——1942. "Xu" 序. In Ji Rongsen 計榮森, ed., *Zhongguo dizhi xuehui gaikuang* 中國地質學會概況 [Overview of the Geological Society of China]. Chongqing: Zhongguo dizhixuehui   1–2.

Wong Wen-hao 1913. "Contribution à l'étude de la porphyrite quartzifère de Lessines," *Mémoires de l'Institut géologique de l'Université de Louvain* 1: 298–325.

Wu, Shellen Xiao 2010. "Underground Empires: German Imperialism and the Introduction of Geology in China   1860–1919." Ph.D. dissertation, Princeton University.

Xia Xiangrong 夏湘蓉 and Wang Genyuan 王根元 1982. *Zhongguo dizhi xuehui shi* 中国地质学会史 [History of the Geological Society of China]. Beijing: Dizhi chubanshe.

"Xiangnan dizhi diaochasuo gaikuang" 湖南地質調查所概況 [Overview of the Hunan Geological Survey] n.d. Manuscript.

Xie Jiarong 謝家榮 1936. "Fakanci" 發刊辤 [Statement on the founding of the journal], *Geological Review* 地質論評 1, 1: 1–2.

Yang Zhongjian 杨钟健 1983. *Yang Zhongjian huiyilu* 杨钟健回忆录 [Memoir of Yang Zhongjian]. Beijing: Dizhi chubanshe.

Yang Zunyi 杨遵仪 1993. *Taoli man tianxia: Jinian Yuan Fuli jiaoshou bainian danchen* 桃李满天下: 纪念袁复礼教授百年诞辰 [Students all over the world: Commemorating the hundredth anniversary of Professor Yuan Fuli's birth]. Wuhan: Zhongguo dizhi daxue chubanshe.

Young, T. I. 楊遵儀 1935. *Bibliography of Chinese Geology up to 1934* 中國地質文獻目錄. Peiping: National Academy of Peiping.

Zhang Hongzhao 章鴻釗 1910–1911. "Shijie geguo zhi dizhi diaocha shiye" 世界各國之地質調查事業 [The workings of the geological surveys of the countries of the world], pts. 1–4, *Earth Sciences Journal* 地學雜誌 1, 3–4: 1–4; 2, 12: 1–5; 2, 13: 1–6; 2, 14: 1–5.

"Zhongguo dizhi xuehui chubanpin" 中國地質學會出版品 [Publications of the Geological Society of China] 1937. *Geological Review* 地質論評 2, 3: 304.

Operatic Escapes

Performing Madness in Neuropsychiatric Beijing

Hugh Shapiro

Abstract

This chapter scrutinizes the experience of one rural woman, an abused housewife who appropriated the cultural authority of the drama, capturing the emotional power of its characters. Three interconnected themes predominate: first, the role of opera in marginalized people's encounter with adversity; second, the ambiguous interplay between healing, theater, possession, gods, actors, and patients; third, the formation of a nascent neuropsychiatry in urban China. In this case, the explanatory power of psychiatry confronts the narrative power of cultural identification, which patients drew from the ubiquitous lore of Peking opera characters and narratives. This case study brings to light a method of resistance among the disempowered, by women suffering domestic violence, women with negligible legal or social recourse. The chapter captures competing vocabulary and modalities for understanding individual experiences, analyzing the problem of explaining and translating a person's interiority into the clinical terms of psychiatry. By drawing on detailed medical case histories from China's preeminent neuropsychiatric ward in the 1930s, this chapter posits a sacrosanct-curative-theatrical zone where performance, religious practice, healing, escape, and anxieties about madness comingled.

On February 7, 1933, Zhang Rangzhen, a thirty-year-old woman, appeared at the outpatient clinic of Beijing's preeminent hospital, the Peking Union Medical College, expressing herself in short, stylized sentences, imitating the style of the opera.[1] She told the attending physician that her brain was "out of order." During the exam he noticed needle marks on her chest. Patient Zhang remarked that her alimentary tract contained needles, and the doctor then referred her to radiology and neuropsychiatry. The following day, Zhang returned to the public outpatient service, where a neurologist examined her; she continued communicating in the language of the opera. During the examination she stood up, walked about, and gestured as if performing onstage. The neurologist, Dr. Wei Yulin,[2] recognized the scripts from traditional country drama.

Born in 1903 to a rural household, Zhang Rangzhen had bound feet and had married into another farming family. Her first three years of marriage were

---

1   *Xiehe yiyuan* (XHYY 協和醫院, Peking Union Medical College), Case No. 39305, pp. 2–28. Hereafter cited as XHYY, 39305:2–28. Pseudonyms are used for all patients.

2   Dr. Wei Yulin, associate professor of neurology, served from 1924 to 1942 (Ferguson 1970, 238).

content. Then, in 1931, began the conflict with the in-laws. Zhang found herself at the center of spiraling family conflicts. She was beaten. When intervention by her natal family failed to help, Zhang tried to kill herself. She survived multiple suicide attempts. Thereafter, Zhang began arranging her speech into short sentences, speaking the language of the stage. When she had no other language, Zhang responded to interlocutors with phrases from internalized scripts.

Observers understood Patient Zhang's singing of folk operas as a sign of physical and psychic disarray, as symptomatic of a disordered state. Her family by marriage viewed the performance as a will toward madness. Then, in the hospital, based on her history of self-injury and ongoing dramatic enactment, the neurologists diagnosed Zhang as "psychosis manic depressive." Zhang's own testimony complicates these judgments.

Her case brings to light the position of opera along a spectrum of responses by women suffering domestic violence, women with negligible legal or social recourse. As subversive repudiation, men, too, invoked operatic characters and language. Marginalized by unemployment or alienated by other personal disasters, men grappled with family strife, workplace conflict, municipal authority, or medical scrutiny by acting out scenes from the stage. First as self-expression within the home or in clashes with authority on the street and then as responses to clinical probing inside the city's new meliorative institutions, people suffering diverse forms of duress mimicked the language and gesture of theatrical performance. During interviews with physicians, patients acted out roles of famous characters, sang arias, narrated plot, invoked renowned actors.

This chapter scrutinizes the experience of one rural woman, an abused housewife who appropriated the cultural authority of the drama, capturing the emotional power of its characters. Three interconnected themes predominate: first, the role of opera in marginalized people's encounter with adversity; second, the ambiguous interplay between healing, theater, possession, gods, actors, and patients; third, the formation of a nascent neuropsychiatry in urban China. Consider first the clinical context that chronicled Zhang Rangzhen's experience.

Neuropsychiatry in 1930s Beijing

The opening of the Peking Union Medical College (PUMC, 協和醫院) in Beijing in 1921 followed several centuries of Sino-Euro-American-Japanese medical exchange.[3] This complex history dates at least to the early seventeenth century, to Jesuitical efforts of employing medicine in the service of ecclesiastical goals, and continuing with diverse medical missionary enterprises of the long nineteenth century.[4] Such cross-cultural medical dialogue was inextricably tied up with the wider ambitions of an emergent West, which initially sought favor with the Great Qing Empire (1644–1911).[5] European, Japanese, Russian, and American aspirations were then played out under the penumbra of declining Chinese power in the second half of the nineteenth century.[6] The PUMC grew out of a concurrent surge of American economic power, which witnessed an unprecedented concentration of capital and a concomitant philanthropic urge, in this case, by the Rockefeller family.[7]

Patron Rockefeller had aimed to create a world-class medical facility purposefully insulated from the immediate locale, "in defiance of the social and economic conditions of the country."[8] Within this environment, sustained efforts by Andrew H. Woods, Wei Yulin (魏毓麟), Xu Yingkui (Hsu Ying-k'uei 許英魁), Feng Yingkun (馮應琨), Bingham Dai (戴秉衡), Theron S. Hill, and especially Richard S. Lyman shaped the PUMC into a dominant force in China's new psychiatry for much of the twentieth century.

The neuropsychiatry ward of the PUMC was built up by Richard Lyman, a 1925 graduate of Johns Hopkins who trained under Swiss-born Adolf Meyer. Among the most influential figures in American psychiatry, Meyer had impatiently critiqued the idea that the mind and the body could be understood separately, savaging the notion of mind-body duality as "medically useless."[9] Severing the psychological from the somatic (and the social) only obfuscated a patient's condition.[10] To Meyer, the lived experience of patients must be read as carefully as a brain lesion or the Wassermann Reaction, the antibody test for syphilis. But both must be read. Meyer's "psychobiology" cast a broad net, probing all the unpredictable forces shaping human experience. The methodology set up by Richard Lyman in the PUMC neuropsychiatric ward closely followed Adolf Meyer's eclectic, "commonsense" approach. The other force

---

3   Beijing's name changed several times during the first half of the twentieth century; "Peking" was adopted both by China's postal administration and by Anglophone users.

4   See Elman 2009.

5   See Xiong 1994.

6   See Rogaski 2004.

7   See Bullock 2011.

8   Bullock 1980, 46.

9   Meyer 1915; Grob 1985.

10   Lief 1948; Lamb 2010.

FIGURE 1   *In 1940, a twenty-three-year-old patient of the PUMC psychiatric ward, a woman
            from a Manchu family in decline, diagnosed with schizophrenia, sketched a
            character from the operatic stage. The headdress, face, image on torso, and the
            sword suggest a woman warrior. The meaning of the two words next to the opera
            figure's head,* 梨替 *li ti, is ambiguous. Reading* 梨 *li as "opera troupe," as in* 梨園
            *liyuan, it might be understood as "opera stands in for" (the patient, reality . . . ?) or
            "substituting in an opera troupe" or "playing the part of 'Pear'" (reading* 梨 *as "pear"
            and here being the name of a character). Another reading, when combined with the
            word for "medicine"* (藥 *yao) written beneath the sword, might be "opera instead of
            medication" or "pears instead of medication"* (梨替藥). *(The days before and after
            she executed this sketch, the patient was given sedatives, barbital and chloral
            hydrate, in standard therapeutic doses). The other words written are "pen, ink, paper,
            inkstone, one piece [or dollar], two pieces . . . [up to] ten pieces [or dollars]"* (筆墨紙
            硯, 一塊, 二塊 . . . 十塊 *bi mo zhi yan, yikuai, erkuai . . . shi kuai). Other
            images include plants, insect-like creatures, wineglasses, the Buddhist swastika* 卐
            *wan, and high-heeled shoes. Attending physicians and nurses recorded in their case
            notes the patient's deep involvement in opera, describing her singing on the wards
            and her expert discourse on all aspects of the theater.* SOURCE: XHYY, 71584:26, 34, 44
            (JUNE 1, 1940).

acting on both the PUMC and on Lyman was the Johns Hopkins University
School of Medicine.

The Hopkins paradigm dictated the totalistic mapping of patient experience. This meant discovering as much as possible about the patient's pread-

mission background, while assiduously recording everything that transpired within the hospital. Quotidian details—food eaten, medications consumed, therapies undergone—and other less tangible aspects such as a patient's foibles or random jottings, all were meticulously captured in the record. Sometimes the night shift nursing staff would record the fitful utterances of a patient talking in her or his sleep. Structured interviews conducted by physicians were recorded in the case histories, often word for word with the support of a stenographer. Within the ward, neuropsychiatrists running the therapeutic interviews usually deployed barbiturates such as sodium amytal to encourage the patient to talk uninhibitedly. By the late 1930s, after the eruption of full-scale war with Japan, sherry became the sedative of choice, replacing expensive or unavailable drugs. At other times, patients were supplied with pen and paper and other materials and were encouraged to write, sketch, or paint to express their ideas. Some patients drew images of themselves as archetypes from the operatic stage. In this sense, the new hospitals of Republican era cities created new rhetorical platforms, affording novel stages for self-expression for patients of diverse backgrounds.

Rural, Foot-Bound, Married, Childless

A physical exam revealed that Zhang Rangzhen experienced irregular menstruation and had had no pregnancies. She was given the sedative Luminal (phenobarbital) and sent for X-rays. Five days later, she was admitted to the PUMC neuropsychiatric ward. Once in the ward, Zhang's singing continued uninterrupted for three days. Depending on the opera she performed, Zhang laughed, grew solemn, or wept. However, when asked questions by hospital staff, she stopped and answered in earnest, revealing her story.[11]

Zhang Rangzhen grew up in the countryside, sixty miles outside Beijing. She had hoped to study yet had not been sent to school.[12] Instead, Zhang sat at the feet of village storytellers, imbibing the historical romances. Zhang also viewed sundry "moving pictures" hired into the village by families holding celebra-

---

11   XHYY, 39305:2–28. This narrative draws on interviews with Zhang by neuropsychiatrists, memos recorded by nighttime staff, and investigations of the Social Service Department.
12   Young women aiming to obtain a modern education, especially when combined with unbinding their feet or refusing to bind them, sometimes met with family violence, as did a young woman in 1907 who was coerced into suicide by her in-laws (Judge 2008, 75–76).

tions.[13] We do not know which films she watched, though the earliest movies shot in China in 1905 contained footage of operatic performance by the luminary Tan Xinpei (譚鑫培).[14]

Eager to participate in work, she helped tend sheep and pigs and learned sewing and cooking. She took over the domestic work at age seventeen when her eldest sister married. Engaged at nineteen, her marriage at twenty-three was modestly hypergynous, into a slightly larger farming family of the same township but different village. At this age she had been extremely healthy, with bound feet, no addictions, and, according the neurological exam, no hereditary predisposition toward mental illness.

Zhang got on well with her husband in the first year of marriage. Serving her elders dutifully, she earned the praise of the household's three generations. As time passed though, she found herself at the center of escalating family conflict. Her husband turned out to be the family "idler," resented by harder-working cousins but protected by a doting mother. The exasperated cousins demanded that the family property be divided (*fenjia* 分家). After the division of assets, Zhang's husband's shortcomings became more apparent and costly, leading to tension between Zhang, her husband, and her mother-in-law. On one occasion, Zhang's mother-in-law beat her in the presence of others, after overhearing her criticize her husband. Zhang's conflicts with her in-laws gradually extended to her sisters-in-law, one of whom had married into Zhang's own village, spreading gossip of the infighting back to her natal family.

When Zhang's brother learned of his sister's predicament he visited the in-laws and quarreled bitterly with them, leading to even worse treatment for Zhang. Relations between the families deteriorated, until recriminations erupted into an unspecified "fight" that resulted in both families appearing in court and in Zhang's brother being imprisoned for two and one-half months.[15] Zhang's situation grew intolerable. She expressed remorse at having involved the men of her family.

---

13   Dikötter 2007, 253. By 1904, film projectors from Japan had made their way into China's interior.

14   Xiao 1998, 5.

15   In winning the suit, the in-laws might have drawn on the experience of the second son, who was twenty-five years old and had been a servant in a government office in Harbin, Manchuria, until the upheaval of the early 1930s. The record is silent on the details of the court case.

Suicide

In October 1931, after four years of marriage, Zhang drove between thirty-five and fifty sewing needles into her neck, chest, and upper abdomen. The suicide attempt failed. X-rays in 1933 revealed fourteen needles in different parts of the abdomen, some broken, the longest being six inches in length. However, sixteen months would transpire before she found her way to a hospital. Over the next year, Zhang attempted suicide five more times:

–   October 1931, wrapped five needles in cabbage leaves and swallowed them; followed by attacks of pain.
–   January 1932, ate bitter almonds; "no acute poisoning; only dizziness."
–   February 1932, drank molten tin; experienced hoarseness of voice for one month.
–   February 1932, cut throat with knife.
–   September 1932, swallowed concentrated salt solution (*yanshui* 鹽水) used in the preparation of bean curd.[16]

Following Zhang's final attempt to take her own life, her behavior changed. During this year of despair, Zhang had repressed her anger, lost sleep, and spoke very little. After September 1932, Zhang adopted a new manner of speaking, and she began quarreling and saying unpleasant things to the sisters-in-law. She began expressing herself in short, stylized sentences, imitating the voice of the opera.

Zhang drew vocabulary from the prose and poems she had heard her nephew read while growing up, but for the most part she turned to the opera, especially the tragedies. She repeated the words throughout the day and continued at night, standing in the courtyard, particularly when the celestial bodies were bright. She generally refused food and began to spit blood. She continued in this way for three months.

Self-Induced Madness

The mother-in-law and Zhang's husband were convinced that by uninterrupted singing she purposefully was trying to render herself insane. In December 1932, they responded by hanging Zhang from her thumbs and beat-

---

16   Of seventy-two fatal or attempted suicides reported in 1919 Hangzhou, sixteen had swallowed salt ("Suicides in Hangchow" 1920).

ing her. Beating those perceived as insane was a therapy reported during this period: striking the body could wake up the diseased mind or drive out the possessing spirit.[17] Though her behavior remained unchanged, Zhang's physical condition continued to deteriorate. Her father-in-law insisted that she be treated in Beijing, and in February 1933, the husband gave his wife seven dollars for traveling money and sent her into the city.

In the hospital Zhang explained that her distress grew from a sense of having failed in being a filial daughter to her parents. The notion of the future made her laugh. She had little to say about the in-laws, expressing hatred only for her husband. After six days of taking sodium amytal, her response to "suggestion" was limited, and her singing disrupted the entire ward. When staff informed her that she would be discharged from the hospital if she continued to disturb the other patients, Zhang quieted down, remarking that she was a hopeless case. After doses of morphine and hyoscine, an opioid and analgesic blend given for the "restraint of excited patients," the singing stopped entirely.[18] Her throat damaged from the suicide attempts, a victim of sustained physical and psychological violence, Zhang gave this explanation for her behavior: she spoke in the language of the opera so it would be "easier for others to share in her sad story."[19]

Self-Injury as Escape

Lacking routes of escape, resistance by women coalesced around self-incapacitation. At one extreme was suicide.[20] Suicide by women in the early twentieth century came to be seen as "an act of despair," protesting "the very conditions of marriage that late imperial womanly virtue celebrated."[21] (Conflict with in-laws would continue to motivate suicide among married women in the 1950s, after the establishment of the PRC.)[22] What remained was a spectrum of social and physical self-injury: fasting, sleep avoidance, illness, feigned madness, and self-mutilation. Self-injury as defiance included inappropriate behavior. In

---

17   McCartney 1926, 623. Madness and possession were complexly intertwined in the popular imagination during this period; discussed below.

18   For "Psychoses," "Morphine and scopolamine hypodermically if necessary" (Dieuaide 1936, 35). Scopolamine is also known as hyoscine.

19   XHYY, 39305:12.

20   S. Lee and Kleinman 2003, 296; Wolf 1975; Wong and Phillips 2009.

21   Mann 2011, 128.

22   Diamant 2000, 115.

Zhang's case, this meant the overt and public articulation of private despair in a manner that disrupted the order of a family, a village, a hospital.

Disempowered, this abused rural housewife appropriated the cultural authority of the drama, capturing the emotional power of its characters. As soon as she discovered her opera voice, Zhang Rangzhen openly defended herself against the sisters-in-law; she became critical and went on the offensive, not with impunity but with force. When the hospital staff instructed her to stop or be expelled from the hospital, she stopped. In Zhang's own testimony, she spoke in the legitimate voice of operatic scripts to reveal her suffering. These transcripts, to use James Scott's words, provided "subordinate groups" with a range of "low-profile forms of resistance that dare not speak in their own name."[23] The memorized scripts empowered Zhang "to speak with authority *without* being authorized to speak."[24]

In interviewing rural women about the 1950s, Gail Hershatter found that "women reached for an older language of performance under duress," language available from "popular tales in local histories, operas, and stories from classical texts, circulating in village performances."[25] In the late 1920s, sociologists in the Mass Education Movement transcribed the arias of fifty traditional folk operas in central Hebei Province   120 miles southeast of Beijing, a region not far from Zhang's village.[26] Within the one thousand copied pages of lyrics, "strife between mother-in-law and daughter-in-law" was "probably the most frequent and obvious generational stress point."[27] In *An'er Brings Rice* (*An'er song mi* 安兒送米), a "wronged wife ... has been driven out of the house by her cruel in-laws and has taken refuge in a nunnery."[28] In *Tumbler Invites a Guest* (*Banbudao qingke* 搬不倒請客), the in-laws beat the wife, and after her suicide attempt, it is the "less hardhearted father-in-law" who eventually sympathizes with the daughter-in-law and enables a temporary retreat from hostilities.[29]

The dramas recited by Zhang provided language to unmask anguish. Marked by traumatic experience, her performances were manifestations that appeared at once self-conscious and spontaneous, controlled and uncontrolled, deliber-

---

23   J. Scott 1990, 19.

24   Butler 1997, 157 (emphasis in original).

25   Hershatter 2011, 29.

26   Dong and Ou 2000, 163, 197.

27   Arkush 1990, 88, 89–90. In contrast to folk opera, the "lifelong struggle between mother and daughter-in-law in the Chinese family" did not feature prominently in classical opera (Hsu and Tseng 1974, 170).

28   Arkush 1990, 90.

29   Arkush 1990, 90.

ate articulations of hopelessness, insensible expressions of despair. The processes by which commonplace, widely recognized cultural forms are internalized
in daily life and then, following the experience of trauma, expressed outwardly
as symptoms are analyzed by Arthur Kleinman and Joan Kleinman. To borrow
their pioneering formulation, this phenomenon suggests "how culture infolds
into the body" and "how bodily processes outfold into social space."[30]
Symptoms "infold the social into the psychobiological"; they "*are* the cultural
forms of lived experience."[31] Librettos and stylized movements became part of
a dormant emotional repertoire, manifesting as symptoms in intense moments
of social disaster or personal despair, uttered with allusion, out of place, dis-
concerting. Expressed within a matrix of sexual, legal, and domestic restraint,
and then subjected to medical analysis, the words themselves became the
symptoms of a disordered state, the gesticulations ominous signs of a disturbed
condition. Zhang's family by marriage judged her singing a willful inducement
of madness, an affront to their moral order, and they beat her for it.

The Symbolic Power of Scripts

What gave the operatic transcript its power? Part of the answer must be the
cultural "centrality" of the form, its sheer ubiquity in life and society.[32] The
"mania for opera" in 1800 Beijing was such that "monks and daoists were
accused of performing Kunqu rather than rituals."[33] During the violence of the
Boxer Uprising circa 1900, combatants widely emulated dramaturgical motifs
witnessed on the stage. The "gods by which the Boxers were possessed were
all borrowed from these operas," explains Joseph Esherick.[34] In 1904, Chen
Duxiu, a founding member of the Chinese Communist Party, criticized theater
that features "gods, fairies, ghosts, or demons," precisely because it can "stir
up the ignorant masses and cause a great deal of trouble," as it did with the
"1900 Boxers" who were "definitely influenced by what they saw in the "stage
plays."[35] In the turmoil of the 1911 revolution, men who opposed the clipping of
their queues "began to comb their hair like the Ming characters from the local

---

30   Kleinman and Kleinman 1994, 710–711.
31   Kleinman and Kleinman 1994, 711, 716 (emphasis in original).
32   Johnson 1990, 41–42, 48, 53–54.
33   Naquin 2000, 635.
34   Esherick 1987, 64–65.
35   Chen 1999, 119.

opera," finding alternative models of self-fashioning on the stage.[36] Sidney Gamble's fifteen-month survey conducted from 1918 to 1919 found "theater" to be "probably the most popular sort of organized recreation in Peking," and "theater-going" and "listening to story-tellers" among the city's most passionate distractions.[37] In quiet times, too, not only during millennial chaos, fluency in opera remained conspicuous in the social imagination of city people.[38] David Strand found that rickshaw men, surveyed during the 1920s, expressed their thoughts with "images and ideas taken from the operatic stage."[39] In the mid-twentieth-century countryside, "the cultivation of cotton, the local practice of weaving, the ubiquity of local opera, and the persistent lack of water ... all shaped household economies and modes of sociality."[40]

Power struggles played across the operatic stage. In 1924 Shanghai, a theatergoer booed a Peking opera star as she sang a famous aria. Enraged, admirers of the performer ferociously beat the heckler. In retaliation, a warlord kidnapped and brutalized a crime boss, triggering a shift in Shanghai's power dynamics and hastening the ascent of legendary gangster Du Yuesheng, who went on to organize the world's first international drug cartel.[41] During wartime, from the 1920s to the 1940s, the Nationalists (GMD), Japan's colonial officials in Taiwan, and the Chinese Communist Party (CCP), all exploited the power of theater.[42] The CCP, for example, organized the performance of patriotic dramas in villages, and then in the early PRC, in the 1950s, it mobilized opera troupes to advocate new policies, such as the Marriage Law.[43] Critics mocked politicians, ridiculing their discourse as "no different than singing Peking opera on stage."[44]

But mockery obscures the gravity of public spectacle. Performance in public threatened power. In the late 1920s and early 1930s, authorities strapped leather harnesses to the heads of the condemned, gagging them to prevent the shouting of antigovernment slogans before their execution.[45] The idea of suppressing possible outbursts was rooted in the past, when the condemned being led to the execution grounds might sing arias from the stage, a notion savaged in

---

36   Harrison 2000, 37.
37   Gamble 1921, 224, 35.
38   Li 1992.
39   Strand 1989, 257.
40   Hershatter 2011, 15.
41   Wakeman 1995, 118–120.
42   Qiu 1992; Merkel-Hess 2012.
43   MacFarquhar 1999; Hung 1994; Gao 2004; Altehenger 2010.
44   Strand 2011, 79.
45   Spence and Chin 1996, 93.

Lu Xun's 1921 short story "Ah Q." The practice of silencing those about to be executed was replayed during the Cultural Revolution (1966–1976), when the jaws of condemned public enemies were dislocated.[46]

Radical transformations in identity, status, and gender also played out on the stage. When the 1911 Revolution liberated actors from their lowly status, the stage became "a platform for their political opinions."[47] The spread of photography, mass-produced images, and entertainment newspapers focused "public attention on outstanding actors," enhancing their social circulation and making them "into national and very public personalities."[48] The rise of women's Yue opera in the 1930s fundamentally reversed centuries of gendered practice, as women began performing men's parts.[49] Children learned about moral paragons from storytellers and from the opera.[50] And the "people who grew up with" these stories, argues David Johnson, "could never unlearn them."[51] Indeed, neurologist Wei Yulin, who had recently returned to Beijing from medical training in Philadelphia when he first examined Patient Zhang, identified straightaway the country drama she performed in the clinic. But when did opera threaten?

Self-Induced Possession and Insanity

For her relentless singing and stylized movements, for her alienating behavior, Zhang's affinal family hung her by the thumbs and beat her. Understanding this assault as anything but wanton and senseless is difficult. Zhang's family, in its evaluation, suspected that she aimed to drive herself mad by inducing a voluntary self-possession through operatic incantation. The violence that the family visited on Zhang was conceived (or justified) as curative, administered to startle her spirit or to pry her loose from the grip of possession. This scenario was not new: physicians in the late imperial era prescribed therapeutic violence for female patients perceived as mad, as Fabien Simonis shows. The cura-

---

46   After King Philip's War (1670s), defeated leader Metacom's skull was displayed at Plymouth, Massachusetts. Cotton Mather "wrenched off and took away Metacom's jawbone, completing his silencing" (Taylor 2002, 202).

47   Goldstein 2007, 106; Karl 2002.

48   Yeh 2003, 16.

49   Jiang 2009.

50   Cohen 2008, 49.

51   Johnson 2009, 2.

tive beatings were designed to purge or suppress the women's behavior that had embarrassed their families.[52]

What makes Zhang's case from 1933 especially distinct is that her family judged her chanting of opera as having induced the possessed state. This idea departs from the analyses of scholarly medicine, which generally understood possession as provoked by exogenous forces, such as ghostly qi, or by internal physiological accumulations—that is, by mucus (*tan* 痰)—an influential idea advocated by Zhu Zhenheng in the early fourteenth century.[53] Nor was Zhang's family punishing her as a case of *xi mi* 戲迷, "opera obsession," though exasperated families in the 1930s did commit kin to the asylum in Beijing when they were unable to endure the nonstop singing of opera at home.[54] Nor was the family reacting to any particularly threatening content of Zhang's recited script, although patients did invoke subversive operas such as *Mei Yu Pei* 梅玉配, in which a woman fakes her own death by burning down the family home, thus avoiding an unwanted betrothal, marrying instead her true love.[55] The rural family's anxiety and its brutal treatment of daughter-in-law Zhang divulge uneasiness about the intertwined nature of possession and madness in the medical-religious imagination. This early twentieth-century perception was in part rooted in the late imperial era, when physicians attributed to possession a constellation of symptoms resembling those of madness: "delirium, strange speech, unprovoked grief, extreme apathy, hallucinations and the desire to injure oneself."[56]

Zhang's family's other idea, that she intentionally provoked the possession, was also a broadly witnessed notion. First, consider that early Tantric Buddhism and certain sects of Daoism theorized that the possessed state could be willed into existence.[57] Volitional possession remained evident in the popular imagination into the twentieth century. In 1920, for example, after conducting hundreds of "mental examinations" in Canton, Robert M. Ross reports that nothing "obsesses the mind of [the] insane" as much as possession.[58] Ross implies that patients self-diagnosed and perhaps provoked the possession. Schipper shows poignant manifestations of this idea in 1960s Taiwan, when "[a]nyone can be

---

52   Simonis 2010, 383, 389.

53   Simonis 2010, 124. Zhu Zhenheng aimed to "naturalize the etiology of acts that commoners, ritual specialists, and probably most physicians, commonly attributed to the action of malevolent forces" (Simonis 2010, 178; on "mucus possession," see passim).

54   XHYY, 43542:2 (February 12, 1934).

55   XHYY, 71584:16 (May 31, 1940). On *Mei Yu Pei*, see Huang and Xu 2001, 414, 536.

56   Simonis 2010, 327.

57   Strickmann 2002, 204–206, passim.

58   Ross 1920, 516.

possessed and become a medium."[59] Second, self-induced possession could be used as a weapon against one's enemies, as a form of "soul-attack."[60] Third, neurologists, psychiatrists, and anthropologists expanded the idea of a possession–madness continuum in the early to mid-twentieth century, positing that passively acquired, self-induced, or even faux possession could eventuate in a lasting insanity or schizophrenia.[61] And, last, this menacing idea: the gods often coerced someone into mediumistic service by first visiting terrible suffering, poverty, disease, and even death on the person, bringing him or her back from the ruined state only to yoke the victim into servitude as a medium.[62]

To gain some understanding of how Zhang's family comprehended her behavior—self-injuring, deliriously strange speaking, singing opera at night under the sky, reciting dramatic scripts, denying food, spitting blood, sleep-walking through choreography witnessed onstage—consider the deeper background blurring distinctions between patient and healer, actor and viewer, and priest and worshiper and the web of interconnections between opera, possession, mediumistic healing, and madness. Ritual performers, spirit mediums, and actors, sharing marginally disreputable roles in society, together inhabited an ambiguous sacrosanct-curative-theatrical space. Inside this ambiguous zone, the sick are healed; yet also, the healthy could be made not right, succumbing to serious illness, perhaps madness.

---

59   Schipper 1993, 46 (though this possession is not necessarily tied to madness). In mid-twentieth-century Sudan, a "shamanistic trance ... may materialize anywhere and at any time," and a "fit or trance ... is often self-induced" (Nadel 1946, 35, 29). See also Seligman 1932, 26, 127 ("purposefully induced dissociation").

60   Strickmann 2002, 195.

61   French neurologist Jean Vinchon observed in 1931 that patients who "have become run down as the result of grief or illness ... come under the category of acquired asthenia." These "griefs and illnesses ... inhibit the psychic control and are largely accountable for mediumistic phenomena," possibly leading to a "permanent dissociation of the personality" and "end in insanity" (Vinchon 1931, 120, 116). Anthropologists reached analogous conclusions: "simulated possessions can end in genuine schizophrenia" (Métraux 1955, 34). French psychiatrists working in colonial Indochina in the 1930s accused local spirit mediums of overheating their patrons, thereby "igniting underlying forms of dementia" and inducing mental illness. Because of their practices, "sorcerers" themselves were described as at "special risk" of succumbing to mental illness (Edington 2013, 750).

62   Shahar 1998, 177.

Priests, Spirit Mediums, Actors, Deities

"The originators of all performance styles were shamans."[63] Scholars broadly support this hypothesis, that the origins of theater in remote antiquity are shamanistic.[64] Much later, in the eleventh and twelfth centuries, we know that in China the "earliest known stages for which there is epigraphic evidence" were "constructed in temples."[65] And the siting of stages in temples was not for expediency or need. Early opera in China was "exorcistic" and had been "integrated into the liturgies of temple festivals."[66] "The stage" was "a sacred area."[67]

Creating religious icons for ritual use on temple stages raises questions: what do gods look like, and how do they behave? Court painters of the Ming puzzled over these problems and found a divine archetype in the image of the Dharma King of Tibetan Buddhism, a sacerdotal model guiding their deification of the emperor.[68] In popular religion, the conduct and iconography of gods emerged from spoken lore and popular literature, from puppet theater and staged drama.[69] Into modern times, "the carved representation of the god" in the "temple" and the "operatic representation of the god on stage" were fungible images; "both had divine potency."[70] Deities "even took roles in some performances."[71] Some spectacles left open the "hint ... that a masked participant in the procession of gods might actually be a god."[72] Theater and shrine, the gods dwelled in both.

Priests, spirit mediums, actors: all performed ritual. The paradox of mediumistic ritual is this: it is both scripted and spontaneous. Anticipatable theatrical sequences were capable of creating "spontaneous cosmological power" (*ling* 靈).[73] To act as a medium, to temporarily host the visitation of a god, first as receptacle and then as instrument, priest and actor both observed ritual preparation. In late 1960s Hong Kong, for example, Daoist priests ritually made ready to "perform religious ceremonies in which the role of a god or bodhisat-

---

63 Werle-Burger 1998, 127.

64 Luo 1956; Mackerras http://books.google.com/books/about/Chinese_theater.html?id=7fx WKB9jijgC1983, 8; Kirby 1974.

65 Johnson 2009, 330.

66 Johnson 2009, 330.

67 Schipper 1993, 45; Ch'iu 1991.

68 Fong 1996, 331.

69 Shahar and Weller 1996.

70 Johnson 2009, 127.

71 Johnson 2009, 134.

72 Johnson 2009, 127.

73 Dean and Zheng 2010, 260.

tva is assumed." During the ceremony, the "actor is still a priest but at the same time regarded ... as the god or bodhisattva."[74] In daily life, the priests did not practice vegetarianism, but when preparing for the role, they abstained from meat and sex. In late imperial times, actors portraying gods on the operatic stage ritually cleansed themselves in analogous fashion, abstaining from food and sex, before performing the role.[75] Actors in 1960s and 1970s Taiwan did the same, observing "dietary and sexual restrictions" when playing characters in dramas with religious-ritual meaning.[76] "Taoist masters," in this same context, "sometimes perform as puppeteers."[77]

"The Posture of a General"

Mediums, in articulating their vision, borrowed vocabulary and imagery from the dramatic stage. One spirit medium took "the posture of a general often seen in Chinese opera" and, in a "dramatic manner," suddenly shouted, shook rhythmically, and monotonously incanted.[78] Theater provided the very voice of the god possessing the medium, godlike utterances that clients or patients knew from the stage: "dressed in the costume of the god he represents, the medium speaks in the shrill voice of the Chinese theater actors."[79]

This interplay, the dialectical influence of the symbols of possession and stagecraft, occurs across cultures. "Every case of possession has its theatrical side," its "histrionics and exhibitionism," writes Alfred Métraux, on Haitian voodoo in the mid-twentieth century. The spirit medium "becomes an actor on the spur of the moment." "Unlike the hysteric, who reveals his anguish and his desires through symptoms," the "ritual of possession must conform to the classic image of a mythical personage." "No modification" of the anticipated performance "is permitted."[80] In his classic study of possession in seventeenth-century France, Michel de Certeau observes that in the "theater of the possessed" there "are no longer any human beings ... there are only roles," only "actors on the stage."[81] All the participants "organize that possession ... as

---

74   Topley 1970, 427.
75   Professor Andrea Goldman, personal communication, January 20, 2012.
76   Schipper 1993, 44–45.
77   Schipper 1993, 44.
78   Tseng 1972, 561.
79   Schipper 1993, 47.
80   Métraux 1955, 33, 24, 29, 25.
81   De Certeau 2000, 85, 88.

though in a theater," each one playing "the role assigned to him."[82] In Malaysia, ritual enactment occurs in a "performance frame,"[83] and in China, religious specialists invoke "standardized methods."[84] The "medium's enactments" in Tibet are "mostly predictable."[85] In mid-1940s Sudan, a "forced falsetto" was "often the accepted spirit voice."[86]

To one veteran spectator in 1909 China, John Macgowan of the London Missionary Society, the "sorcerer" is a charlatan, feigning possession. "Repeating certain formulae," he "passes into a state of frenzy"; he "leaps and flings himself about as though he were the most veritable madman that ever escaped a lunatic asylum. During his paroxysms, he pretends that he is inspired by the god."[87] However, despite the broad semantic, visual, and symbolic overlay between theater and possession, between histrionics and religious experience, dramaturgy did not obstruct religious power.

To the contrary, the efficacy of mediumistic healing may very well depend on the client's or patient's ability to read predictable signs of the performance.[88] To be sure, devout spectators recognized the dramaturgical element of possession, and they required authentication. By mutilating the flesh, spirit mediums proved their authenticity.[89] By climbing a ladder with rungs of sharp steel blades or grinding glass in the mouth, mortified flesh is shown impervious to pain, verifying the brief presence of the divine. Still, corporeal proofs are merely part of the agreed-upon sacrament. The audience reads the embodied lexicon, anticipates choreographed liturgy, and responds to its impeccable execution. This participatory element cannot be separated out from the power of ritual healing. Devotional participation boosts therapeutic effect.[90]

Shaman and patient, medium and crowd, actor and audience: electrified, dynamic exchanges pass between them. In contemporary Fujian Province, the "crowd moving as one body expresses the *intercorporeality* of ritual-events."[91] This "intercorporeality of the crowd," Kenneth Dean argues, "enables the rapid transfer of affect/intensity between and through individual bodies." Movements performed by the medium are transmitted as "waves of affect" and are "physi-

---

82   Oughourlian 1991, 78.

83   DeBernardi 2006, 10.

84   AvRuskin 1988, 288.

85   Sidky 2011, 93.

86   Nadel 1946, 35.

87   Macgowan 1909, 100.

88   R. Lee (1989, 262–263) also suggests this idea.

89   Jordan 1972, 79–80.

90   Kleinman, Eisenberg, and Good 1978.

91   Dean 1998, 33 (emphasis in original).

cally acted out by mass movements of the crowd."[92] Sometimes there occurs the "spontaneous transmission of trance states through a crowd."[93] Suspicious of this type of theomorphic power, authorities suppressed mediums, as did the colonial bureaucracy in early twentieth-century Taiwan.[94] Authorities distrusted the medium's ability to create eruptions of emotional experience, traverse social space, spontaneously induce trance, stir the crowd, divert attention from pain, heal. Theater shares this power. It is the "power of drama to shape experience,"[95] the capacity to bring religious ceremony to life, making the words and actions of the sacrament meaningful.

Ritual and theater are inseparable. As Clifford Geertz theorizes, drama plays out as an "enactment of mass ritual," as "great collective gestures, mass enactments of elite truths."[96] In China, too, we witness an "interpenetration of opera and ritual" by which performances take on the semblance of "large-scale ritual prayers."[97] And spectator involvement enhanced ritual, as in a 1979 exorcistic drama in rural Shanxi.[98] At "the climax of the performance the entire audience joined in, first falling on their knees and then giving three great shouts with the actors and singing the 'Song of Subjugating the Demons.'"[99] As in Fujian and Shanxi, in Bali the "throng of lookers-on and joiners-in" turned it "into a kind of choreographed mob scene," giving "the *negara* an expressive power that neither palaces as copies of the cosmos, nor kings as icons of divine authority, could themselves produce."[100] This "enveloping movement of the whole drama on the soul"[101] and the mesmeric power of performers stirred anxieties.

Sophie Volpp explicates the unease troubling Confucian literati in the mid-seventeenth century. That a "rogue storyteller" could "manipulate those under his thrall," amazing "the people of the markets," driving audiences to "lose control of their bodies," so that their "hair will stand on end, and their tongues will cleave to [the] roofs of their mouths."[102] The "trickster" storyteller induced listeners to lose "themselves as though they were in a trance," to "lose control of

---

92   Dean 1998, 33.
93   Dean and Zheng 2010, 260.
94   Lin 1953, 323.
95   Geertz 2000, 30.
96   Geertz 1980, 13, 116.
97   Johnson 1990, 41–42, 48, 53–54; 2009, 2.
98   In the 1930s, troupes improvised "in reaction to audience responses" (Jiang 2009, 8).
99   Johnson 2009, 78.
100   Geertz 1980, 116.
101   Fallon 1945, 523, 524, citing Charles Morgan.
102   Volpp 2011, 215, 216, 219, 220.

their minds."[103] (The profound empathy experienced by audiences irked power holders: earlier, during the mid-Ming, local elites tried to purge dissenting values from local drama.)[104] The raconteur's "power to mesmerize" provoked the aficionado's dread, who "fears losing mastery over the self, fears being manipulated by the actor."[105] Even the most erudite viewers could be "bewitched" by performers, cautioned the late Ming connoisseur Zhang Dai.[106]

This intensity of audience experience exposed opera troupes to danger. In the early twentieth century, writer Lu Xun, a harsh critic of opera, observed a rural performance in which the captivated audience reacted to onstage fiction with genuine violence. Viewers in the audience, enraged by a stage murder, behaved "as if they were 'intoxicated or deranged'" and became "part of the drama" and "chased actors deep into the surrounding countryside."[107]

The shared quality of sacramental and dramaturgical contagion, in which devotees and viewers could be infected by ritual or dramatic events—with rage, with ecstasy, with prescience, or, more privately, with a diminution of pain—goes to the source of the family's attack on Patient Zhang. Anyone can become a spirit medium, any time, any place; the possessed are sometimes actors, priests, audience members, bystanders, or victims of abuse who are near death; and this state of self-induced temporary possession can be directed maliciously outward and, if not blocked or ameliorated, can lead to insanity. For these reasons it became impossible for Zhang to remain safely in her husband's home, a conclusion reached by the father-in-law, who, like the character in *Tumbler Invites a Guest* who helps the daughter-in-law get away, instructed his son to provide seven dollars of traveling money and send her into the city, to the hospital.

Feigned Madness of the Stage

There is an added convolution to patients emulating operatic characters in 1930s China. Patients imitated operatic figures that, onstage, in character, acted out madness. Yet within the scripted narrative, this stage madness was often explicitly feigned. Feigned madness itself dates to late antiquity and is not a

---

103   Volpp 2011, 220.
104   Tanaka 1985, 160.
105   Volpp 2011, 220, 221, 222. Loss of self-possession could also weaken the body (Kuriyama 1993, 54).
106   Cass 2012, 18.
107   Johnson 1990, 41–42, 48, 53–54.

surprising gambit to encounter on the stage, or in life, to wit, Jizi's 箕子 "undo-ing his hair to feign madness" (*beifa yangkuang* 被髮佯狂), by which he survived the chaotic Shang-Zhou transition in the eleventh century BCE.[108] Arresting our attention are patients in the asylum or in the neuropsychiatric ward who emulated characters that in the drama were themselves feigning madness.[109]

The feigned insanity of Miss Zhao in *Beauty Defies Tyranny* (*Yu zhou feng* 宇宙鋒) was among the great Mei Lanfang's 梅蘭芳 (1894–1961) most celebrated roles during the 1930s.[110] Miss Zhao feigns madness to evade an unwanted marriage. The ruse works. The Manchu patient who executed the sketch of an opera character preferred the work of Shang Xiaoyun 尚小雲, who performed *Driven Mad by the Lost Son* (*Shizi jingfeng* 失子驚瘋), in which a mother's infant son is kidnapped, driving her insane; when the grown son returns, her sanity is restored.[111]

Knowledgeable audiences looked to stage insanity as a test of the artist's virtuosity. Jacobean actors garnered ideas from the inmates of Bethlehem Hospital ("Bedlam") in London.[112] Players on the Victorian stage assiduously mimicked the symptoms observed in the "wards of our lunatic hospitals," as did a Miss Lingard, who, transcending normal "stage-madness," with its insufferable "bundle of inconsistencies provocative of laughter or disgust," realized an exacter personification of "genuine melancholia attonita."[113] "Mad scene" choreography in China called up different signifiers. Blankly staring eyes, "mad eyes" (*fengyan* 瘋眼), expressed feigned madness.[114] Audiences listened for the symptoms of feigned madness in an exact falsetto or understood psychic distress from specific facial expressions or onstage costume changes that revealed torn clothing. To persuade her father of the impossibility of marriage, to extinguish its very idea, Miss Zhao raised her voice to a high pitch and "tears down her black cloud" (her hair).[115] She then executes a celebrated feigned-madness sequence requiring virtuosity, the "*lan hua*" (蘭花 orchid): laughing three times, Miss Zhao grasps her father's beard and pretends to yank out the whiskers. She lies on the floor and babbles gibberish, hallucinating cow-headed, horse-faced demons. Miss Zhao is then presented to the emperor, who disbe-

108 Sima 1983.

109 While Patient Zhang did not invoke characters that feigned madness, others did.

110 Goldstein 2007, 188.

111 XHYY, 71584:16 (May 31, 1940).

112 Reed 1952, 6; Neely 2004.

113 "A Stage-Study of Madness" 1884.

114 Idema 2007, 84n53.

115 On disordered hair signifying female transgression in opera, see Goldman 2012, 175.

lieves the reports of her lunacy. By refusing to kneel, and by other outrageous antics, she displays an insane disregard for the imperial presence. Emulating a deep tradition of political critique, Miss Zhao utters words that only a crazy person would utter: you laugh at me because I'm mad; I laugh at you because you are an immoral emperor. This kingdom belongs to the people; it is not your possession.[116]

Symbolic gestures that on the stage were understood to represent madness were sometimes viewed by the police as markers of genuine madness. Just as Miss Zhao insanely tore down the "black cloud" of her hair, lectures delivered at the Beijing Police Academy instructed cadets to recognize aberrant behavior in women this way. "If you see a woman running with hair all disheveled, her behavior is suspect ... Normal women comb and wash their hair."[117] Operatic symbols of illness or madness, in other words, overlapped with broadly construed ideas of psychic instability, notions that were used by the police in detaining suspects or remanding patients to the asylum.[118]

"The Outward Sense Is Gone, the Inward Essence Feels"[119]

That patients of the neuropsychiatric ward and the asylum emulated characters witnessed on the stage is not unique to China. The imitation of figures read about in fiction, heard about from storytellers, observed in daily life, or, later, watched in film or on television occurs cross-culturally and diachronically. Even in neuropsychological conditions that appear shaped by genetic predisposition, transformations in the cultural environment can explicitly impinge upon expressions of distress.[120] The proliferation of television channels is a vivid example. For people diagnosed with multiple personality disorder in Europe and North America, Ian Hacking posits a connection between the spread of TV remote controls and the "rapid switching of characters" gleaned from television programs.[121]

The physician, too, offered the patient models for imitation. Jean-Martin Charcot's celebrated lectures on hypnotism at Salpêtrière, in the 1880s, show

---

116   On details of *Beauty Defies Tyranny*, see A. Scott 1958, 78–80.

117   Strand 1989, 84, "Rules of Police Work."

118   On Nationalist (GMD) police imprisoning political suspects, who were then admitted to asylums, in late 1940s Shanghai, see Perry 2005, 141.

119   Fallon 1945, 524.

120   Guo and Kleinman 2011; Phillips 2004.

121   Hacking 1995, 32, 40.

the influence of inadvertent cues by the physician on patient behavior. When Charcot's staff sounded a gong, the archetypal "hysterical" patient Blanche "instantaneously" fell "into catalepsy," assuming "plastic poses" previously ascertained from her physician.[122] Earlier, in the eighteenth century, we witness this type of psychological contagion in the dramaturgy of the animal magnetists, in Gassner and then in Mesmer.[123] And earlier, in antiquity, physicians in Greece exploited the power of dramatic suggestion. Theaters were a standard feature in Greek healing temples, and in the great medical centers of Pergamon, the theater had a distinct therapeutic function.[124] In the Asclepeion, an institution of healing, people seeking treatment consumed drugs and then traversed an underground passageway. As they walked the subterranean corridor in a reverie, healers called down the voices of the gods from openings in the ceiling, whispering suggestively to the patients. Modern techniques hint at provocative continuities, such as the administration of barbiturates to enhance the patient's suggestibility and to provoke self-revelation during interviews.

Disempowered people in Zhang's position aimed to appropriate the cultural authority of the drama. By capturing the emotional power of its characters, they might express themselves with some impact, in a way that attracted the attention that had otherwise eluded them. Perhaps their private calamity, their wildly unjust circumstance, might be noticed—if not understood, at least witnessed. Once in the hospital, their narratives were encouraged, probed, and recorded in detail, until the patients' voices or conduct disrupted the ward, and they were compelled to stop.

Pharmacology thus also functioned as a restraint. The opioid and analgesic injections given to Zhang, combined with the threat of expulsion from the hospital, stopped her singing. Once Zhang grew quiet, the focus of investigation turned to her acute abdominal discomfort and to the fever and nausea that she began to suffer. An abscess formed on her arm, where it was discovered that needles had also been stabbed. The surgery consult determined that "nothing" was wrong with her abdomen and ruled against removing the needles surgically. Chief neuropsychiatrist Richard Lyman strongly protested. In light of Zhang's aggravated symptoms, he was appalled by Surgery's nonchalance. Yet the service had made its determination, putting Lyman in the position of advocate, of defending the need to remove the needles from Zhang's body. Required then to eliminate any psychological basis for her complaint, Lyman acrimoni-

---

122   Ellenberger 1981, 95. On Paul Regnard, "Catalepsy Provoked by the Abrupt Sound of a Tam-Tam," in *Les maladies endémiques de l'esprit* (1887), see Didi-Huberman 2004, 209.

123   Winter 2000; Ellenberger 1981, 53–81.

124   Mitchell-Boyask 2007.

ously wrote: it is almost an "absolute rule in my experience that a happy hypo-manic patient will never seek to produce nor will exaggerate serious physical symptoms." "Frankly," Lyman continued, "I am *not satisfied* with any statement that there is 'nothing' wrong in the abdomen."[125] The "internal wandering of needles is highly likely and could easily produce [an] increase of pathology."[126] The needles were not removed.

Zhang discharged herself AMA, against medical advice, on February 26, 1933, several days after Lyman's appeal to remove the needles failed, and nine-teen days after she had first presented at the PUMC outpatient clinic. A hospi-tal social worker on March 2 paid a follow-up visit to a cousin's Beijing address where Zhang stayed, only to discover that she had left the city and returned to the village.[127]

Bibliography

Altehenger, Jennifer E. 2010. "Love, Law and Legality: Marriage Law Campaigning in the Early People's Republic of China." Ph.D. dissertation, University of Heidelberg.

Arkush, David 1990. "The Moral World of Hebei Village Opera," in Paul A. Cohen and Merle Goldman, eds., *Ideas across Cultures: Essays on Chinese Thought in Honor of Benjamin I. Schwartz*. Cambridge, MA: Harvard University Asia Center   87–107.

AvRuskin, Tara L. 1988. "Neurophysiology and the Curative Possession Trance: The Chinese Case," *Medical Anthropology Quarterly*, n.s., 2, 3: 286–302.

Bullock, Mary Brown 1980. *An American Transplant, the Rockefeller Foundation and Peking Union Medical College*. Berkeley: University of California Press.

_____ 2011. *The Oil Prince's Legacy: Rockefeller Philanthropy in China*. Stanford, CA: Stanford University Press.

Butler, Judith 1997. *Excitable Speech: A Politics of the Performative*. New York: Routledge Press.

Cass, Victoria B. 2012. "Female Performance Cultures in the Ming and the Grammar of Jiangnan Identity." Manuscript.

Chen, Duxiu 1999. "On Theater," in Faye Chunfang Fei, ed. and trans., *Chinese Theories of Theater and Performance from Confucius to the Present*. Ann Arbor: University of Michigan Press   117–120.

Ch'iu, K'un-liang 1991. *Les aspects rituels du théâtre chinois* [Ritual aspects of the Chinese theater]. Paris: Collège de France, Institut des hautes études chinoises.

---

125   XHYY, 39305:22 (italics is underlining in original).
126   XHYY, 39305:23.
127   XHYY, 39305:(social service)3.

Cohen, Paul A. 2008. *Speaking to History, the Story of King Goujian in Twentieth-Century China*. Berkeley: University of California Press.

de Certeau, Michel 2000. *The Possession at Loudun*. Chicago: University of Chicago Press.

Dean, Kenneth 1998. *Lord of the Three in One: The Spread of a Cult in Southeast China*. Princeton, NJ: Princeton University Press.

Dean, Kenneth, and Zhenman Zheng 2010. *Ritual Alliances of the Putian Plain*. Vol. 1. Leiden: E. J. Brill.

DeBernardi, Jean Elizabeth 2006. *The Way That Lives in the Heart: Chinese Popular Religion and Spirit Mediums in Penang, Malaysia*. Stanford, CA: Stanford University Press.

Diamant, Neil 2000. *Revolutionizing the Family: Politics, Love, and Divorce in Urban and Rural China 1949–1968*. Berkeley: University of California Press.

Didi-Huberman, Georges 2004. *Invention of Hysteria: Charcot and the Photographic Iconography of the Salpêtrière*, translated by Alisa Hartz. Cambridge, MA: MIT Press.

Dieuaide, Francis R. 1936. *Manual for the Medical Services of the Peiping Union Medical College Hospital*. 5th ed., revised by the staff of the Department of Medicine. Beijing: n.p.

Dikötter, Frank 2007. *Exotic Commodities: Modern Objects and Everyday Life in China*. New York: Columbia University Press.

Dong, Xiaoping 董晓萍 and Ou Dawei 欧达伟 [David Arkush] 2000. *Xiangcun xiqu biaoyan yu zhongguo xiandai minzhong* 乡村戏曲表演与中国现代民众 [The performance of village opera and China's modern populace]. Beijing: Beijing shi fan daxue chubanshe.

Edington, Claire 2013. "Going in and Getting out of the Colonial Asylum: Families and Psychiatric Care in French Indochina," *Comparative Studies in Society and History* 55, 3: 725–755.

Ellenberger, Henri 1981. *The Discovery of the Unconscious: The History and Evolution of Dynamic Psychiatry*. New York: Basic Books.

Elman, Benjamin 2009. *A Cultural History of Modern Science in China*. Cambridge, MA: Harvard University Press.

Esherick, Joseph W. 1987. *The Origins of the Boxer Uprising*. Berkeley: University of California Press.

Fallon, Gabriel 1945. "Dramatic Illusion," *Irish Monthly* 73, 870: 523–528.

Ferguson, Mary E. 1970. *China Medical Board and Peking Union Medical College, a Chronicle of Fruitful Collaboration 1914–1951*. New York: China Medical Board.

Fong, Wen C. 1996. "Imperial Portraiture of the Ming Dynasty," in Wen C. Fong and James C. Y. Watt, eds., *Possessing the Past, Treasures from the National Palace Museum, Taipei*. New York: Metropolitan Museum of Art; Taipei: National Palace Museum 327–333.

Gamble, Sidney D., assisted by John Stewart Burgess 1921. *Peking, a Social Survey*. New York: George H. Doran.

Gao, James Z. 2004. *The Communist Takeover of Hangzhou: The Transformation of City and Cadre   1949–1954*. Honolulu: University of Hawai'i Press.

Geertz, Clifford 1980. *Negara: The Theater State in Nineteenth-Century Bali*. Princeton, NJ: Princeton University Press.

____ 2000. "Blurred Genres, the Refiguration of Social Thought," in Clifford Geertz, *Local Knowledge: Further Essays in Interpretive Anthropology*. 3rd ed. New York: Basic Books   19–35.

Goldman, Andrea 2012. *Opera and the City: The Politics of Culture in Beijing   1770–1900*. Stanford, CA: Stanford University Press.

Goldstein, Joshua 2007. *Drama Kings: Players and Publics in the Re-creation of Peking Opera   1870–1937*. Berkeley: University of California Press.

Grob, Gerald N. 1985. *The Inner World of American Psychiatry   1890–1940*. New Brunswick, NJ: Rutgers University Press.

Guo, Jinhua, and Arthur Kleinman 2011. "Stigma, HIV/AIDS, Mental Illness, and China's Non-persons," in Arthur Kleinman et al., eds., *Deep China, the Moral Life of a Person*. Berkeley: University of California Press   237–262.

Hacking, Ian 1995. *Rewriting the Soul: Multiple Personality and the Sciences of Memory*. Princeton, NJ: Princeton University Press.

Harrison, Henrietta 2000. *The Making of the Republican Citizen: Political Ceremonies and Symbols in China   1911–1929*. Oxford: Oxford University Press.

Hershatter, Gail 2011. *The Gender of Memory: Rural Women and China's Collective Past*. Berkeley: University of California Press.

Hsu, Jing and Wen-Shing Tseng 1974. "Family Relations in Classic Chinese Opera," *International Journal of Social Psychiatry* 20: 159–171.

Huang Jun 黄钧 and Xibo Xu 徐希博 2001. *Jing ju wen hua ci dian* 京剧文化词典 [Dictionary of Beijing opera culture]. Shanghai: Hanyu da cidian chuban, Shiji chuban.

Hung, Chang-tai 1994. *War and Popular Culture: Resistance in Modern China   1937–1945*. Berkeley: University of California Press.

Idema, Wilt L. 2007. "Madness on the Yuan Stage," *Horin, Comparative Studies in Japanese Culture* 14: 65–86.

Jiang, Jin 2009. *Women Playing Men: Yue Opera and Social Change in Twentieth-Century Shanghai*. Seattle: University of Washington Press.

Johnson, David 1990. "Scripted Performances in Chinese Culture: An Approach to the Analysis of Popular Literature," *Chinese Studies* [*Han Hsueh Yen Chiu* 漢學研究] 8, 1: 37–55.

____ 2009. *Spectacle and Sacrifice: The Ritual Foundations of Village Life in North China*. Cambridge, MA: Harvard University Asia Center.

Jordan, David K. 1972. *Gods, Ghosts, and Ancestors: The Folk Religion of a Taiwanese Village*. Berkeley: University of California Press.

Judge, Joan 2008. *The Precious Raft of History: The Past, the West, and the Woman Question in China*. Stanford, CA: Stanford University Press.

Karl, Rebecca E. 2002. *Staging the World: Chinese Nationalism at the Turn of the Twentieth Century*. Durham, NC: Duke University Press.

Kirby, E. T. 1974. "The Shamanistic Origins of Popular Entertainments," *Drama Review: TDR* 18, 1: 5–15.

Kleinman, Arthur, Leon Eisenberg, and Bryon Good 1978. "Culture, Illness, Care: Clinical Lessons from Anthropologic and Cross-Cultural Research," *Annals of Internal Medicine* 88, 2: 251–258.

Kleinman, Arthur, and Joan Kleinman 1994. "How Bodies Remember: Social Memory and Bodily Experience of Criticism, Resistance, and Delegitimation following China's Cultural Revolution," *New Literary History* 25, 3: 707–723.

Kuriyama, Shigehisa 1993. "Concepts of Disease in East Asia," in Kenneth F. Kiple, ed., *The Cambridge World History of Human Disease*. Cambridge: Cambridge University Press   52–59.

Lamb, Susan D. 2010. "Pathologist of the Mind: Adolf Meyer, Psychobiology and the Phipps Psychiatric Clinic   1908–1917." Ph.D. dissertation, Johns Hopkins University.

Lee, Raymond L. M. 1989. "Self-Presentation in Malaysian Spirit Seances: A Dramaturgical Perspective on Altered States of Consciousness in Healing Ceremonies," in Colleen A. Ward, ed., *Altered States of Consciousness and Mental Health: A Cross-cultural Perspective*. Newbury Park, CA: Sage Publications   251–266.

Lee, Sing, and Arthur Kleinman 2003. "Suicide as Resistance in Chinese Society," in Elizabeth J. Perry and Mark Selden, eds., *Chinese Society: Change, Conflict and Resistance*   2nd ed. London: RoutledgeCurzon   289–311.

Li Hsiao-t'i 李孝悌 1992. *Qingmo de xiaceng shehui qimeng yundong* 清末的下層社會啟蒙運動 [Lower-class enlightenment in the late Ch'ing period   1901–1911]. Nankang: Academia Sinica.

Lief, Alfred 1948. *The Commonsense Psychiatry of Dr. Adolf Meyer: Fifty-Two Selected Papers*. New York: McGraw-Hill.

Lin, Tsung-yi 1953. "A Study of the Incidence of Mental Disorder in Chinese and Other Cultures," *Psychiatry* 16, 4: 315–335.

Luo Jintang 羅錦堂 1956. *Zhongguo sanqu shi* 中國散曲史 [History of Chinese verse]. Taibei: Zhonghua wenhua chuban.

MacFarquhar, Roderick 1999. *The Origins of the Cultural Revolution*. Vol. 3. New York: Columbia University Press.

Macgowan, John 1909. *Lights and Shadows of Chinese Life*. Shanghai: North China Daily News & Herald.

Mackerras, Colin 1983. *Chinese Theater: From Its Origins to the Present Day.* Honolulu: University of Hawai'i Press.

Mann, Susan 2011. *Gender and Sexuality in Modern Chinese History.* Cambridge: Cambridge University Press.

McCartney, J. Lincoln 1926. "Neuropsychiatry in China: A Preliminary Observation," *Chinese Medical Journal* 40, 7: 617–626.

Merkel-Hess, Kate 2012. "Acting Out Reform: Theater and Village in the Republican Rural Reconstruction Movement," *Twentieth-Century China* 37, 2: 161–180.

Métraux, Alfred 1955. "Dramatic Elements in Ritual Possession," translated by James H. Labadie, *Diogenes* 3, 18: 18–36.

Meyer, Adolf 1915. "Objective Psychology or Psychobiology with Subordination of the Medically Useless Contrast of Mental and Physical," *Journal of the American Medical Association* 65, 10: 860–863.

Mitchell-Boyask, Robin 2007. "The Athenian Asklepieion and the End of the Philoctetes," *Transactions of the American Philological Association* 137, 1: 85–114.

Nadel, S. F. 1946. "A Study of Shamanism in the Nuba Mountains," *Journal of the Royal Anthropological Institute of Great Britain and Ireland* 76, 1: 25–37.

Naquin, Susan 2000. *Peking: Temples and City Life   1400–1900.* Berkeley: University of California Press.

Neely, Carol Thomas 2004. *Distracted Subjects: Madness and Gender in Shakespeare and Early Modern Culture.* Ithaca, NY: Cornell University Press.

Oughourlian, Jean-Michel 1991. *The Puppet of Desire, the Psychology of Hysteria, Possession, and Hypnosis.* Stanford, CA: Stanford University Press.

Pearson, Veronica 1991. "The Development of Modern Psychiatric Services in China   1891–1949," *History of Psychiatry* 2: 133–147.

Perry, Elizabeth J. 2005. *Patrolling the Revolution: Worker Militias, Citizenship, and the Modern Chinese State.* Lanham, MD: Rowman & Littlefield.

Phillips, Michael P. 2004. "Suicide and the Unique Prevalence Pattern of Schizophrenia in Mainland China: A Retrospective Observational Study," *Lancet* 364, 9439: 1062–1068.

Qiu Kunliang 邱坤良 1992. *Rizhi shiqi Taiwan xiju zhi yanjiu   1895–1945: Jiuju yu xinju* 日治時期臺灣戲劇之研究, 1895–1945: 舊劇與新劇 [Research on drama in Taiwan during the period of Japanese rule   1895–1945: Old drama and new drama]. Taipei: Zili wanbao she wenhua chubanbu.

Reed, Robert Rentoul, Jr. 1952. *Bedlam on the Jacobean Stage.* Cambridge, MA: Harvard University Press.

Rogaski, Ruth 2004. *Hygienic Modernity, Meanings of Health and Disease in Treaty-Port China.* Berkeley: University of California Press.

Ross, Robert M. 1920. "The Insane in China: Examinations Hints," *Chinese Medical Journal* 34, 5: 514–518.

Schipper, Kristofer 1993. *The Taoist Body*, translated by Karen C. Duval. Berkeley: University of California Press.

Scott, A. C. 1958. *An Introduction to the Chinese Theatre*. Singapore: Donald Moore.

Scott, James C. 1990. *Domination and the Arts of Resistance: Hidden Transcripts*. New Haven, CT: Yale University Press.

Seligman, C. G. 1932. *Pagan Tribes of the Nilotic Sudan*. London: Routledge.

Shahar, Meir 1998. *Crazy Ji, Chinese Religion and Popular Literature*. Cambridge, MA: Harvard University Asia Center.

Shahar, Meir, and Robert P. Weller, eds. 1996. *Unruly Gods, Divinity and Society in China*. Honolulu: University of Hawai'i Press.

Sidky, Homayun 2011. "The State Oracle of Tibet, Spirit Possession, and Shamanism," *Numen* 58: 71–99.

Sima Qian 司馬遷 1983. *Guice lie zhuan* 龜策列傳, 128 *juan* 卷 [Biographies of the turtle-shell diviners], in *Shiji* 史記 [Historical records]. Taipei: Hongshi shuju.

Simonis, Fabien 2010. "Mad Acts, Mad Speech, and Mad People in Late Imperial Chinese Law and Medicine," Ph.D. dissertation, Princeton University.

Spence, Jonathan D., and Annping Chin 1996. *The Chinese Century: A Photographic History of the Last Hundred Years*. New York: Random House. "A Stage-Study of Madness" 1884. *British Medical Journal* 1, 1223: 1100.

Strand, David 1989. *Rickshaw Beijing: City People and Politics in the 1920s*. Berkeley: University of California Press.

_____ 2011. *An Unfinished Republic: Leading by Word and Deed in Modern China*. Berkeley: University of California Press.

Strickmann, Michel 2002. *Chinese Magical Medicine*, edited by Bernard Faure. Stanford, CA: Stanford University Press.

"Suicides in Hangchow" 1920. *Chinese Medical Journal* 34, 5: 560.

Tanaka, Issei 1985. "The Social and Historical Context of Ming-Ch'ing Local Drama," in David Johnson, Andrew J. Nathan, and Evelyn S. Rawski, eds., *Popular Culture in Late Imperial China*. Berkeley: University of California Press 143–160.

Taylor, Alan 2002. *American Colonies: The Settling of North America*. New York: Penguin Books.

Topley, Marjorie 1970. "Chinese Traditional Ideas and the Treatment of Disease: Two Examples from Hong Kong," *Man*, n.s., 5, 3: 421–437.

Tseng, Wen-Shing 1972. "Psychiatric Study of Shamanism in Taiwan," *Archives of General Psychiatry* 26, 6: 561–565.

Vinchon, Jean 1931. "The Devil and the Psychoneurotics," in M. Laignel-Lavastine, *The Concentric Method in the Diagnosis of Psychoneurotics*. New York: Harcourt, Brace 112–130.

Volpp, Sophie 2011. *Worldly Stage: Theatricality in Seventeenth-Century China*. Cambridge, MA: Harvard University Asia Center.

Wakeman, Frederic, Jr. 1995. *Policing Shanghai   1927–1937*. Berkeley: University of California Press.

Werle-Burger, Helga 1998. "Interactions of the Media: Storytelling, Puppet Opera, Human Opera and Film," in Vibeke Børdahl, ed., *The Eternal Storyteller, Oral Literature in Modern China*. New York: Routledge   122–136.

Winter, Alison 2000. *Mesmerized: Powers of Mind in Victorian Britain*. Chicago: University of Chicago Press.

Wolf, Margery 1975. "Women and Suicide in China," in Margery Wolf and Roxane Witke, eds., *Women in Chinese Society*. Stanford, CA: Stanford University Press   111–141.

Wong, Susan Y., and Michael R. Phillips 2009. "Nonfatal Suicidal Behavior among Chinese Women Who Have Been Physically Abused by Their Male Intimate Partners," *Suicide and Life-Threatening Behavior* 39, 6: 648–658.

Xiao, Zhiwei 1998. "Chinese Cinema," in Yingjin Zhang and Zhiwei Xiao, eds., *Encyclopedia of Chinese Film*. New York: Routledge   3–30.

Xiehe yiyuan 協和醫院 (XHYY), Peking Union Medical College.

Xiong Yuezhi 熊月之 1994. *Xixue dongjian yu wan Qing shehui* 西學東漸与晚清社會 [The dissemination of Western learning and late Qing society]. Shanghai: Shanghai People's Publishing.

Yeh, Catherine Vance 2003. "A Public Love Affair or a Nasty Game? The Chinese Tabloid Newspaper and the Rise of the Opera Singer as Star," *European Journal of East Asian Studies* 2, 1: 13–51.

Index

Academia Sinica:
  establishment of   32, 216
  Research Institute of Geology   282
academic disciplines:
  "art" (*shu*) distinguished from "science"
      (*xue*)   12, 96, 104–105, 107, 109–111
  and the "Chinese Society for the History of
      Science"   56
  and the civil examination system   67, 101
  Comte's ordering of   95, 101
  globalization of   67–70, 85–87
  and the historiography of science   39–40
  kagaku/kexue
      as "academics divided into disciplines"
        7, 23, 24–25, 100
      and "science"   12, 93–100, 94n5, 103–110
  in Nishi Amane's *Hyakugaku renkan*   95
  and the "physical sciences" of Yan Fu   108,
      110
  and traditional Chinese science   68, 86–87
  Western-inspired taxonomy of Liu Shipei's
      Zhuomo xueshushi xu   67–68, 70–76,
      77–78, 84–86
  See also education; "science"; Yan Fu
Adolph, William Henry   51
alphabetic writing systems:
  and China's disadvantaged entry into global
      communication   10, 10, 153, 155–158,
      164–170
  scientific thinking associated with   5,
      120–121
  Wang Zhao's *guanhua zimu* ("Mandarin
      alphabet")   122, *123fi*
  See also Chinese characters; Chinese-
      language typewriter; phonetization
      and alternative writing schemes;
      shorthand
Ancient Greece:
  and early China   5
  healing temples of   318
  and Liu Shipei's "History of Psychology"
      77–78

ancient learning. *See* Chinese classical
      learning; Zhou period intellectual
      history
Anderson, Johan Gunnar   272
Anderson, Paul A. (1898–1990)   253
Arabic numerals:
  Chinese characters encoded in   159, 161,
      162ffi–2, 164–165, 172, 173, 175
  and Chinese numerals for pagination of the
      Geological Bulletin   272
"art" (*shu*, Japanese *jutsu*), distinguished from
      "science" (*xue*)   12, 96, 104–105, 107,
      109–111
astronomy:
  Department of Mathematics and
      Astronomy at the Beijing School of
      Foreign Languages   21
  Herschel's *The Outline of Astronomy*   20
  Hobson's *Summary of Astronomy*   17, 18
  in Japan   25
  and Jesuit astronomers   15–16, 19, 44
  and the "nine arts" (*jiu shu*) of Jartoux   54
  and Qing Self-Strengthening reforms
      16–17, 23
authenticity:
  and commodification   186–188, 191,
      206–207
  and copying (*fangzao*) of foreign technolo-
      gies   201–205, 206
  counterfeiting of industrial products   193
  identification of fake scientists   234–237
  and the National Products Movement
      200–205, 200–201n34
  pirating (*fanyin*)   186, 191–193
  and scientific truth   222n45, 235, 238, 270
  and *shiyan* (investigating through practice)
      187, 193, 197–198, 197n28, 199–200, 206
  See also experimentation

Ba Jin   223–224
Bacon, Francis:
  Chinese origins of Baconian logic   83
  and the "three Great Inventions"   47

---

*   Page numbers in *italic* indicate figures.

Printed in the United States
by Baker & Taylor Publisher Services